Sample Survey Methods and Theory

Sample Survey Methods and Theory

VOLUME II · THEORY

MORRIS H. HANSEN

Assistant Director for Statistical Standards
Bureau of the Census

WILLIAM N. HURWITZ

Chief Statistician
Bureau of the Census

WILLIAM G. MADOW

Chairman, Statistical Research Laboratory
University of Illinois

Wiley Classics Library Edition Published 1993

A Wiley-Interscience Publication
JOHN WILEY & SONS, INC.
New York • Chichester • Brisbane • Toronto • Singapore

Wiley Classics Library Edition Published 1993.

This text is printed on acid-free paper.

ISBN 0-471-30966-4 (Vol. II pbk.)

ISBN 0-471-00628-9 (Set)

Library of Congress Catalog Card Number 53-8112

Printed in the United States of America

10 9 8 7 6 5 4 3 2 1

To

MILDRED, HANNAH, and LILLIAN

Preface

THIS VOLUME CONTAINS THE FUNDAMENTAL THEORY ON WHICH sampling methods are based together with derivations of the formulas and proofs of statements made in Volume I. Volume I gives the principles and methods of sampling and their applications to various types of problems, and states without proof the formulas appropriate to the methods presented. The two volumes combined are an attempt to provide a comprehensive presentation of both sampling theory and practice.

The first three chapters of this volume present the fundamental theorems on probability, expected values, and variances that are needed in the development of the sampling theory in the remaining chapters. Chapters 4 through 11 contain derivations, proofs, and some extensions of theory for the corresponding chapters of Volume I, and provide a convenient summary of sampling formulas. These chapters have been designed primarily to facilitate reference from Volume I, and therefore suffer somewhat in continuity. They do not contain a discussion of the application of the results derived; for this the reader is referred to Volume I. Chapter 12 discusses some of the practical implications of the treatment of response errors in surveys, and develops a theory for the methods described. Applications of recent developments in decision theory have not been included.

Readers desiring only the ability to understand the derivations of sampling formulas can apply the theorems of Chapters 2 and 3, without proof, as they are introduced in the proofs given in Chapters 4 through 12. Many will wish to have the fuller command of the methods and the ability to extend them that comes through understanding the theory in Chapters 2 and 3, where proofs of the theorems are given.

For the most part, the mathematical background assumed for this volume is college algebra, although some calculus is used for a few of the proofs.

For textbook purposes, the following suggestions may serve as guides in the organization of courses in which proofs are given. For courses with proofs either volume can serve as the text, the choice depending on the emphasis desired; the other can serve as a reference book and provide supplemental material for the teacher. A one-year course may begin with Chapter 1 of either volume, the development of selected theorems from Chapters 2 and 3 of Volume II as indicated in the introductions to those chapters, followed by Chapters 4 through 6 and selected materials from Chapters 8 through 12 of either volume. The appropriate additional theorems of Chapters 2 and 3 may be developed as they are needed. For a one-semester course, material from Chapters 1 through 6 and 11 may be sufficient. Throughout Chapters 4 through 11 of this volume some of the sections are footnoted "May be deferred." For these, in particular, it may be convenient to read the theorems but omit the proofs if only selected materials are to be covered.

Most of the theory of sample survey design is an immediate consequence of statistical theory that has been developed and extended by many persons over a long period of years. No attempt has been made to trace or give credit in the text to original sources of sampling theory, except for quite recent developments. Over and above the specific credits noted in the text we are highly indebted to Miss Blanche Skalak and Dr. Margaret Gurney for their assistance in preparing this volume. Miss Skalak developed a number of the proofs in Chapters 4 through 11, wrote up most of the proofs, and reviewed and made numerous helpful suggestions on the entire manuscript. Dr. Gurney reviewed the manuscript, made many helpful suggestions, and prepared the index.

MORRIS H. HANSEN
WILLIAM N. HURWITZ
WILLIAM G. MADOW

May, 1953

Contents

INTRODUCTION

FUNDAMENTAL THEORY

DERIVATIONS, PROOFS, AND SOME EXTENSIONS OF THEORY FOR CORRESPONDING CHAPTERS OF VOLUME I

A THEORY FOR RESPONSE ERRORS

Sample Survey Methods and Theory

CHAPTER 1

Introduction and Definitions

1. The scope of the theory of sample surveys. The contents of this chapter. The theory of sample surveys is concerned with developing, analyzing, and improving methods of formulating the information wanted from a sample survey, selecting the sample, obtaining the information from the sample, translating that information into statements relating to the purposes of the survey, and evaluating the accuracy of those statements.

As a preliminary to considering the parts of the theory of probability and mathematical expectation that are needed for the effective study of the statistical characteristics of sample surveys, we shall, in Sec. 2–5 of this chapter, introduce some of the more important definitions. In Sec. 6 we shall discuss why probability methods of selecting samples should be used.

2. Definitions of *population*, *element*, and *list*. By a finite *population*, $\mathscr{A}$, is meant any well-defined set or class containing a finite number of *elements*, $A_1, A_2, \cdots, A_N$. These elements may be plants, farms, persons, blocks, counties, businesses, electric light bulbs, insects, and so on. The population will then consist of certain of these elements: the plants of a certain kind in a specified field, the farms of over a specified size, the unemployed persons in the United States, the blocks in a specified city, the counties in which coal is mined, the grocery stores in a specified state, the United States income tax returns for a stated year, the electric light bulbs produced in a given plant during a stated period of time, the insects in a given field. Thus, to define a population we must be able to state the kind of elements of which it consists and to give rules for including or excluding any particular element. These rules may take the form of an enumeration of the elements of the set or may be a statement of the conditions the elements must satisfy.

When the elements of a population have been numbered or otherwise identified, we call that population together with its identification system a *list*. (In some other studies of sample surveys the word *frame* is used as we use the word *list*.) For example, if the population consists of the blocks of a specified city, we may obtain a map of that city on which the blocks and streets are outlined and assign numbers to the blocks. The map and numbers constitute a list. If the blocks were numbered in a different order, we would have a different list. Similarly, if the population consists of the cards in a file, the list is determined by a particular arrangement of the cards. For two lists to be the same they must consist of the same elements with the same identifications; consequently, a single population may yield many lists.

For some methods of selecting samples the order in which the population is arranged in a list will not affect the precision of the information obtained, whereas for others the order of arrangement may influence the precision considerably.

We have introduced the term *list* because the recommended methods of selecting samples will all involve the use of some form of list. With such lists, it is possible to select a sample of elements from the population with *known probabilities* of selection, a prerequisite of the sampling theory to be considered. We will ordinarily select a sample of elements by selecting the numbers that identify them. Thus, if we were selecting 2 elements out of a population of 10 elements, we might identify these 10 elements by the numbers 1, 2, · · ·, 10, select 2 of the 10 numbers, and say that the sample consisted of the elements identified by these numbers.

Sometimes the list is such that any of several numbers in the list identify the same element of the population. Suppose that we want to select a sample of law firms and use as a list the names given in a register of lawyers. Then, if a law firm has 10 lawyers in the register, any of 10 numbers will identify that firm.

The numbering of the elements of the population need not be simply 1, · · ·, N, but may for various administrative and other reasons be more complicated. For example, a given household in a city may have 2 numbers that jointly identify it—the number assigned to the block on which the household is located, and the number assigned to the household within the block.

3. Definitions of *characteristic*, *elementary unit*, and *population of analysis*. By a *characteristic* of a population is meant any quantity or relationship relating to the population.

Illustration 3.1. Illustrations of characteristics.

POPULATION	SOME CHARACTERISTICS
All persons in a city.	Average height; total income; per cent of income spent on food at different levels of income; number of females; distribution of total income among families by size of income; attitude towards taxes; intentions to purchase refrigerators; relation of education and amount of crime.

POPULATION	CHARACTERISTICS
Electric light bulbs produced on a given date in a given plant.	Number of defective bulbs; average duration of burning of bulbs; variation in length and intensity of burning of bulbs; relationship of all these quantities to the order of production of the bulbs for each machine producing them.

We mean to give to *characteristic* the broadest possible meaning. It may or may not be numerical. It is anything we may wish to learn about the population.

Sometimes it will be convenient to speak of the characteristics of the elements of the population, i.e., the total income of all persons in an area, the height of the person, whether the bulb or the group of bulbs is defective, and so on.

Often, particular interest will center on a particular class of elements for which frequency distributions or averages of the characteristic are desired. The particular elements for which such distributions or averages are desired will be referred to as the *elementary units*. A characteristic of an elementary unit may be an *attribute*, e.g., a person is male or female; or a value of a variable, e.g., the income of an individual or of a family. Since the same survey may yield different characteristics, it may also refer to several elementary units, e.g., the person and the family. Usually the objectives of the survey will determine both the elementary units and the population consisting of them. The population whose elements are the elementary units is called the *population of analysis*.

Illustration 3.2. Suppose that the population of analysis consists of all families in the District of Columbia. Then we may say that X_i, Y_i, Z_i are the respective characteristics—food expenditures, rent expenditures, and income in a specified period for the ith family, A_i, in the population which consists of all N families $A_1, \cdots, A_N$ in the District of Columbia.

We will then say that X_i is the value of the characteristic X for the ith element, A_i, of the population. For example, Z_i is the value of the characteristic Z, total income, for the ith family in the District of Columbia.

Exercises

3.1. Suppose that the population of analysis consists of N electric light bulbs and the characteristic is the proportion of defective bulbs. Define a numerical characteristic for each of the bulbs such that, if you know the value of this characteristic for each element of the population, you can compute from these values and the size of the population the characteristic of the population.

3.2. State a population of analysis and define two characteristics of the population. Also define characteristics of the elements of the population so that you can compute the characteristics of the population from those of the elements.

3.3. Suppose that the population of analysis consists of 5 families, A_1, A_2, $\cdot \cdot \cdot$, A_5, and that the values of X, Y, and Z, designating, respectively, food expenditures, rent expenditures, and total income for these families, are as follows:

Family	X	Y	Z
1	10	20	110
2	15	15	200
3	13	18	80
4	18	25	90
5	24	16	70

Compute the following characteristics: total expenditures for food and rent, total income, and per cents of income spent on food and rent.

4. Definitions of terms such as *sample*, *sampling plan*, and *sampling unit*. A *sample* is a subset of a population selected to obtain information concerning the characteristics of the population, i.e., if the population consists of the N elements A_1, A_2, $\cdot \cdot \cdot$, A_N, then the sample will consist of some of these elements, e.g., A_1, A_3, A_7, $\cdot \cdot \cdot$. We shall be concerned only with *probability samples*, i.e., all elements of the population have a *known* probability of being included in the sample. (Probability is discussed in Sec. 4, Ch. 2.) We select a sample and obtain certain information for the elements of the sample, say the values of the characteristics X, Y, Z, and combine this information in such a way that we shall have useful information concerning certain characteristics of the population.

Sometimes the population used for selecting the sample is the population of analysis, but often a different population is defined. For example, the elementary units may be people but the populations from which we select a sample are the populations of blocks and families. Thus, some populations are defined as a result of the problem we are investigating; others are defined to help us select the sample.

For assistance in selecting a sample we may define several populations and methods of selecting samples. By the *sampling plan* we shall mean all the steps we take in selecting the sample once the population of analysis is defined.

Illustration 4.1. To select a sample of the families of New York City we might first define a population consisting of all the blocks in New York City and select a sample of those blocks. Then we might define the families living on the selected blocks to be a population and select a sample of them. Thus, the final sample would consist of families, but we would have first selected a sample of blocks to help us select the sample of families. In picking the first sample we might use as a list a map on which the streets were shown and the blocks numbered. In picking the second sample, i.e., the sample of families, we might construct a list by sending people to the blocks selected in the first sample and having these people list the addresses of all families living on the sample blocks. Each line on the listing would identify the corresponding family. (There may be two or more lines for one address, if there is more than one family at a given address.)

This method of obtaining a sample of families is an illustration of *cluster sampling* where the cluster is a city block and the family is an elementary unit. In general, by *cluster sampling* we mean that, for purposes of selecting a sample, we have defined a population whose elements are groups or clusters of elementary units, and that we will select a sample of the clusters from that population. Clusters may be natural, like the blocks of a city, or they may be constructed, as in Illustration 4.3 below.

The elements of the population from which we select the sample are called *sampling units*. If, as in Illustration 4.1 of this section, the elements of the sample selected initially are clusters, and a subsample is taken from the selected sampling units, we often refer to the clusters as *primary sampling units* or as *first-stage sampling units*, and to the elements of which the second sample consists as *second-stage sampling units*. The sample selected in Illustration 4.1 was selected in two stages, but we sometimes continue the process to three or more stages. The sample selected at the second stage is a *subsample* of that selected at the first, i.e., the sample obtained in the first selection is defined to be a population from which a sample is selected at the second stage, and so on. We refer to all such sampling plans as *multi-stage* sampling plans.

Illustration 4.2. In order to pick a sample of people living in cities, we might begin by selecting a sample of counties (the primary sampling units are counties), then select a sample of cities within the counties selected in the first stage (the second-stage sampling units are cities), then select a sample of blocks within the selected cities (the third-stage sampling units are blocks), then select a sample of families within the selected blocks (the fourth-stage sampling units are families), and finally select a sample of people within the selected families. This is an example of a five-stage sampling plan. Counties, cities, blocks, and families are all clusters of elementary units and are also elements of populations from which samples are selected. Thus, the counties may be considered to be a cluster of cities, of blocks, of families, or of people, but also may be considered elements of a population which has counties as elements. The important thing is the flexibility with which the words *element* and *cluster* are used, i.e., cluster simply stands for a grouping of elements that is convenient for the selection of the sample, but once the clusters are defined they are themselves the elements of a population from which we select a sample.

In order to avoid confusion in using the terms *population*, *element*, and *sample*, it is sufficient to state exactly what the elements and population are for each stage of the selection of the sample. It may be remarked that, although the population of analysis and, sometimes, the elementary units that make it up are uniquely determined by the purposes of the survey, ordinarily there are many ways in which populations may be defined for selecting the sample to obtain information concerning any specified population of analysis. Whenever we select a sample, there will be many possible samples of which we select one to be "the" sample. We can say that these possible samples are the elements of a population from which one element is selected to be "the" sample.

Illustration 4.3. Suppose that we are selecting a 10 per cent sample of the families in a city from a list by first selecting 1 of the first 10 families, $A_1, A_2, \cdots, A_{10}$, and every tenth family afterwards. Then, if A_1 is selected, the sample consists of $A_1, A_{11}, A_{21}, \cdots$; if A_2 is selected, the sample consists of $A_2, A_{12}, A_{22}, \cdots$, and so on. Thus, there are 10 possible samples, of which we select 1 to be "the" sample. Each of these possible samples is a cluster but not a natural cluster.

Exercises

4.1. Suppose that a population consists of 5 families, A_1, A_2, A_3, A_4, A_5, where A_i contains i persons, $i = 1, 2, \cdots, 5$. List all possible samples consisting of 2 families and show the number of persons in each sample.

4.2. A population consists of 10 families living on 2 blocks B_1 and B_2, 5 families living on each block. The 5 families living on block B_1 are denoted by $A_{11}, \cdots, A_{15}$, where the first subscript identifies the block, and the second identifies the family within the block, and the families consist of 3, 2, 1, 3, 1 persons, respectively; those living on block B_2 are denoted by $A_{21}, A_{22}, \cdots, A_{25}$, and consist of 1, 6, 2, 3, 7 persons, respectively. The sampling plan is a two-stage plan; the list for the first-stage sample consists of B_1 and B_2, of which 1 is to be selected. If B_1 is selected in stage 1, we select 3 elements from the list $A_{11}, \cdots, A_{15}$; if B_2 is selected in stage 1, we select 3 elements from the list $A_{21}, A_{22}, \cdots, A_{25}$. List all possible samples, showing the number of persons in each sample.

4.3. State a three-stage sampling plan for estimating a specified characteristic of a population you define.

4.4. Is there only one population that is "correct" to estimate the values of certain characteristics for the population? If we wish to estimate total personal income, can we regard people, families, blocks, or file cards listing personal income as elementary units of the population?

5. Definitions of *estimate, sample design, precision, true value, accuracy, survey design.* After a sample has been selected we prepare *estimates* based on the sample for specified characteristics of the population of analysis. Often, as in estimating the average personal income in the United States, the estimate may be a number, e.g., the average income is $2350; but also the estimate may consist of an interval of numbers, e.g., the average income is more than $2200 and less than $2600; and often the estimates consist of several numbers, intervals, and functions. The *sample design* will consist of the sampling plan and method of estimation.

Although, as we shall see later, there are many estimation equations, we shall for the present consider only one, the arithmetic mean.

Suppose that the population consists of the five families listed in Ex. 3.3, and that when questioned they give the data listed in the table in that exercise. Suppose also that we wish to estimate $\bar{Z}$, the average family income. If we knew the data of Ex. 3.3, we would calculate $\bar{Z} = \dfrac{110 + 200 + 80 + 90 + 70}{5} = \dfrac{550}{5} = 110$. But suppose that we do not know the data for the population, and in order to estimate the unknown value, $\bar{Z}$, we decide to select a sample of 2 families, learn the values of Z for these families, and use the average income of these 2 families as an estimate of $\bar{Z}$. There are 10 possible samples of 2 of the elements of the population. These possible samples, the corresponding values of Z, and the estimates of $\bar{Z}$ are given in Table 1. Each of the 10 possible samples will yield an estimate of $\bar{Z}$.

Table 1. Samples of 2 from a population of 5 elements

Possible samples	Values of Z	Estimates of $\bar{Z}$
A_1, A_2	100, 210	155
A_1, A_3	100, 80	90
A_1, A_4	100, 90	95
A_1, A_5	100, 70	85
A_2, A_3	210, 80	145
A_2, A_4	210, 90	155
A_2, A_5	210, 70	145
A_3, A_4	80, 90	85
A_3, A_5	80, 70	75
A_4, A_5	90, 70	80

By a measure of *precision* of the estimate we shall mean a measure of how close the set of possible sample estimates for a particular sample design may be expected to come to $\bar{Z}$. Such measures of precision, to be useful, must be approximately known from other information, perhaps an earlier survey, or it must be possible to estimate them from the sample.

Ordinarily, we would use an approximate value of the measure of precision in designing the sample; and then, if we had a reasonably large sample, estimate the measure of precision from the data obtained from the sample itself.

It is not enough, however, to consider measures of precision. Was the information that would have been obtained and listed in Ex. 3.3 correct? Did some families understate their incomes? Did the interviewer forget to ask some families about income from pensions and similar sources? Did the interviewer make a mistake and substitute family A_3 for family A_1, which was supposed to be in the sample? By the *true value* of the characteristic we shall mean the value that would be obtained if no errors were made in any way in obtaining the information or computing the characteristic. By a measure of *accuracy* of an estimate we mean a measure of how close the estimate may be expected to come to the true value of the characteristic. Thus, even a complete enumeration may not be entirely accurate, but, according to our definition of precision, a complete enumeration is precise.

By the *survey design* will be meant the sample design together with the questionnaire and the method of obtaining the information from the sample, or, more generally, the method of measurement. Thus, the survey design includes the plans for all the parts of the survey except the statement of the objectives. It includes:

(*a*) The questionnaire,
(*b*) Decision on method of observation or interview,
(*c*) Sample design,
(*d*) Choice and training of interviewers,
(*e*) Assignments of interviewers,
(*f*) Decisions on treatment of noninterviews,
(*g*) Estimation equations,
(*h*) Processing of questionnaires,
(*i*) Preparation of tables,
(*j*) Studies of precision and accuracy of information,

as well as instructions and methods followed for carrying through these operations. It will be seen that each of these parts affects the accuracy of the information to be obtained. Since the objective of survey design is to maximize the accuracy of the estimate (or, more generally, to minimize, in some sense, the losses that may result from the fact that the estimate will almost certainly not equal the true value), it follows that the expenditures for the different parts of the survey should be allocated with this objective in mind.

6. Why probability methods should be used in selecting samples. There are many possible methods of selecting a sample from a population. Some of these depend on the judgment of people who claim to know the population; others merely consist in defining the sample to be the part of the population that is most conveniently available; others (and it is these that we shall study in the following pages) are based on the use of the theory of probability. In applying these probability methods, the following two points must be kept in mind as the justification of their use and as the condition of their valid application:

(*a*) Methods of selecting samples based on the theory of probability are the only general methods known to us which can provide a measure of precision. Only by using probability methods can objective numerical statements be made concerning the precision of the results of the survey.

(*b*) It is necessary to be sure that the conditions imposed by the use of probability methods are satisfied. It is not enough to hope or expect that they are. Steps must be taken to meet these conditions by selecting methods that are tested and are demonstrated to conform to the probability model.

It should be obvious that we are not presenting methods based on probability theory as just one more means of selecting samples. Rather we assert that, with rare exceptions, the precision of estimates not based on known probabilities of selecting the samples cannot be predicted before the survey is made, nor can the probabilities or precision be estimated after the sample is obtained. If we know nothing of the precision, then we do not know whether to have much faith in our estimates, even though highly accurate measurements are made on the units in the sample. Hence, when the information to be obtained is of real importance, it will be desirable to choose methods based on the use of the theory of probability.

It is sometimes argued that any sample is selected by probability methods; that the interviewers who select a sample by approaching people they meet in certain localities are using probability methods when they do their jobs as they should, not selecting more than designated numbers from one specified group or another. Such probabilities are, however, unknown; they may vary from enumerator to enumerator, and they may vary over time. It is not adequate to test to see whether the probabilities seem to be what they should, because the conditions existing at the next application of the procedure will differ from what they were during the test. Considerable evidence exists that only carefully tested methods of selection which are capable of being repeated can be depended upon to yield either equal probabilities or any other specified probabilities of selection.

We ask for greater care in selecting samples than is customary in many other applications of statistics. Yet in these other branches of statistics we would ask for such care if it were physically, administratively, and economically possible as it usually is in survey design. One of the most important facts about the selection of samples is that, if a list can be constructed and if proper methods of selection are used, we need not guess, we can *determine* the probabilities of selecting the possible samples. Furthermore, the increases in precision and the ability to measure precision that thus become available will in general more than repay the cost of applying these methods.

The uses to which some survey results will be put are sufficiently crude so that almost any method for selecting the sample will yield satisfactory information. Obviously, the cheapest method that meets the purposes of the survey should be chosen. In some instances probability sampling methods may not be feasible. Probability statements concerning the precision of surveys should, however, not be made unless probability methods have been employed. It may be very misleading to apply probability statements to nonprobability surveys. In fact, the need for such statements might be taken as the test of whether, if feasible, the sample should be selected on a probability basis.

Exercises

6.1. Assume that a sample of families is selected by means of probability methods of selection; questionnaires are sent to them by mail, and the information is then obtained only for those families that return the questionnaire. Is this a probability method of sampling in the sense that there are known probabilities for specified families being in the sample? Can a population be defined from which it is a probability sample?

6.2. Sometimes a sample of people is selected by first estimating the proportions of the population that are in certain classes, e.g., 52 per cent female, 48 per cent male; then distributing the sample in those same proportions; and finally asking the interviewers to find specified numbers of people in each of these categories, these specified numbers being so selected that the total sample is selected in the correct proportions. What are the problems with this kind of sampling? Under what conditions would it be good or bad? Is it a probability method of sampling?

CHAPTER 2

Fundamental Theory—Operations, Events, and Probability

1. Introduction. In this chapter, we shall give the minimum introduction to the theory of probability that permits us to develop the theorems that we need for the theory of expected values. Of these, Theorems 3 and 5 of Sec. 6 (p. 28) on the probabilities of "product" and "sum" events, respectively, will be found exceedingly useful.

The notions of probability and the operations of selecting an element from a population, discussed in Sec. 3 and 4, must be clearly understood and related to sampling. If not, much of the remainder of this chapter will appear to be a formal treatment unrelated to reality.

As a preliminary we introduce summation notation in Sec. 2. This notation is used throughout the volume.

2. Summation notation—why we study summation notation. A knowledge of summation notation is very useful in statistics in the following ways:

(*a*) It provides a convenient shorthand for expressions that would otherwise be very cumbersome.

(*b*) By proving theorems concerning the notation itself we obtain results that would otherwise need to be proved in many special cases.

The integers. We will refer to the positive integers 1, 2, · · ·, the negative integers -1, -2, · · ·, and zero as integers.

Summation notation. The symbol Σ is the Greek capital letter "sigma" and is the notation used to indicate summation. The expression $\sum_{i=1}^{M} f(i)$ stands for $f(1) + f(2) + \cdots + f(M)$, and the expression

$$\sum_{i=1}^{M} \sum_{j=1}^{N_i} f(i, j) = f(1, 1) + \cdots + f(1, N_1) + f(2, 1) + \cdots + f(2, N_2) + \cdots + f(M, 1) + \cdots + f(M, N_M)$$

whatever may be the functions $f(i)$ and $f(i, j)$.

Whenever the summation is from 1 to some indicated value, the "= 1" is omitted below the summation sign. Thus $\sum_{i}^{N}$ has the same meaning as $\sum_{i=1}^{N}$. If the lower limit of summation is different from 1, it will be indicated.

Illustration 2.1. The heights in inches of 5 people are measured and turn out to be 67, 72, 63, 68, 70. They are denoted by x_1, x_2, x_3, x_4, and x_5. The mean or average height is calculated by

$$\frac{67 + 72 + 63 + 68 + 70}{5} = \frac{340}{5} = 68$$

which may be more concisely written

$$\bar{x} = \frac{\sum_{i}^{5} x_i}{5} = 68$$

In general, the arithmetic mean $\bar{x}$ is defined by the equation

$$\bar{x} = \frac{1}{N} \sum_{i}^{N} x_i$$

if there are N observations. It is important to realize that $\bar{x}$ is not a function of i. It would have the same value if $\bar{x}$ were written:

$$\bar{x} = \frac{1}{N} \sum_{g}^{N} x_g$$

Exercises

2.1. Show that

$$\sum_{i}^{N} p_i x_i = \sum_{j}^{N} p_j x_j = \sum_{k}^{N} p_k x_k$$

2.2. Evaluate $\sum_{i}^{N} (x_i - \bar{x})$, where $\bar{x} = \frac{1}{N} \sum_{i}^{N} x_i$.

2.3. If $x_i = a$, $i = 1, 2, \cdots, N$, evaluate $\sum_{i}^{N} x_i$, i.e., $\sum_{i}^{N} a$.

2.4. Write out $\sum_{i}^{N} i^2 x_i$.

2.5. If $f(i) = a + bx_i + cy_i$, evaluate $\sum_{i}^{5} f(i)$ in terms of $\bar{x}$ and $\bar{y}$.

2.6. If $z_i = xp_i + yq_i$, where $\sum_{i}^{N} p_i = \sum_{i}^{N} q_i = 1$, evaluate $\sum_{i}^{N} z_i$.

2.7. If $f(i) = (x_i - \bar{x})^2$, where $i = 1, 2, 3$, and $x_1 = 0$, $x_2 = 4$, $x_3 = 2$, then evaluate $\sum_{i}^{3} f(i)$.

2.8. Show that

$$\sum_{i}^{N-1} x_i = \sum_{i}^{N} x_i - x_N$$

Illustration 2.2. Summation with respect to two subscripts. Let us evaluate the product $\sum_{i}^{M} x_i \sum_{j}^{N} y_j$. This is, by definition, equal to

$$(x_1 + \cdots + x_M)(y_1 + \cdots + y_N)$$

or

$$x_1(y_1 + \cdots + y_N) + x_2(y_1 + \cdots + y_N) + \cdots + x_M(y_1 + \cdots + y_N)$$

or

$$\begin{aligned} &x_1y_1 + x_1y_2 + \cdots + x_1y_N \\ &+ x_2y_1 + x_2y_2 + \cdots + x_2y_N \\ &+ \cdots \\ &+ x_My_1 + x_My_2 + \cdots + x_My_N \end{aligned}$$

which may be written, according to the definition of summation notation, as

$$\sum_{i}^{M} \sum_{j}^{N} x_i y_j$$

Hence we have proved that, whatever may be the terms x_i and y_j, we have

$$\sum_{i}^{M} \sum_{j}^{N} x_i y_j = \sum_{i}^{M} x_i \sum_{j}^{N} y_j \tag{2.1}$$

For example, if $x_i = \log i$ and $y_j = j^2$, we have

$$\sum_{i}^{M} \sum_{j}^{N} j^2 \log i = \sum_{i}^{M} \log i \sum_{j}^{N} j^2$$

If $M = N$ and $y_j = x_j$, then Eq. 2.1 becomes

$$\sum_{i}^{M} \sum_{j}^{M} x_i x_j = \left(\sum_{i}^{M} x_i\right)^2 \tag{2.2}$$

and it is easy to verify from Eq. 2.2 that

$$\sum_{i}^{M} x_i^2 + \sum_{i \neq j}^{M} x_i x_j = \left(\sum_{i}^{M} x_i\right)^2 \tag{2.3}$$

where $\sum_{i \neq j}^{M}$ means that the summation extends over all possible values of i and j, excluding those for which $i = j$, and

$$\sum_{i}^{M} x_i^2 + 2 \sum_{i<j}^{M} x_i x_j = \left(\sum_{i}^{M} x_i\right)^2 \tag{2.4}$$

where $\sum_{i<j}^{M}$ means that the summation extends over all possible values of i and j for which $i < j$, so that

$$\sum_{i\neq j}^{M} x_i x_j = 2 \sum_{i<j}^{M} x_i x_j$$

For example, if $M = 3$, we have

$$\sum_{i\neq j}^{3} x_i x_j = x_1x_2 + x_1x_3 + x_2x_1 + x_2x_3 + x_3x_1 + x_3x_2$$

and

$$\sum_{i<j}^{3} x_i x_j = x_1x_2 + x_1x_3 + x_2x_3$$

Exercises

2.9. Prove Eq. 2.2 and 2.3 of this section.

2.10. Prove that

$$\sum_{i}^{M}\sum_{j}^{N} x_{ij} = \sum_{j}^{N}\sum_{i}^{M} x_{ij} = \sum_{i}^{M} y_i = \sum_{j}^{N} z_j$$

where $y_i = \sum_{j}^{N} x_{ij}$ and $z_j = \sum_{i}^{M} x_{ij}$, by writing out the summations and matching terms. (This is called inverting the order of summation.)

2.11. Prove that

$$\sum_{i}^{M}\sum_{j}^{i} x_{ij} = \sum_{j}^{M}\sum_{i\geq j}^{M} x_{ij}$$

by writing out both summations and matching terms.

2.12. Show that in general

$$\sum_{i}^{M} x_i y_i \neq \sum_{i}^{M} x_i \sum_{i}^{M} y_i$$

although equality may occur for specified values of the x's and y's.

2.13. Show that

$$\sum_{i}^{M}(x_i + y_i) = \sum_{i}^{M} x_i + \sum_{i}^{M} y_i$$

2.14. Show that in general

$$\frac{\sum_{i}^{2} x_i}{\sum_{i}^{2} y_i} \neq \frac{1}{2}\sum_{i}^{2}\frac{x_i}{y_i}$$

although equality may occur for specified values of the x's and y's. (This result is also true when 2 is replaced by M.)

2.15. Show that

$$\sum_{i}^{M}(ax_i + by_i)^2 = a^2\sum_{i}^{M} x_i^2 + 2ab\sum_{i}^{M} x_i y_i + b^2\sum_{i}^{M} y_i^2$$

2.16. Show that

$$\sum_i^M (a_0 + a_1 x_{1i} + a_2 x_{2i} + \cdots + a_K x_{Ki})$$
$$= Ma_0 + a_1 \sum_i^M x_{1i} + a_2 \sum_i^M x_{2i} + \cdots + a_K \sum_i^M x_{Ki}$$
$$= Ma_0 + a_1 x_{1\cdot} + \cdots + a_K x_{K\cdot}$$

where

$$x_{1\cdot} = \sum_i^M x_{1i}, \cdots, x_{K\cdot} = \sum_i^M x_{Ki}$$

3. The notion of probability. Experience suggests that many an operation is such that, *when the operation is carried out under suitably controlled conditions, it is impossible to predict exactly which of the several possible results of the operation will occur on a particular performance*, even if one has complete information about the outcomes of preceding performances. Nevertheless, experience also indicates that *we may expect a high degree of stability in the proportion of times a particular result will occur in a sufficiently long series of performances of the operation.* Illustrations may be found in such operations as tossing a coin or in the production of goods after the process of production is in a state of control.

Our ability to use data obtained from a sample in order to make inferences about a larger universe from which the sample has been drawn depends upon our ability to select the sample by means of an operation having the above-mentioned properties. Experience has indicated that such operations may be based on the tables of random numbers, and it is for that reason that these tables are important in sampling practice.

In order to develop the properties of such an operation quantitatively to the extent necessary for their application to sampling work, it is convenient to associate numerical values with the possible outcomes of the operation, in the following manner: If the operation has K possible outcomes $A_1, \cdots, A_K$ and n_i/n is the proportion of times that the outcome A_i is observed in a series of n trials, and if we can expect the proportion n_i/n to be arbitrarily close to a number P_i independent of n provided n is sufficiently large, we shall say that the probability* of A_i is P_i. The numbers P_i are clearly non-negative and their sum is 1, since $n_1 + \cdots + n_K = n$.

We shall often refer to a possible result as an *elementary event*. Thus, if $A_1, \cdots, A_K$ are the possible results of an operation, then they also are K elementary events, one and only one of which will occur when the operation is performed. If A_i occurs when the operation is performed,

* For a more rigorous development of the theory of probability, the reader should consult W. Feller (2).

we say that the elementary event A_i has occurred. We shall denote the set of all possible results or elementary events associated with an operation by $\mathscr{A}$.

By an event, $A^\star$, we shall mean a subset of the set, $\mathscr{A}$, of elementary events. Thus, $A^\star$ may consist of, say, A_2, A_4, and A_7. We shall say that the event $A^\star$ occurs if and only if one of its constituent elementary events occurs. In the above illustration, $A^\star$ would occur if and only if A_2 or A_4 or A_7 occurred. The event $A^\star$ may be $\mathscr{A}$ itself, or a specific A_i, or any subset of $\mathscr{A}$. The *complementary event* to $A^\star$ is designated by $\bar{A}^\star$ and consists of all the elementary events of $\mathscr{A}$ that are not in $A^\star$ and none of the elementary events of $\mathscr{A}$ that are in $A^\star$.

Abstraction from experience suggests the following definition of the probability of an event, since in n performances of the operation the relative frequency of occurrence of $A^\star$ will be the sum of the relative frequencies of occurrence of the elements that constitute $A^\star$.

Definition. If $A^\star$ is an event consisting of the elementary events A_{i_1}, $\cdots$, A_{i_k}, then we define the probability of $A^\star$ to be

$$Pr(A^\star) = P_{i_1} + P_{i_2} + \cdots + P_{i_k}$$

where $i_1, \cdots, i_k$ are k of the K integers $1, \cdots, K$.

Thus, if the set, $\mathscr{A}$, consists of K possible results all of which have equal probability, i.e., $P_i = 1/K$, $i = 1, \cdots, K$, and if the set $A^\star$ consists of k of these elementary events (possible results), then, by the definition of probability

$$Pr(A^\star) = \frac{k}{K}$$

Since $A^\star$ and $\bar{A}^\star$ include all elements of $\mathscr{A}$ and since $Pr(\mathscr{A}) = 1$, it follows that

$$Pr(A^\star) + Pr(\bar{A}^\star) = 1$$

4. Probability selection methods and the equal probability selection method. The operation that is basic in sampling is the selection of one element from a population in such a way that each element has a known probability of being selected. To do this, it must be possible to assign the appropriate probabilities to all possible results of the operation. If the elements of a population are $A_1, A_2, \cdots, A_K$, then the selection of one element from that population has one of the possible results: A_1 is selected, or A_2 is selected, or, $\cdots$, or A_K is selected. If each of the possible results of the selection has a known probability, we say we have a probability selection method. More specifically, by a *probability selection method* we shall mean an operation applied to the elements of a population such that when the operation is performed one and only one element of the population is selected; the probability that the element A_i is selected from the population consisting of $A_1, \cdots, A_K$ is P_i, $i = 1, \cdots, K$, $P_i \geq 0$, $P_1 + \cdots + P_K = 1$, where the P_i are known numbers. A selection method is called an *equal probability selection method* or *epsem* if $P_1 = \cdots = P_K = 1/K$, i.e., if all elements are equally likely to be selected. The sample itself may be selected by first selecting one element, then another, and so on, or by defining an auxiliary population consisting of all the possible samples and selecting one of the elements of this population to be the sample. A widely accepted practice of assigning probabilities to the results of a selection method is the use of a table of random numbers. (See Sec. 8 for detailed description of the properties of tables of random numbers and illustrations of their use. Chapter 4 of Vol. I gives some further illustrations.)

Illustration 4.1. Assume that the elements of the population $\mathscr{A}$ are A_1, A_2, A_3, and that we wish to select two elements from the population $\mathscr{A}$ with equal probability. We can select the first element with equal probability, and then select the second element with equal probability from the remaining two. In this selection method, the possible results of the first selection are

$$A_1, A_2, A_3$$

The possible results of the second selection are

A_2 or A_3—if A_1 is selected on the first selection,

A_1 or A_3—if A_2 is selected on the first selection, and

A_1 or A_2—if A_3 is selected on the first selection.

Hence, the possible results of first selecting an element by an epsem, and then making the second selection by an epsem from among the remaining two elements in the population, are

$$B_1 = A_1, A_2 \qquad B_4 = A_2, A_3$$
$$B_2 = A_1, A_3 \qquad B_5 = A_3, A_1$$
$$B_3 = A_2, A_1 \qquad B_6 = A_3, A_2$$

Instead of selecting the first element from the original population and then selecting the second element from the remaining elements of the population, we can also select two elements by defining another population, $\mathscr{B}$, whose elements are the pairs listed above. We can then select one of the elements B_1, B_2, B_3, B_4, B_5, or B_6. If, say, B_3 is the element selected from $\mathscr{B}$, then the sample consists of A_2 and A_1. Thus, the elements of $\mathscr{B}$ are the possible samples of elements of $\mathscr{A}$. Note that, since each pair is selected with equal probability, the probability of selecting one of the elements of $\mathscr{B}$ is $\frac{1}{6}$.

The selection method illustrated here is a special case of simple random sampling without replacement, which is defined later in this section.

Illustration 4.2. Two coins are tossed. The elements of $\mathscr{A}$ are A_1, A_2, A_3, and A_4, where A_1 is HH, A_2 is HT, A_3 is TH, and A_4 is TT, and HH stands for two heads; HT stands for the first coin a head, the second coin a tail; TH stands for the first coin a tail and the second coin a head; and finally TT stands for two tails. In this illustration, the operation consists of tossing two coins; the results of the operation are given by A_1, A_2, A_3, and A_4 above. Finally, the probability of each of the results for "true" coins is $\frac{1}{4}$.

Illustration 4.3. Two coins are tossed. The results of this operation, A_1, A_2, A_3, and A_4, are as defined in Illustration 4.2. Also assume that the coins are "true" so that the probability that one of the results A_1, A_2, A_3, or A_4 occurs is equal to $\frac{1}{4}$. Let us now find the probability of obtaining exactly one head. Thus, $A^\star$ in this illustration consists of HT and TH. By the definition of the probability of an event

$$Pr(A^\star) = P_{\text{HT}} + P_{\text{TH}} = \tfrac{1}{4} + \tfrac{1}{4} = \tfrac{1}{2}$$

Similarly, for finding the probability of at least one head, the event $A^\star$ consists of HT, TH, and HH. Hence $Pr(A^\star) = \frac{3}{4}$.

Illustration 4.4. Three blocks B_1, B_2, and B_3 contain 3, 8, and 1 houses. To select one block by the method called selection with probability proportionate to size we proceed as follows: select one of the integers $1, 2, \cdots, 12$ by an epsem; if the selected integer is 1, 2, or 3, then B_1 is selected; if the selected integer is $4, 5, \cdots, 11$, then B_2 is selected; and if the selected integer is 12, then B_3 is selected. What are the probabilities of selection for each block?

Since the selection of integers is by an epsem, it follows that $Pr(i) = \frac{1}{12}$, $i = 1, \cdots, 12$. Hence, by the definition of probability,

$$Pr(\text{selecting } B_1) = \tfrac{3}{12}$$

$$Pr(\text{selecting } B_2) = \tfrac{8}{12}$$

$$Pr(\text{selecting } B_3) = \tfrac{1}{12}$$

For further theory and application of probability proportionate to size see Ch. 8 (Vol. I and II) and Ch. 9 (Vol. I and II).

a. Definition of sampling without replacement. If a population consists of N elements, and if a sample of n elements is obtained by first selecting one of the N elements, and, without replacing it, selecting one of the remaining $N - 1$ elements, and, without replacing the two selected elements, selecting one of the remaining $N - 2$ elements, and so on, so that at the nth selection, there are $N - n + 1$ elements, then we say that the sample has been selected without replacement. Since there are N possible results of the first selection, $N - 1$ possible results of the second selection, $\cdots$, and $N - n + 1$ possible results of the nth selection, it follows that there are $N(N - 1) \cdots (N - n + 1)$ possible results of the n selections. (This evaluation of the number of possible results follows from Theorem A.2, p. 37, of the Appendix to this chapter.)

As a special case, if we let $N = n$, we have $n! = n(n - 1)(n - 2) \cdots 1$ as the number of possible orders in which a specified set of n elements may be selected without replacement. Each possible selection is then simply an arrangement of the n elements in the order of selection.

b. Definition of simple random sampling without replacement. If the method of selecting a sample of n elements from N elements is such that each of the possible C_n^N n-combinations of elements is equally likely to be selected, then the sampling plan is called a simple random sampling plan without replacement or, simply, simple random sampling. It will be assumed that the term *simple random sampling* applies to sampling without replacement unless otherwise qualified. The symbol C_n^N is the number of combinations of N things taken n at a time, and is referred to as the number of n-combinations. (Combinations and permutations are discussed in the Appendix to this chapter, which should now be read by the student who is not already acquainted with the subject.)

Thus, a simple random sampling plan is such that each of the C_n^N combinations has probability $1/C_n^N$ of being the sample actually selected. Also, if $A^\star$ is an event occurring if any of a specified $N_{A^\star}$ combinations (where $N_{A^\star}$ is the number of elements of $A^\star$) is selected, then by the definition of probability (Sec. 3 of this chapter)

$$Pr(A^\star) = \frac{N_{A^\star}}{C_n^N}$$

For example, if the elements of the population are $A_1, A_2, \cdots, A_N$, and if $A^\star$ is the event "the sample of size n contains A_i," then there are C_{n-1}^{N-1} of the C_n^N possible samples of n elements containing the element A_i, $N_{A^\star} = C_{n-1}^{N-1}$, and

$$Pr(A^\star) = \frac{C_{n-1}^{N-1}}{C_n^N} = \frac{n}{N}$$

If $A^\star$ is the event "the sample of size n contains $A_1, A_2, \cdots, A_m$ (or any other *specified* m elements)," then

$$Pr(A^\star) = \frac{C_{n-m}^{N-m}}{C_n^N} = \frac{n(n - 1) \cdots (n - m + 1)}{N(N - 1) \cdots (N - m + 1)}$$

Theorem 1. *In sampling without replacement, if each of the n-permutations of N elements has equal probability of being selected to be the sample, then each n-combination of N elements has probability* $1/C_n^N$ *of being selected.*

Proof. There are $N!/(N-n)!$ possible n-permutations of N elements so that if they have equal probability then each has probability $(N-n)!/N!$. Each of the n-permutations is an element of $\mathscr{A}$. If $A^\star$ is the event occurring if a specified n-combination is selected, then there are $n!$ elements of $\mathscr{A}$ in $A^\star$. Hence, by the definition of the probability of an event (Sec. 3),

$$Pr(A^\star) = n!\,\frac{(N-n)!}{N!} = \frac{1}{C_n^N}$$

Thus, one way to select a simple random sample of n elements is to give each n-permutation an equal chance of being selected.

The probability is $1/N$ that, in a sample selected by simple random sampling, the element A_i is the jth element selected, since there are $N!/(N-n)!$ n-permutations and the number of possible selections of j elements such that A_i is not selected before the jth selection and is selected at the jth selection is, by Theorem A.2 (p. 37) of the Appendix, $(N-1)\cdots(N-n+1)$, no matter what the value of j. Then

$$Pr(A_i \text{ is the } j\text{th selected element}) = \frac{(N-1)\cdots(N-n+1)}{N(N-1)\cdots(N-n+1)} = \frac{1}{N}$$

Note that this is equivalent to saying that if n selections are made without replacement, then the probability of A_i being selected at any one of the n selections is equal to $1/N$. (This probability is also evaluated using conditional probability in Sec. 7*a*.)

Similarly,

$$Pr(A_1, \cdots, A_M \text{ are the first } M \text{ selected elements}) = \frac{(N-M)(N-M-1)\cdots(N-n+1)}{N(N-1)\cdots(N-M+1)(N-M)(N-M-1)\cdots(N-n+1)} = \frac{1}{N(N-1)\cdots(N-M+1)}$$

(This probability is also evaluated using conditional probability in Sec. 7*a*.)

c. Definition of sampling with replacement. Assume that the population consists of N elements. If a sample of n elements is obtained by first selecting one of the N elements, replacing it, then making a second selection and replacing the element before making a third selection, etc., until n

selections are made, then we say that the sample is selected with replacement. Since there are N possible results in each of the selections, the number of possible results of two selections is N^2, of three selections, N^3, and of n selections, N^n. Note that there is no restriction to the number of times a particular element may be included in the sample.

d. Definition of simple random sampling with replacement. If the sampling is done with replacement, and each element has probability $1/N$ of being selected at each selection, then we call the system of selection simple random sampling with replacement. Now let us find $Pr(A^\star)$, where $A^\star$ is the event "the sample of size n contains the element A_i at least once"; we have

$$N^n = \text{number of samples}$$

$$(N-1)^n = \text{number of samples that do not contain } A_i$$

$$N^n - (N-1)^n = \text{number of samples that contain } A_i \text{ at least once}$$

$$\frac{N^n - (N-1)^n}{N^n} = Pr(A^\star)$$

Illustration 4.5. Let a population consist of L sets of elements such that each element of the population is in one and only one of these sets. Let the ith of these L sets consist of N_i elements, $i = 1, \cdots, L$, where $N = N_1 + \cdots + N_L$. If a simple random sample of n elements is selected, let us find $Pr(A^\star)$, where $A^\star$ is the event "the n elements are so distributed that n_1 will fall in class 1, n_2 in class 2, $\cdots$, n_L in class L."

The number of ways in which n_i elements can be selected from N_i is $C_{n_i}^{N_i}$. Hence the number of samples in which there will be exactly n_1 in class 1, n_2 in class 2, $\cdots$, n_L in class L will be $C_{n_1}^{N_1}C_{n_2}^{N_2}\cdots C_{n_L}^{N_L}$. Since there are C_n^N possible samples of n elements, it follows that

$$Pr(A^\star) = \frac{C_{n_1}^{N_1}C_{n_2}^{N_2}\cdots C_{n_L}^{N_L}}{C_n^N}$$

Thus, if a simple random sample of 10 elements is selected from a population consisting of 20 elements of one kind and 80 elements of another, then the probability that the sample contains 4 elements of the first kind and 6 of the second is

$$\frac{C_4^{20}C_6^{80}}{C_{10}^{100}}$$

e. Definition of systematic sampling. Let us suppose that a population consists of the elements $A_1, A_2, \cdots, A_N$ arranged in some fixed order and that the possible samples from the population are defined to be the subsets $A_1^\star, A_2^\star, \cdots, A_K^\star$ of the population, where the elements of $A_i^\star$

are A_i, A_{K+i}, A_{2K+i}, $\cdots$, $A_{(n-1)K+i}$, $i = 1, \cdots, K$. Then, if one of the possible samples $A_1^\star, \cdots, A_K^\star$ is selected, either by an epsem or with probability proportionate to the number of elements it contains or by any other means, we say that a systematic sampling plan is being used.

We shall now suppose for simplicity that $N = Kn$. Then, since each A_j occurs in one and only one of $A_1^\star, \cdots, A_K^\star$, it follows that

$$Pr(A_j \text{ is in the sample}) = Pr(A_i^\star \text{ containing } A_j \text{ is the sample})$$
$$= \frac{1}{K} = \frac{n}{N}$$

just as in simple random sampling.

Also, note that, whereas there are C_n^N possible simple random samples of n elements from a population of N elements, there are only $K = N/n$ possible systematic samples of n elements from a population of N elements.

Exercises

4.1. It is desired to select a sample of *law firms*. For selecting this sample an up-to-date register of *lawyers* is available. There are N lawyers listed in this register. To obtain a law firm, it is planned that 1 lawyer be selected by an epsem, and the firm to which this lawyer belongs will be in the sample.

Does this procedure select law firms by an epsem? What is the probability of selecting a specified firm containing 1 lawyer, 3 lawyers, k lawyers?

4.2. A city contains 1000 blocks of which 10 are vacant. A block is selected by an epsem. What is the probability of selecting a vacant block?

4.3. In a town of 1000 families, 100 consist of 1 person, 300 of 2 persons, 500 of 3 persons, and 100 of 4 persons. One family is selected by an epsem. What is the probability that it consists of 2 persons? Two families are selected so that all pairs of families are equally probable. What is the probability that both consist of 3 persons?

Ten families are selected by simple random sampling. What is the probability that 3 of them consist of 1 person, 4 of 2 persons, and 3 of 4 persons?

4.4. If simple random sampling with replacement is used, then what is the probability that neither A_1 nor A_2 is obtained in a sample of size n from $A_1, \cdots, A_N$?

5. Product events. Independence. Conditional probability. Let $\mathscr{A}$ be the set of possible results $A_1, \cdots, A_K$ of an operation and let these possible results have probabilities $P_1, P_2, \cdots, P_K$, $P_i \geq 0$, $\sum_i^K P_i = 1$. Let $A^\star$ and $B^\star$ be two events; i.e., $A^\star$ consists of certain of the elements of $\mathscr{A}$, and $B^\star$ consists of certain of the elements of $\mathscr{A}$.

Let $A^\star B^\star$ be the product event, i.e., the event that occurs if both $A^\star$ and $B^\star$ occur when the operation is performed. In other words, the event $A^\star B^\star$ consists of all elementary events common to $A^\star$ and $B^\star$ and occurs if and only if one of these common elementary events occurs. Consequently, if $A_{i_1}, \cdots, A_{i_k}$ (where $k \leq K$) are the only elementary events common to $A^\star$ and $B^\star$, we have, by the definition of probability (Sec. 3),

$$Pr(A^\star B^\star) = P_{i_1} + \cdots + P_{i_k} \tag{5.1}$$

Illustration 5.1. Population $\mathscr{A}$ consists of 4 elements, A_1, A_2, A_3, A_4. A simple random sample of 2 elements is selected from this population. The possible samples of 2 elements are

$$A_1, A_2;\ A_1, A_3;\ A_1, A_4;\ A_2, A_3;\ A_2, A_4;\ A_3, A_4$$

Let $A^\star$ consist of the possible samples containing A_1, and let $B^\star$ consist of the possible samples containing A_2. Then, $Pr(A^\star) = Pr(B^\star) = \frac{1}{2}$, and $Pr(A^\star B^\star)$ is the probability of possible samples containing both A_1 and A_2 and is equal to $\frac{1}{6}$.

Illustration 5.2. Two coins are tossed. The event $A^\star$ is defined to be "at least 1 head occurs." The event $B^\star$ is defined to be "at least 1 tail occurs." Then, $\mathscr{A}$ consists of the 4 elements HH, HT, TH, and TT; $A^\star$ consists of HH, HT, and TH; and $B^\star$ consists of HT, TH, and TT. Hence, $Pr(A^\star) = Pr(B^\star) = \frac{3}{4}$. The product event $A^\star B^\star$ then consists of HT and TH, so that

$$Pr(A^\star B^\star) = P_{\text{HT}} + P_{\text{TH}} = \tfrac{1}{2}$$

If $A_1^\star, \cdots, A_M^\star$ are M events, then the product event $A_1^\star \cdots A_M^\star$ consists of the elements common to *all M* events.

Definition of independent events. Two events $A^\star$ and $B^\star$ are called *independent* if and only if

$$Pr(A^\star B^\star) = Pr(A^\star)\, Pr(B^\star)$$

Illustration 5.3. Assume that the population consists of the 4 elements A_1, A_2, A_3, A_4, and that a simple random sample of 2 is selected from this population with replacement. The following are the possible samples of 2 to be selected from this population where the element on the left of each pair listed below represents the element drawn on the first selection and the element on the right represents the element drawn on the second selection:

A_1A_1	A_2A_1	A_3A_1	A_4A_1
A_1A_2	A_2A_2	A_3A_2	A_4A_2
A_1A_3	A_2A_3	A_3A_3	A_4A_3
A_1A_4	A_2A_4	A_3A_4	A_4A_4

Let $A^\star$ be the event that A_1 be drawn on the first selection of a sample of 2. Let $B^\star$ be the event that A_3 be drawn on the second selection of a sample of 2. $Pr(A^\star) = Pr(B^\star) = \frac{1}{4}$, since, in 4 of the 16 pairs listed above, A_1 is the first selection, and similarly in 4 of the 16 pairs A_3 is the second selection. Now $A^\star B^\star$ is the event that A_1 be drawn on the first selection and A_3 on the second selection. Hence, $Pr(A^\star B^\star) = \frac{1}{16}$. By the definition of independence $A^\star$ and $B^\star$ are independent events, since $Pr(A^\star B^\star) = Pr(A^\star)\, Pr(B^\star) = \frac{1}{16}$.

Illustration 5.4. Let a dime and a quarter be tossed and let the result of tossing the dime be listed first. Let the tossing be such that all 4 possible results have equal probability $\frac{1}{4}$. Define $A^\star$ to be the event "heads occurs when the dime is tossed" and $B^\star$ the event "heads occurs when the quarter is tossed." Then

$$Pr(A^\star) = P_{\mathrm{HH}} + P_{\mathrm{HT}} = \tfrac{1}{2}$$

$$Pr(B^\star) = P_{\mathrm{HH}} + P_{\mathrm{TH}} = \tfrac{1}{2}$$

and

$$Pr(A^\star B^\star) = P_{\mathrm{HH}} = \tfrac{1}{4}$$

Hence, $A^\star$ and $B^\star$ are independent, since $Pr(A^\star B^\star) = Pr(A^\star)\, Pr(B^\star)$.

If the events $A^\star$ and $B^\star$ were defined as in Illustration 5.2, then we would have

$$Pr(A^\star) = Pr(B^\star) = \tfrac{3}{4}$$

and

$$Pr(A^\star B^\star) = \tfrac{1}{2}$$

so that $A^\star$ and $B^\star$ would not be independent.

Exercise 5.1. Show that $A^\star$ and $B^\star$ in Illustration 5.1 are not independent.

Definition of conditional probability. Let $A^\star$ and $B^\star$ be two events. If $Pr(B^\star) > 0$, then the conditional probability of the occurrence $A^\star$ subject to the occurrence of $B^\star$ is defined to be

$$Pr(A^\star | B^\star) = \frac{Pr(A^\star B^\star)}{Pr(B^\star)}$$

Conditional operation. We shall now consider the following operation, which we call a *conditional operation.* Let $B^\star$ be an event, $0 < Pr(B^\star) \leq 1$, and suppose that the only change in the original operation is that all results possible originally which are not contained in $B^\star$ become impossible. Physically, this could be accomplished by skipping any performance in which $B^\star$ does not occur. Thus, the possible results of the conditional operation are the elements of $B^\star$. We shall denote their probabilities by $Pr(A_i|B^\star)$. Then by our previous discussion of the notion of an operation we have $Pr(A_i|B^\star) \geq 0$ and

$$\sum_{A_i \varepsilon B^\star} Pr(A_i|B^\star) = 1$$

where by $\sum_{A_i \varepsilon B^\star}$ we mean the summation over all elements A_i that are in $B^\star$. Note that $A_i \varepsilon B^\star$ is shorthand for "A_i is an element of $B^\star$." Also from the definition of conditional probability it follows that

$$\begin{aligned} Pr(A_i|B^\star) &= \frac{P_i}{Pr(B^\star)} && \text{if } A_i \text{ is an element of } B^\star \\ &= 0 && \text{if } A_i \text{ is not an element of } B^\star \end{aligned}$$

Illustration 5.5. Assume that the population consists of 5 elements, A_1, A_2, A_3, A_4, and A_5. Assume that a simple random sample of 2 elements is selected without replacement from this population. We shall denote the result of the first selection by a_1 and of the second by a_2. Then the possible samples are given in the following table, where a dash

$a_1 \diagdown a_2$	A_1	A_2	A_3	A_4	A_5
A_1	—	√	√	√	√
A_2	√	—	√	√	√
A_3	√	√	—	√	√
A_4	√	√	√	—	√
A_5	√	√	√	√	—

(—) indicates that the sample is impossible and a check (√) indicates that the sample consists of the elements shown in row and column headings. It is clear from the above table that the possible selections for a_2 depend on which element is selected for a_1. In fact

$$\begin{aligned} Pr(a_2 = A_i|a_1 = A_i) &= 0 \\ Pr(a_2 = A_j|a_1 = A_i) &= \tfrac{1}{4} \qquad \text{if } i \neq j \end{aligned}$$

where $Pr(a_2 = A_j|a_1 = A_i)$ is read "the conditional probability that $a_2 = A_j$ given that (or subject to the condition that) $a_1 = A_i$." Note that a_1 and a_2 are random events (discussed more fully in Chapter 3).

Illustration 5.6. Suppose that a population consists of 2 blocks B_1 and B_2 on the first of which are located 3 houses A_{11}, A_{12}, A_{13}, while on the second are located 2 houses A_{21} and A_{22}. We first select 1 block by an epsem, and then from the selected block select 1 house by an epsem.

Denote the result of selecting a block by b and the result of selecting a house by a. Then

$$\begin{aligned} Pr(a = A_{i2}|b = B_j) &= 0 \qquad \text{if } i \neq j \\ &= \tfrac{1}{3} \qquad \text{if } i = j = 1 \\ &= \tfrac{1}{2} \qquad \text{if } i = j = 2 \end{aligned}$$

where we read $Pr(a = A_{i2}|b = B_j)$ as the "conditional probability that the selected house is the second on block i subject to the condition that the selected block is the jth block." Here a and b are random events (see Chapter 3 for fuller discussion).

Illustration 5.7. Suppose that, of the N persons in a population, $N_{A\star} > 0$ have incomes over \$3000, and $N_{A\star B\star}$ have incomes over \$3000 and expenditures under \$2500. An epsem is used to select one of the N persons. Then from the definition of the probability of an event we see that

$$Pr(\text{the selected person has over \$3000 income}) = \frac{N_{A\star}}{N}$$

$$Pr(\text{the selected person has over \$3000 income and expenditures under \$2500}) = \frac{N_{A\star B\star}}{N}$$

$$Pr(\text{the selected person has expenditures under \$2500 if the selected person has over \$3000 income}) = \frac{N_{A\star B\star}}{N_{A\star}}$$

Exercises

5.2. A population consists of M blocks, the ith of which contains $N_i > 0$ houses. A house is selected by first selecting a block by an epsem, and then selecting a house from the selected block by an epsem. What is the probability of selecting a specific household on block j? What would be the probability of selecting that house if one house were selected from all the houses by an epsem?

5.3. Suppose now that, instead of selecting a block by an epsem, we select a block so that the probability of selecting the ith block is N_i/N, $i = 1, \cdots, M$, $N = N_1 + \cdots + N_M$. From the selected block a house is then selected by an epsem. Answer the two questions asked in connection with Ex. 5.2.

5.4. Suppose that a simple random sample of 2 elements is selected with replacement from the population given in Illustration 5.5. Show that

$$Pr(a_2 = A_i|a_1 = A_i) = \tfrac{1}{5}$$
$$Pr(a_2 = A_j|a_1 = A_i) = \tfrac{1}{5} \qquad \text{for } i \neq j$$

6. Some theorems on probabilities. Let $A_1^\star, \cdots, A_K^\star$ be events associated with a particular operation. Then by $A_1^\star + \cdots + A_K^\star$, the sum event, we mean the event that occurs if at least one of the $A_j^\star$ occurs. By $A_1^\star A_2^\star \cdots A_K^\star$, the product event, we mean, as earlier defined in Sec. 5, the event that occurs if $A_1^\star, A_2^\star, \cdots$, and $A_K^\star$ all occur.

Illustration 6.1. Let us suppose that $N = 10$ and that $A_1, \cdots, A_{10}$ are the 10 spots in Fig. 1. As shown by the figure, A_1, A_3, and A_4 are

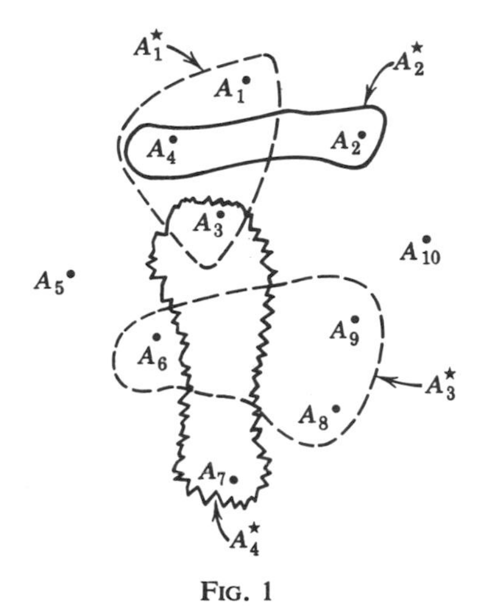

FIG. 1

the elements of $A_1^\star$; i.e., if A_1, A_3, or A_4 is selected, then $A_1^\star$ is selected. Similarly, A_2 and A_4 are the elements of $A_2^\star$; A_6, A_8, and A_9 are the elements of $A_3^\star$; and A_3 and A_7 are the elements of $A_4^\star$. Then the elements of $A_1^\star + A_2^\star + A_3^\star + A_4^\star$ are $A_1, \cdots, A_4, A_6, \cdots, A_9$, since if any element but A_5 and A_{10} is selected at least one of the events $A_1^\star$, $A_2^\star$, $A_3^\star$, and $A_4^\star$ will occur. On the other hand, $A_1^\star A_2^\star A_3^\star A_4^\star$ can never occur, since no element is an element of all four of them. We call $A_1^\star A_2^\star A_3^\star A_4^\star$ a *null event*, i.e., an event that cannot occur. The product event $A_1^\star A_2^\star$ will occur if and only if A_4 occurs, while the product event $A_1^\star A_2^\star A_4^\star$ is a null event. The complementary event to $A_1^\star + \cdots + A_4^\star$ is $\overline{A_1^\star + \cdots + A_4^\star}$, which consists of A_5 and A_{10} and is *not* the same as $\bar{A}_1^\star + \bar{A}_2^\star + \bar{A}_3^\star + \bar{A}_4^\star$.

Theorem 2. *If $A_1^\star, \cdots, A_K^\star$ are events such that $Pr(A_1^\star A_2^\star \cdots A_K^\star) > 0$ then $Pr(A_1^\star A_2^\star \cdots A_h^\star) > 0$ for any integer $h < K$.*

Proof. By assumption, there is at least one elementary event A_i with probability $P_i > 0$ common to all the events $A_1^\star, \cdots, A_K^\star$. Any such event A_i will necessarily be common to $A_1^\star, \cdots, A_h^\star$, so that the sum of the P_i over the elementary events common to $A_1^\star, \cdots, A_h^\star$ will be at least as large as the sum of the P_i over the elementary events common to $A^\star, \cdots, A_K^\star$.

Theorem 3. *If* $A_1^\star, \cdots, A_K^\star$ *are events, with* $Pr(A_1^\star A_2^\star \cdots A_K^\star) > 0$, *then*

$$Pr(A_1^\star A_2^\star \cdots A_K^\star) = Pr(A_1^\star)\, Pr(A_2^\star | A_1^\star) \cdots Pr(A_K^\star | A_1^\star \cdots A_{K-1}^\star).$$

Proof. If $K = 1$, then each side of the equation is $Pr(A_1^\star)$. Suppose now that the theorem is true for $K = j$. Then we shall show the theorem to be true for $K = j + 1$, which will complete the proof by induction. This means that the theorem is proved for any positive integral value of K by the following steps: The theorem is true if $K = 1$. Then, putting $j = 1$, we show that it follows for $K = 2$. Then the theorem is true if $K = 2$, and, putting $j = 2$, we show that it follows for $K = 3$, and so on.

Now, by the definition of conditional probability, treating $A_1^\star A_2^\star \cdots A_j^\star$ as one event, we have

$$Pr(A_1^\star A_2^\star \cdots A_{j+1}^\star) = Pr(A_{j+1}^\star | A_1^\star \cdots A_j^\star)\, Pr(A_1^\star \cdots A_j^\star)$$

By the hypothesis of the induction, Theorem 3 is true for $K = j$. Substituting for $Pr(A_1^\star A_2^\star \cdots A_j^\star)$, we see that Theorem 3 is also true for $K = j + 1$, which completes the proof.

Several events $A_1^\star, \cdots, A_K^\star$ are said to be *mutually exclusive* if no element is common to 2 or more of them and are said to be *exhaustive* if $A_1^\star + \cdots + A_K^\star =$ the entire population.

Thus, if $A^\star$ is an event, then $A^\star$ and $\bar{A}^\star$ (the complementary event) are both mutually exclusive and exhaustive.

Illustration 6.2. Let us define the event $A_5^\star$ to be $\overline{A_1^\star + \cdots + A_4^\star}$. Then, in Illustration 6.1, $A_1^\star, \cdots, A_5^\star$ are exhaustive but not mutually exclusive. $A_1^\star$ and $A_3^\star$ are mutually exclusive but not exhaustive. The five events $A_1^\star$, $A_2^\star \bar{A}_1^\star$, $A_3^\star$, $A_4^\star \bar{A}_1^\star$, and $A_5^\star$ are both exhaustive and mutually exclusive.

Theorem 4. *If* $A^\star$ *is a subset of* $\mathscr{A}$ *then*

$$Pr(A^\star \mathscr{A}) = Pr(A^\star)$$

Proof. Every element in $A^\star$ is also in $\mathscr{A}$, and hence the elements in $A^\star \mathscr{A}$ are exactly those in $A^\star$.

The following theorem is of considerable importance in calculating probabilities.

Theorem 5. *Let* $A_1^\star, \cdots, A_K^\star$ *be mutually exclusive events. Then, if* $A^\star = A_1^\star + \cdots + A_K^\star$, *it follows that* $Pr(A^\star) = Pr(A_1^\star) + \cdots + Pr(A_K^\star)$.

Proof. Since $A^\star = A_1^\star + \cdots + A_K^\star$, and since $A_1^\star, A_2^\star, \cdots, A_K^\star$ are mutually exclusive, the theorem follows immediately if we substitute for $Pr(A^\star)$ and $Pr(A_1^\star) + \cdots + Pr(A_K^\star)$ the sum of the probabilities of the elementary events of which they consist, using the definition of the probability of an event.

Corollary 1. Let $A_1^\star, \cdots, A_K^\star$ be mutually exclusive, and let $B^\star$ and $C^\star$ be any events. Then, if $A^\star = A_1^\star + \cdots + A_K^\star$, it follows that

$$Pr(A^\star C^\star | B^\star) = Pr(A_1^\star C^\star | B^\star) + \cdots + Pr(A_K^\star C^\star | B^\star)$$

Proof. Since $A_1^\star, \cdots, A_K^\star$ are mutually exclusive, it follows that $A_1^\star C^\star, \cdots, A_K^\star C^\star$ are mutually exclusive. Furthermore, the set $A^\star C^\star$ consists of exactly the same elements as the set $A_1^\star C^\star + \cdots + A_K^\star C^\star$; or

$$A^\star C^\star = A_1^\star C^\star + \cdots + A_K^\star C^\star$$

and similarly

$$A^\star C^\star B^\star = A_1^\star C^\star B^\star + \cdots + A_K^\star C^\star B^\star$$

where $A_1^\star C^\star B^\star, \cdots, A_K^\star C^\star B^\star$ are mutually exclusive, since $A_1^\star$, $A_2^\star, \cdots, A_K^\star$ are mutually exclusive. Then

$$Pr(A^\star C^\star | B^\star) = \frac{Pr(A^\star C^\star B^\star)}{Pr(B^\star)}$$

$$= Pr(A_1^\star C^\star | B^\star) + Pr(A_2^\star C^\star | B^\star) + \cdots + Pr(A_K^\star C^\star | B^\star)$$

An alternative proof for the special case where an epsem is being used is as follows: Suppose that $A_i^\star$ consists of N_i elements, $i = 1, \cdots, K$. Of these N_i elements suppose that N_{ic} are also in $C^\star$. Then $A_i^\star C^\star$ consists of N_{ic} elements none of which can be in $A_j^\star C^\star (i \neq j)$, since the N_{ic} elements are in $A_i^\star$, and no element of $A_i^\star$ is also in $A_j^\star$. Furthermore, $A^\star C^\star$ consists of the $N_{1c} + N_{2c} + \cdots + N_{Kc}$ elements which are both in one of the $A_i^\star$ and in $C^\star$. Hence, from the definition of the probability of an event, we have

$$Pr(A^\star C^\star) = Pr(A_1^\star C^\star) + \cdots + Pr(A_K^\star C^\star)$$

There are several other interesting corollaries to Theorem 5. Of these we give the two that follow:

Corollary 2. If $A_1^\star, \cdots, A_K^\star$ are exhaustive as well as mutually exclusive, then $Pr(A^\star C^\star | B^\star) = Pr(C^\star | B^\star)$ and we have

$$Pr(C^\star | B^\star) = Pr(A_1^\star C^\star | B^\star) + \cdots + Pr(A_K^\star C^\star | B^\star)$$

Corollary 3. If $A_1^\star, \cdots, A_K^\star$ are mutually exclusive, then

$$Pr(A^\star C^\star | B^\star) = Pr(A_1^\star | B^\star C^\star)\, Pr(C^\star | B^\star) + \cdots + Pr(A_K^\star | B^\star C^\star)\; Pr(C^\star | B^\star)$$

$$= Pr(C^\star | A_1^\star B^\star)\, Pr(A_1^\star | B^\star) + \cdots + Pr(C^\star | A_K^\star B^\star)\; Pr(A_K^\star | B^\star)$$

7. Some illustrations of the uses of theorems on independent and conditional probability. *a. Evaluation of some probabilities associated with simple random sampling.* We have previously seen that the probability of any specified possible sample is $1/C_n^N$ if simple random sampling is being used. Let us now prove that result and others by conditional probability.

(1) First, let $B_i^\star$ be the event "A_i is in the sample." Then the product event $B_1^\star B_2^\star \cdots B_n^\star$ is the event "$A_1, A_2, \cdots, A_n$ are in the sample" since the product event occurs if and only if all the component events occur. By Theorem 3

$$Pr(B_1^\star B_2^\star \cdots B_n^\star) = Pr(B_1^\star)\, Pr(B_2^\star | B_1^\star) \cdots Pr(B_n^\star | B_1^\star \cdots B_{n-1}^\star)$$

Now

$$Pr(B_1^\star) = \frac{n}{N}$$

Also $Pr(B_2^\star | B_1^\star) = (n-1)/(N-1)$, since the condition that A_1 is in the sample reduces the problem to one of selecting $n-1$ elements by simple random sampling from a population consisting of $A_2, \cdots, A_N$.

Similarly

$$Pr(B_i^\star | B_1^\star \cdots B_{i-1}^\star) = \frac{n-i+1}{N-i+1}$$

so that

$$Pr(\text{sample consists of } A_1, \cdots, A_n) = \frac{n(n-1)\cdots 1}{N(N-1)\cdots(N-n+1)} = \frac{1}{C_n^N}$$

(2) The probability that A_i is the Kth element selected for the sample is obtained similarly. Let $B_i^\star$ be the event "A_j is not obtained at the ith selection," $i = 1, \cdots, K-1$, and let $B_K^\star$ be the event "A_j is obtained at the Kth selection." Then we want to evaluate the probability of $B_1^\star B_2^\star \cdots B_K^\star$, and by Theorem 3

$$\begin{aligned} Pr(B_1^\star B_2^\star \cdots B_K^\star) &= Pr(B_1^\star)\, Pr(B_2^\star | B_1^\star) \cdots Pr(B_K^\star | B_1^\star \cdots B_{K-1}^\star) \\ &= \frac{N-1}{N}\,\frac{N-2}{N-1} \cdots \frac{N-K+1}{N-K+2}\,\frac{1}{N-K+1} \\ &= \frac{1}{N} \end{aligned}$$

b. Definition of stratified simple random sampling. Suppose that the N elements of a population are classified into L strata, the ith of which contains N_i elements, $i = 1, \cdots, L$. Suppose also that simple random samples are independently selected from each of the strata. Then we say that a stratified simple random sampling plan is used.

It is instructive to compare the numbers of possible samples of size n when simple random sampling and stratified simple random sampling are used.

There are C_n^N possible samples of size n from a population of size N if simple random sampling is used.

Let $n = n_1 + \cdots + n_L$. Then, from Theorem A.2 of the Appendix, there are $C_{n_1}^{N_1} C_{n_2}^{N_2} \cdots C_{n_L}^{N_L}$ possible stratified simple random samples consisting of n_1 elements from stratum 1, n_2 elements from stratum 2, $\cdots$, n_L elements from stratum L.

Clearly, since each possible stratified simple random sample of $n = n_1 + \cdots + n_L$ elements is also a possible simple random sample of n elements, it follows that

$$C_{n_1}^{N_1} C_{n_2}^{N_2} \cdots C_{n_L}^{N_L} < C_n^N$$

and, in fact, since every sample of size n will, for some values of $n_1, \cdots, n_L$, consist of n_1 elements from stratum 1, n_2 elements from stratum 2, $\cdots$, n_L elements from stratum L, we have

$$\Sigma C_{n_1}^{N_1} C_{n_2}^{N_2} \cdots C_{n_L}^{N_L} = C_n^N$$

where the summation is over all $n_1, \cdots, n_L$ such that $n_i \geq 0$ and $n_1 + \cdots + n_L = n$.

We have already seen that if a simple random sample of size n is selected then the probability that some specified element, say A_i, is in the sample is n/N. If that element is classified into the jth stratum, and if n_j elements are selected from that stratum in the stratified simple random sample, then the probability that A_i is in the sample becomes n_j/N_j.

Furthermore, if proportionate sampling is used, i.e., if the sampling ratio $n_j/N_j = f, j = 1, \cdots, L$, then $n_j = fN_j$ and hence

$$f = \frac{n_j}{N_j} = \frac{n_1 + \cdots + n_L}{N_1 + \cdots + N_L} = \frac{n}{N}$$

Thus, if proportionate stratified simple random sampling is used, then the probability of a specified element being in the sample is the same as when a simple random sample of the same size is selected; whereas, if the sampling is disproportionate, then, of course, this will not be so in any stratum for which $n_j/N_j \neq n/N$.

c. Cluster sampling. By a *cluster* we mean a set of elements that are treated as a single element for purposes of selecting a sample. For example, the people within a family constitute a cluster if we select families, the people or families within a block constitute a cluster if we select blocks, the people or families or farms within a county constitute a cluster if we select counties. Thus, clusters may be small or large. They are treated as units in selecting a sample.

We have already derived the probabilities needed for simple random sampling, stratified simple random sampling, and systematic sampling of clusters. Once the clusters have been selected, however, the selected clusters may themselves be treated as a population from which a sample is selected. When this is done, we are using cluster sampling with *subsampling*.

Let the clusters be denoted by $A_1, A_2, \cdots, A_M$ and let the cluster A_i consist of the elements $A_{i1}, A_{i2}, \cdots, A_{iN_i}$, $i = 1, \cdots, M$. Suppose that we select a simple random sample of m clusters. Then the probability that we select, say, $A_1, \cdots, A_m$ is $1/C_m^M$. Once we have selected $A_1, \cdots, A_m$ we may treat these clusters as a single population and select a simple random sample or systematic sample from it, or we may treat the selected clusters as substrata and select a stratified simple random sample or stratifield systematic sample from the strata. We then use Theorem 3 (p. 28) to calculate the probability of obtaining a specified sample of elements. This probability is obtained by multiplying $1/C_m^M$ by the conditional probability of obtaining a sample of elements from the selected clusters.

Suppose that m of the M clusters are selected by simple random sampling and that from the selected clusters a subsample of n_j elements is selected, again by simple random sampling. Then the probability of selecting a specified element is

$$\frac{m}{M}\frac{n_j}{N_j}$$

where N_j is the total number of elements in the jth cluster. In the special case where $n_j/N_j = \bar{n}/\bar{N}$, the probability of selecting a specified item is $m\bar{n}/M\bar{N} = n/N$, the same probability of selecting a specified element as existed for simple random sampling and for proportionate stratified sampling. The probability of obtaining a specified pair of elements differs, and it is this difference which has a bearing on the relative accuracy of the three sampling designs.

For simple random sampling the probability of obtaining a pair of elements is $\dfrac{n(n-1)}{N(N-1)}$. For stratified random sampling the probability depends on whether both elements are in the same or different strata even if proportionate sampling is used. If both elements are in the jth stratum, the probability is $\dfrac{n_j(n_j-1)}{N_j(N_j-1)}$, whereas if they are in the jth and kth strata, the probability is $\dfrac{n_j}{N_j}\dfrac{n_k}{N_k}$. For cluster sampling, if both elements are in the same cluster, the probability is $\frac{m}{M}\frac{n_j(n_j-1)}{N_j(N_j-1)}$, where N_j is the size of cluster and n_j the size of sample selected from that cluster, whereas if the elements are in the jth and kth clusters, the probability is $\frac{m(m-1)}{M(M-1)}\frac{n_j}{N_j}\frac{n_k}{N_k}$.

8. Methods of achieving probability (or measurable) sampling plans. The table of random numbers. By a probability (or a measurable) sampling plan we shall mean a sampling plan where the elements are selected with known probabilities. We have already indicated the importance of a list in selecting samples; i.e., when the list is known, each of the elements of the population is identified by a number and if that number is selected then the corresponding element of the population is in the sample. For this reason, in sampling, we often want to select a sample of numbers. If we can use a probability selection method for obtaining a sample of numbers, then our sampling plan is a probability plan for obtaining a sample from the list or population represented by it. The most important and frequently used tool in obtaining a probability sampling plan is the table of random numbers.

Tests have been performed on the tables of random sampling numbers in common use that make it reasonable to assume that the following statements are true for all practical purposes:

(1) Each number is the result of performing an epsem.

(2) The selection operations are independent.

It follows from the above two statements that each possible pair of numbers in the table may be interpreted as the result of performing an epsem to select one number from 1, · · ·, 100, where 100 is said to be selected if and only if we obtain 00.

The results of 100 performances of the operation can also be used to obtain the results of 100 tosses of a true coin by referring to an odd number as heads and to an even number or zero as tails.

Problems on the table of random numbers

(1) Select an integer from 1, · · ·, 113 by an epsem. Two methods are in common use.

Method a. Choose any 3 columns of a table of random numbers* (p. 117, Vol. I), say columns 5, 6, and 7. As we go down these columns, the numbers are 422, 044, so that 44 is the selected integer since it is the first integer found between 1 and 113 inclusive.

Method b. Divide the first number found by 113. Now $422/113 = 3\frac{83}{113}$.

* M. G. Kendall and B. Babington Smith, *Tracts for Computers*, No. XXIV, Cambridge University Press, second edition, 1946, p. 8.

Then 83 is the selected integer. One precaution must be taken in using method *b*. The basis of this method is that, since all the numbers 001, · · ·, 999, 1000 (1000 is denoted by 000) are equally likely in columns 5, 6, and 7, the remainders yielded by these numbers will occur with equal probability. Yet, we would obtain 113 if any of the 8 numbers 113, 226, · · ·, 904 occurred, whereas 1 could occur if any of the 9 numbers 1, 114, 227, · · ·, 905 occurred. Thus, the numbers 1, · · ·, 113 would not be equally probable. We can avoid this difficulty by passing by the numbers 905, · · ·, 999 until a smaller number is reached. In this way, while preserving the epsem we reduce considerably the number of numbers to be passed by.

(2) Select three integers from 1 to 113 by an epsem. Let us first note that the statement of the problem needs clarification. Shall we permit an integer to occur more than once among the three selected? Since an epsem was defined in terms of selecting one element, how shall we interpret the selection of more than one element? Let us answer the second of these questions first. If we wish to select several elements from one population we can either explicitly or implicitly define a second population whose elements are the possible selections of several elements from the first. Then we select one element from the second population by a selection method that gives to each of its elements the desired probability that it be selected. Another method is that of choosing one number at a time until the sample is obtained. Suppose, now, that we do not wish any of the integers from 1 to 113 to be selected more than once. Then after selecting 44 on the basis of columns 5, 6, 7 we continue on down and reach 9, and continuing in columns 8, 9, 10 we reach 67. Thus, these are the three selected numbers. If we had been willing to permit a number to be selected more than once, and if we had come upon 44 again as the second number as well as the first, we would have selected it. But, if we did not wish to include the same number more than once, we would pass over 44 the second time we came upon it just as though it were a number greater than 113.

Exercises

8.1. Toss 2 coins 100 times. For each toss define the possible results to be 0, 1, or 2 heads. Compute the relative frequencies with which each of these possible results occurs in the first K tosses, $K = 1, 2, \cdots, 100$. Is there any tendency of these relative frequencies to become stable? What probabilities would you associate to each of the 3 possible results?

8.2. Open a book 100 times, trying to do so in such a way that you do not know at what page you will open the book. If the *next to the last digit* (for pages 1, 3, 5, 7, 9 we assume this number to be 0) of the right-hand page number is 0, 2, 4, the result is 0, otherwise it is 1. Calculate the relative frequencies of 0 and 1 for each of the first K performances of this operation, $K = 1, \cdots, 100$. Is there any tendency of these relative frequencies to stabilize? About what values? If the relative frequencies do not seem to stabilize what conclusions would you draw?

8.3. Use a table of logarithms to 4 or more places. If the last digit is odd, say that the result 1 has occurred. Otherwise the result 0 has occurred. Calculate the relative frequencies of 0 and 1 for each of the first K performances, $K = 1, \cdots, 100$. Is there any tendency of these relative frequencies to stabilize? About what value? Repeat this exercise, using the next to the last digits of the logarithms.

8.4. Which of the following ways would you regard as most closely approximating an epsem: (*a*) by opening a book at various points and using the page numbers; (*b*) by using a table of logarithms; or (*c*) by writing the numbers on identical-appearing cards or chips, shuffling them very well, and selecting blindfolded? What are the dangers of employing such methods instead of a table of random numbers?

8.5. Suppose now that in using a city directory you number all the listed dwelling units and use an epsem to select one of these dwelling units. Is this the same as using an epsem to select one of the dwelling units in the city?

8.6. Suppose that, on the basis of an up-to-date and complete listing of the population, a sample of say 10,000 is selected by means of an epsem, and questionnaires are sent to these 10,000 by mail. Of these 10,000 questionnaires, 3000 are returned immediately, 1500 more after one follow-up letter, and 500 after an additional follow-up letter, or 5000 in all. Suppose that these 5000 are kept in a file in the order in which they are received. Several tests are made comparing the results of these 5000 questionnaires and their serial numbers to those of the entire 10,000 questionnaires. None of these tests contradict the assumption that the 5000 questionnaires were obtained by using an epsem. Should these 5000 questionnaires be treated as though obtained by means of an epsem from the same population as that from which the 10,000 elements were selected? What is the danger that the 5000 who returned their questionnaires were essentially selected by an epsem from those who felt strongly about the subject matter of the questionnaire? Can we by an essentially internal analysis of a sample ever obtain satisfactory evidence that it was selected by an epsem from a specified population? Discuss the following: Suppose that a population is expected to have somewhat more males than females and we want to estimate the proportion of males in the population by choosing 100 persons by means of an epsem and using the proportion of males among them. After the 100 pieces of data are recorded someone loses them, and, being afraid to admit it, replaces the data by tossing 100 times a coin having probability of heads equal to .51 and recording male when heads occurs, female when tails occurs.

APPENDIX

Combinations and Permutations

Combinations and permutations are introduced so that we can have convenient tools for specifying certain sets of elements and for counting the numbers of elements in those sets. The discussion is brief since these topics will be known to many of the readers of this volume.

Theorem A.1. *Suppose that one operation has K possible results $A_1, \cdots, A_K$. Suppose that if A_i occurs when the first operation is performed, then a second operation has M_i possible results $B_{i1}, \cdots, B_{iM_i}$, $i = 1, \cdots, K$. Then, the number of possible results of performing these two operations is $M_1 + M_2 + \cdots + M_K$.*

Proof. The possible results of performing these two operations may be denoted by:

$$A_1B_{11}, \quad A_1B_{12}, \cdots, A_1B_{1M_1}$$
$$A_2B_{21}, \quad A_2B_{22}, \cdots, A_2B_{2M_2}$$
$$\cdots$$
$$A_KB_{K1}, A_KB_{K2}, \cdots, A_KB_{KM_K}$$

where the ith row above lists all the M_i possible results of both operations such that A_i occurs when the first operation is performed. Counting the number of possible results listed, we see that there are $M_1 + \cdots + M_K$ possible results.

Corollary. If $M_1 = M_2 = \cdots = M_K = M$, i.e., if the *number* of possible results of performing the second operation is the same, whatever the result of the first operation, then the total number of possible results of performing both operations is KM.

The proof consists in replacing M_i by M in Theorem A.1.

Illustration A.1. (*a*) Suppose that a population consists of 3 blocks, A_1, A_2, A_3, and that A_1 contains 2 families, B_{11}, B_{12}; A_2 contains 3 families, B_{21}, B_{22}, B_{23}; and A_3 contains 1 family, B_{31}. Then there are 6 possible results of performing the 2 operations of first selecting a block and then selecting a family from the selected block. These possible results are:

$$A_1B_{11}, A_1B_{12}, A_2B_{21}, A_2B_{22}, A_2B_{23}, \text{ and } A_3B_{31}$$

(*b*) Suppose that a population consists of 10 elements and that first one element is selected and then from the 9 remaining elements a second element is selected. There are 10 possible results of the first operation, so that $K = 10$. Whatever result occurs for the first operation, there are 9 possible results of the second operation, so that $M_1 = \cdots = M_{10} = 9$. Hence, there are 90 possible results of the pair of operations.

It will simplify the notation if, in the generalization of Theorem A.1 above to N operations, we assume that whatever may be the results of the first $i - 1$ operations, the ith operation has a constant number K_i of possible results. A similar theorem holds when the number of possible results of the ith operation depends on the actual results of performing the first $i - 1$ operations ($i = 1, \cdots, N$).

Theorem A.2. *Let us suppose that the ith of N operations has K_i possible results no matter what may be the results of performing the first $i - 1$ operations ($i = 1, \cdots, N$). Then the number of possible results of performing all operations in a specified order is $K_1K_2 \cdots K_N$.*

Proof. If $N = 2$, this theorem has already been proved in the corollary to Theorem A.1. Then the results of performing the first two operations may be considered to be those of performing one (complex) operation with K_1K_2 results so that we obtain $(K_1K_2)K_3 = K_1K_2K_3$ possible results for performing the first three operations. Continuing, the results of the first $N - 1$ operations may be considered to be those of performing a complex operation having $K_1 \cdots K_{N-1}$ possible results, so that by the corollary to Theorem A.1 the number of possible results of performing all N operations is $(K_1K_2 \cdots K_{N-1})K_N = K_1K_2 \cdots K_N$.

Definition of permutation. Let a set consist of N elements $A_1, A_2, \cdots, A_N$. Then an n-permutation consists of any n of those elements arranged in a specified order. An N-permutation of N elements is called a permutation.

Thus, two n-permutations of $A_1, \cdots, A_N$ will differ either if they contain different elements or if they contain the same elements arranged in different orders. In Illustration A.1 we have proved the following theorem:

Theorem A.3. *The number of n-permutations of N elements is*

$$N(N-1)\cdots(N-n+1) = \frac{N!}{(N-n)!}$$

Definition of combination. Let a set consist of N elements $A_1, A_2, \cdots, A_N$. Then an n-combination consists of any n of these elements.

Thus, two n-combinations of $A_1, \cdots, A_N$ will differ only if they do not contain exactly the same elements.

Illustration A.2. (*a*) The number of 2-permutations of A, B, C is 6. They are AB, AC, BA, BC, CA, CB.

(*b*) The number of 2-combinations of A, B, C is 3. They are AB, AC, BC. The permutations AB and BA are distinct permutations, but they are two ways of stating the same combination.

Theorem A.4. *Let n elements be selected without replacement from N elements and let two selections be considered different if they are distinct n-combinations. Then the number of possible distinct selections, or n-combinations of N elements, is*

$$C_n^N = \frac{N(N-1)\cdots(N-n+1)}{n!} = \frac{N!}{n!(N-n)!}$$

Proof. We have already seen (Illustration A.1) that there are $N(N-1)\cdots(N-n+1)$ possible selections of n elements from N elements without replacement. Consider a specific selection of n elements. There will be $n!$ possible arrangements or n-permutations of these n elements, each of which will occur among the $N(N-1)\cdots(N-n+1)$ possible selections or n-permutations of N elements, i.e., each distinct combination gives rise to $n!$ permutations. Thus,

$$\begin{aligned} n! \cdot (\text{Number of combinations}) &= \text{Number of selections} \\ &= N(N-1)\cdots(N-n+1) \end{aligned}$$

or

$$\text{Number of combinations} = \frac{N(N-1)\cdots(N-n+1)}{n!} = C_n^N$$

which completes the proof.

REFERENCES

(1) H. Cramér, *Mathematical Methods of Statistics*, Princeton University Press, 1946.

(2) W. Feller, *An Introduction to Probability Theory and Its Applications*, Vol. I, John Wiley & Sons, New York, 1950.

CHAPTER 3

Fundamental Theory—Random Variables, Expected Values, Variances, Covariances, and Convergence in Probability

1. Introduction. In this chapter we establish the theorems we shall use in deriving the expected values, variances, and covariances that are needed for the proofs given in the subsequent chapters. Of these the more important are the following: (1) The expected value of a sum of random variables is the sum of their expected values (Theorem 5, Sec. 3, p. 48). More generally, the expected value of a linear combination of random variables is the same linear combination of their expected values (Theorem 6, Sec. 3, p. 49). (2) The expected value of a random variable is the expected value of the conditional expected value of that random variable (Theorem 14, Sec. 5, p. 61).

Theorems 11 and 12 of Sec. 4 (pp. 56 and 57) on the variances and covariances of linear combinations of random variables greatly simplify the problem of deriving variances in a number of practical sampling designs. Theorems 15, 16, and 17 of Sec. 6 (pp. 65 and 68) on conditional variances simplify greatly the development of the variance for a multi-stage sampling design in terms of its components.

2. Random variables—mathematical expectation. *a. Definition of random variable.* The values of one or more variables are usually associated with the elements of a population. For example, if a population consists of N families $A_1, \cdots, A_N$, then associated with the family A_i are values of such variables as the age of the head of the family, the number of children under 18 years of age, and the annual family income. If the elements $A_1, \cdots, A_N$ of the population are farms, then associated with A_i, the farm identified by i, are such variables as the number of acres, yields of different crops, income, and expenditures. If the elements $A_1, A_2, \cdots, A_{N^\star}$ of the population are the $N^\star$ possible samples according to some sampling plan (see Ch. 2, Sec. 4), then with each A_i is associated the value of the estimate or estimates that will be obtained if that possible sample is selected to be the sample. The value of any variable, U, that is associated with A_i will be called U_i. Thus, in the above discussion U might stand for age of the head of the family and U_i would be the age of the head of the family identified by i, or U might stand for estimate of average family income of this population and U_i might be the value of that estimate for A_i, the ith possible sample. Then, if the probability of selecting A_i is P_i, and if we denote by u the value of U that we obtain when one of the elements $A_1, \cdots, A_N$ is selected, it follows that u has possible values $U_1, \cdots, U_N$ and

$$Pr(u = U_i) = Pr(A_i) = P_i \qquad i = 1, \cdots, N$$

where $Pr(u = U_i)$ is read "the probability that $u = U_i$." It should be noted that some of the values $U_1, U_2, \cdots, U_N$ may be the same numerically but they are distinguishable by the different subscripts that associate them with different elements of the population.

In a case such as that described above, we call u a *random variable*, i.e., u is called a random variable if it has a finite number of possible values $U_1, U_2, \cdots, U_N$ and if with each possible value U_i there is associated a probability

$$Pr(u = U_i) = P_i \qquad i = 1, \cdots, N$$

where

$$P_i \geq 0 \quad \text{and} \quad P_1 + \cdots + P_N = 1$$

We can treat any real single-valued function defined on a finite population as a random variable; the probabilities of the possible values of the function are the probabilities of selecting the corresponding elements of the population.

It will be noted that we do not require that the probabilities P_i all be positive. Thus, some of the so-called "possible values" of u may have zero probability of being assumed.

Illustration 2.1. Suppose that $N = 5$ and that to A_1, A_2, A_3, A_4, and A_5 correspond the values $U_1 = 2$, $U_2 = 2$, $U_3 = 7$, $U_4 = 6$, $U_5 = 3$.

Then

$$Pr(u = U_i) = Pr(A_i), \quad \text{and} \quad Pr(u = 2) = Pr(A^\star)$$

where $A^\star$ consists of A_1 and A_2 so that

$$Pr(u = 2) = P_1 + P_2$$

Also

$$Pr(u \leq 4) = Pr(A^\star)$$

where $A^\star$ consists of A_1, A_2, and A_5 so that

$$Pr(u \leq 4) = P_1 + P_2 + P_5$$

Suppose that, in addition to the values of u above, the values $V_1 = 8$, $V_2 = 3$, $V_3 = 5$, $V_4 = 1$, $V_5 = 3$ are also associated with the A_i. Then, if $u \leq 4$, it follows that $A^\star$ occurs, where $A^\star$ consists of A_1, A_2, A_5; if $v \leq 6$, it follows that $B^\star$ occurs, where $B^\star$ consists of A_2, A_3, A_4, and A_5; if both $u \leq 4$ and $v \leq 6$, then $A^\star B^\star$ occurs, where $A^\star B^\star$ consists of only A_2 and A_5, since only if A_2 or A_5 occurs will we have both $u \leq 4$ and $v \leq 6$. Hence,

$$Pr(u \leq 4, v \leq 6) = Pr(A^\star B^\star) = P_2 + P_5$$

Also we can evaluate the conditional probabilities that these random variables assume certain possible values. For example,

$$Pr(u \leq 4 | v \leq 6) = Pr(A^\star | B^\star) = \frac{Pr(A^\star B^\star)}{Pr(B^\star)} = \frac{P_2 + P_5}{P_2 + P_3 + P_4 + P_5}$$

We now need to define *independent random variables*.

Suppose that $u_1, \cdots, u_k$ are k random variables and that u_d has possible values $U_{d1}, U_{d2}, \cdots, U_{dm_d}$, $d = 1, \cdots, k$. Then the random variables $u_1, \cdots, u_k$ are said to be independent, if and only if,

$$\begin{aligned} Pr(u_1 &= U_{1i_1}, u_2 = U_{2i_2}, \cdots, u_k = U_{ki_k}) \\ &= Pr(u_1 = U_{1i_1})\, Pr(u_2 = U_{2i_2}) \cdots Pr(u_k = U_{ki_k}) \end{aligned}$$

for all possible values of $u_1, \cdots, u_k$, where U_{di_d} is any one of the possible values of u_d, $d = 1, \cdots, k$.

Suppose now that two elements are being selected from a population consisting of $U_1, \cdots, U_N$. Denote by u_1 and u_2 the results of the two selection operations. Now, if we select at random with replacement (see Ch. 2, Sec. 4*c*), it follows that

$$\begin{aligned} Pr(u_1 = U_i, u_2 = U_j) &= Pr(u_1 = U_i)\, Pr(u_2 = U_j) \\ &= \frac{1}{N^2} \qquad i, j = 1, \cdots, N \end{aligned}$$

whereas if we select at random without replacement (see Ch. 2, Sec. 4*a*), then

$$\begin{aligned} Pr(u_1 = U_i, u_2 = U_j) &= 0 & \text{if } i = j \\ &= \frac{1}{N(N-1)} & \text{if } i \neq j \end{aligned}$$

so that in the latter case

$$Pr(u_1 = U_i, u_2 = U_j) \neq Pr(u_1 = U_i)\, Pr(u_2 = U_j)$$

These results hold not only for two but also for n selections so that, whereas the random variables that are the results of the individual selections at random with replacement are independent, those that are the results of the individual selections at random without replacement are dependent.

In Feller (2), p. 87, and Cramér (1), p. 162, will be found simple examples showing that even if the random variables $u_1, \cdots, u_k$ are *pairwise independent*, i.e., u_i and u_j are independent for all $i \neq j$, it may still be true that the random variables $u_1, \cdots, u_k$ are *dependent*. However, from the definition of independent random variables, it is not difficult to prove that if $u_1, \cdots, u_k$ are independent then any subset, say $u_1, \cdots, u_h$, $h < k$, of $u_1, \cdots, u_k$ are independent. To see this for, say, $k = 3$ we note that by Ch. 2, Sec. 6, Theorem 5 (p. 28) it follows that, if u_1, u_2, and u_3 are independent, and if U_{1i_1} and U_{3i_3} are particular values of u_1 and u_3, and u_2 takes on possible values U_{2j}, $j = 1, 2, \cdots, m_2$, then

$$\begin{aligned} Pr(u_1 = U_{1i_1}, u_3 = U_{3i_3}) &= \sum_j^{m_2} Pr(u_1 = U_{1i_1})\, Pr(u_2 = U_{2j})\, Pr(u_3 = U_{3i_3}) \\ &= Pr(u_1 = U_{1i_1})\, Pr(u_3 = U_{3i_3}) \end{aligned}$$

since

$$\sum_j^{m_2} Pr(u_2 = U_{2j}) = 1$$

A generalization of the term *random variable* that will be helpful is the term *random event*.

If, when an operation is performed, one of the K exhaustive and mutually exclusive events (see Ch. 2, Sec. 6) $A_1^\star, \cdots, A_K^\star$ must occur, we call the result of that operation a random event, and denote it by $a^\star$. We call $A_1^\star, \cdots, A_K^\star$ the *possible states* of the random event.

If $Pr(A_i^\star) = P_i$ we will write $Pr(a^\star = A_i^\star) = P_i$, where, if $a^\star$ is a random event and $A_i^\star$ is one of its possible states, the statement "$a^\star = A_i^\star$" is to be read "$a^\star$ takes on the state $A_i^\star$."

Thus the result of performing a selection operation to select one of N elements $A_1, \cdots, A_N$ of a population is a random event $a^\star$ having possible states $A_1, \cdots, A_N$ and

$$Pr(a^\star = A_i) = P_i$$

where P_i is the probability that A_i is obtained when the selection operation is performed.

Illustration 2.2. (*a*) One of 5 blocks $C_1, \cdots, C_5$ is to be selected. The result of that selection we call a random event with 5 possible states; these are "C_1 is selected," $\cdots$, "C_5 is selected."

(*b*) Two of the 5 blocks are selected by an epsem. Then the result is a random event having 20 possible states each of probability $\frac{1}{20}$.

(*c*) A sample is selected. Numerical information is obtained for the sample and an estimate is calculated. Then, the result of the selection is a random event, and, as previously mentioned, the estimate is a random variable.

The definition of random event includes that of random variable if by the states of the event we mean the taking on by the random variable of its possible values.

Let us consider how $a^\star$, the random event, and u, the value of some variable for the event, are related.

First, although $a^\star$ need not be numerically valued, u must be, according to the definitions we have given. The random event may be the selection of a block or a house or a person or a set of them; but the random variable must refer to something like the number of families on the block, the income of all persons living in the house, or the age of the person.

Second, a random event may and often will determine the values of several random variables. If we denote by U_i the number of persons, by V_i the total income, and by W_i the total expenditures of the ith family A_i, then the selection of the ith family implies that

$$u = U_i, \quad v = V_i, \quad \text{and} \quad w = W_i$$

so that

$$Pr(u = U_i, v = V_i, w = W_i) = Pr(a^\star = A_i) = P_i$$

Third, sometimes the elements of the population may be taken to be numbers. If we wish to estimate the average size of family, then we may say that the elements of the population are $U_1, U_2, \cdots, U_N$, where U_i is the number of persons in the ith family. However, it should be kept in mind that the probabilities of selection are unaffected by choosing to represent an element by the values of certain variables for that element.

Fourth, just as we are interested in the probability of a random event which consists of several of the elements of $\mathscr{A}$, so we are concerned with the probability that u takes on one of its possible numerical values. Let $A_u^\star$ be a subset of the possible values of u. Then, since the values of u are determined by which of the A_i are selected, it is clear that

$$Pr(u \varepsilon A_u^\star) = Pr(a^\star \varepsilon A^\star) \tag{2.1}$$

where ε is read "is an element of," and $A^\star$ consists of all elements of $\mathscr{A}$ such that if one of these elements is selected then u takes on one of the values contained in $A_u^\star$.

Let us note that if $u_1, \cdots, u_k$ are random variables, then functions such as those given below will also be random variables. For example, if

$$\bar{u} = \frac{u_1 + \cdots + u_k}{k}$$

and

$$s^2 = \frac{\sum_i^k (u_i - \bar{u})^2}{k - 1}$$

then $\bar{u}$, $c\bar{u}$ (where c is a constant), s^2, s, $s/\bar{u}$ are all random variables. To see this, let us recall that to define a random variable we need only state its possible values and their probabilities. To find the possible values of a function of the k random variables $u_1, \cdots, u_k$ we must give to $u_1, \cdots, u_k$ all their possible values. The possible value of the function that occurs when $u_1 = U_{1i_1}$, $u_2 = U_{2i_2}$, $\cdots$, $u_k = U_{ki_k}$ then has probability $Pr(u_1 = U_{1i_1}, \cdots, u_k = U_{ki_k})$. Usually, Eq. 2.1 is used to evaluate the probabilities of the possible values of functions of random variables.

Illustration 2.3. Let u_1 have the possible values 0, 1, let u_2 have the possible values 1, 2, 3, and let $Pr(u_1 = i, u_2 = j) = P_{ij}$, $i = 0, 1, j = 1, 2, 3$. Then, if we define $u = u_1 + u_2$ it follows that u has possible values $0 + 1, 0 + 2, 0 + 3, 1 + 1, 1 + 2$, and $1 + 3$, or 1, 2, 3, 2, 3, 4, or more concisely 1, 2, 3, 4. Also $Pr(u = 1) = P_{01}$, $Pr(u = 2) = P_{02} + P_{11}$, $Pr(u = 3) = P_{03} + P_{12}$, and $Pr(u = 4) = P_{13}$. As an exercise, determine the possible values and probabilities of u_1u_2, and u_1/u_2.

b. Definition of mathematical expectation. Before discussing the intuitive meaning of mathematical expectation let us define it and train ourselves in its computation.

If u is a random variable with possible values $U_1, \cdots, U_N$ and probabilities $P_1, \cdots, P_N$, then the mathematical expectation or expected value of u is

$$Eu = P_1U_1 + P_2U_2 + \cdots + P_NU_N = \sum_i^N P_iU_i$$

Thus, to calculate the expected value of u, we multiply each possible value by its probability and add the products thus obtained.

Illustration 2.4. Suppose that u can take on the values 1 and 0 with probabilities $\frac{1}{4}$ and $\frac{3}{4}$. Then

$$Eu = 1(\tfrac{1}{4}) + 0(\tfrac{3}{4}) = \tfrac{1}{4}$$

Illustration 2.5. Suppose that u_1 and u_2 are independent, u_1 takes on the values 0, 1 with probabilities $\frac{1}{4}$ and $\frac{3}{4}$, u_2 takes on the values 1, 2, and 3 with probabilities $\frac{1}{6}$, $\frac{1}{2}$, and $\frac{1}{3}$, and $u = u_1 + u_2$. Then, as shown in

Illustration 2.3, u has possible values 1, 2, 3, and 4; and, using the assumption of independence, we have

$$Pr(u = 1) = \tfrac{1}{4}\tfrac{1}{6} = \tfrac{1}{24}$$
$$Pr(u = 2) = \tfrac{1}{4}\tfrac{1}{2} + \tfrac{3}{4}\tfrac{1}{6} = \tfrac{1}{4}$$
$$Pr(u = 3) = \tfrac{1}{4}\tfrac{1}{3} + \tfrac{3}{4}\tfrac{1}{2} = \tfrac{11}{24}$$
$$Pr(u = 4) = \tfrac{3}{4}\tfrac{1}{3} = \tfrac{1}{4}$$

Hence

$$Eu = 1(\tfrac{1}{24}) + 2(\tfrac{1}{4}) + 3(\tfrac{11}{24}) + 4(\tfrac{1}{4}) = 2\tfrac{11}{12}$$

As exercises compute Eu_1u_2 and $E(u_1/u_2)$ for this illustration by determining possible values and probabilities, and using the definition of independence.

Illustration 2.6. A population contains M elements, of which M_i have the value U_i, $i = 1, \cdots, K$; $M_1 + \cdots + M_K = M$. One element of the population is selected by an epsem (Sec. 4, Ch. 2). Let us find the mathematical expectation of the value associated with the selected element. Since an epsem is used, the probability of selecting an element that has value U_i associated with it is M_i/M. Hence, if u denotes the value associated with the selected element, it follows that

$$Eu = \sum_i^K \frac{M_i}{M} U_i$$

If the element of the population is selected by a method that gives probability P_{ij} to the jth of the M_i elements that have the value U_i, then

$$Eu = \sum_i^K P_i U_i$$

where

$$P_i = \sum_j^{M_i} P_{ij}$$

since P_i is the probability that u takes on the value U_i (Ch. 2, Sec. 6, Theorem 5, p. 28).

We now consider the intuitive meaning of expected value. Suppose that an operation is performed N times and that U_i occurs N_i times, $N_1 + \cdots + N_K = N$. Then, the average value of u in these N performances is

$$\bar{u} = \frac{N_1U_1 + \cdots + N_KU_K}{N} = \frac{N_1}{N} U_1 + \cdots + \frac{N_K}{N} U_K$$

where N_i/N is the relative frequency of the occurrence of U_i in the N performances. As the number, N, of performances increases, we expect the relative frequencies N_i/N to come closer to the probabilities P_i so that $\bar{u}$ will come close to Eu as we have defined it. Since we will be selecting samples by methods, such as the epsem, that assign known probabilities to each of the possible values, it follows that we can calculate Eu and thus learn to what value the average result of selecting samples by this same method would be expected to tend. (This result is obtained in Sec. 7.)

Let us prove two simple theorems about expected values.

Theorem 1. *If the possible values of the random variable u are non-negative, i.e., if* $U_i \geq 0$, $i = 1, \cdots, N$, *the expected value of u is non-negative.*

Proof. Since $P_1, \cdots, P_N$ are non-negative, it follows that

$$Eu = \sum_i^N P_i U_i$$

is a sum of non-negative terms, and hence

$$Eu \geq 0$$

Theorem 2. *If f(u) and g(u) are two functions of u, where the possible values of u are* $U_1, \cdots, U_N$, *such that*

$$f(U_i) \leq g(U_i) \qquad i = 1, \cdots, N \tag{2.2}$$

then

$$Ef(u) \leq Eg(u)$$

Proof. Since

$$Eg(u) - Ef(u) = \sum_i^N P_i[g(U_i) - f(U_i)]$$

the theorem follows from Eq. 2.2 and the fact that probabilities are non-negative.

c. Biased and unbiased estimates. Let u be an estimate and let U be the quantity we wish to estimate by u. Then, since the value u takes on is determined by the particular sample that is selected, it follows that u is a random variable. If $Eu = U$, we call u an *unbiased* estimate of U. Otherwise we call u a *biased* estimate of U, and refer to the difference, $Eu - U$, as the bias.

Then, if u has possible values $U_1, \cdots, U_N$ with probabilities $P_1, \cdots, P_N$, it will follow that u is an unbiased estimate of $\sum_i^N P_i U_i$.

It is not necessarily an advantage to use unbiased estimates, since it has been found that often they are not as good as biased estimates in the sense that the biased estimates may tend to come close to the quantities that one wishes to estimate with higher probability than the unbiased estimates. We delay any further discussion of biased and unbiased estimates to Sec. 7 of this chapter.

Exercises

2.1. (*a*) A population consists of the 5 elements A_1, A_2, A_3, A_4, A_5 having the values 1, 1, 3, 5, 10, respectively (say A_i represents the ith family and the associated values are sizes of family, or the A_i are businesses and the associated values are incomes). One of the elements is selected by an epsem and the associated value is taken to be an estimate, u. What is the expected value of u?

(*b*) Select 100 samples of size 1 from this same population, using a table of random numbers. Compare the average value of the first 10, first 25, first 50, and all 100 samples with the expected value.

2.2. Repeat both parts of Ex. 2.1 when the element is selected so that the probabilities of selecting A_1, A_2, A_3, A_4, and A_5 are .2, .4, .1, .1, and .2, respectively. Of what quantity is u an unbiased estimate?

2.3. Let u have possible values 1 and 0 with $Pr(u = 1) = P$, and $Pr(u = 0) = Q = 1 - P$. Calculate Eu, Eu^2, Eu^3, and Eu^j, where $j > 0$.

2.4. Let u have possible values 1, 2, 3, 4, 5 with probabilities $\frac{1}{5}$, $\frac{1}{10}$, $\frac{1}{10}$, $\frac{1}{2}$, $\frac{1}{10}$. Calculate Eu. Of what quantity is u an unbiased estimate?

2.5. Let u have possible values $-1, 0, 1$ with probabilities $\frac{1}{4}$, $\frac{1}{2}$, $\frac{1}{4}$. Calculate Eu.

2.6. In Ex. 2.4 calculate Eu^2 and Eu^3.

2.7. In Ex. 2.5 calculate Eu^2 and Eu^3. Of what quantities are u^2 and u^3 unbiased estimates?

3. Some theorems on mathematical expectation. The following theorems are helpful in computing expected values, since they reduce the calculation of the expected values of relatively complicated random variables to the calculation of the expected values of simpler random variables.

Theorem 3. *If c and d are constants and u is a random variable, then*

$$E(cu + d) = cEu + d$$

Note that this theorem implies that $Ed = d$.

Proof. If the possible values of u are $U_1, \cdots, U_N$ with probabilities $P_1, \cdots, P_N$, then $cu + d$ has possible values $cU_i + d$, $i = 1, \cdots, N$ with probabilities $P_1, \cdots, P_N$ so that

$$E(cu + d) = \sum_i^N P_i(cU_i + d) = c\sum_i^N P_iU_i + d\sum_i^N P_i$$

$$= cEu + d$$

Let u and v be random variables, u having possible values $U_1, \cdots, U_N$, v having possible values $V_1, \cdots, V_M$, and $Pr(u = U_i, v = V_j) = P_{ij}$, $i = 1, \cdots, N, j = 1, \cdots, M$. We consider the occurrence of both U_i

and V_j to define an elementary event A_{ij} with probability P_{ij}, $i = 1, \cdots, N$, $j = 1, \cdots, M$, so that there are NM elementary events, some of which may have probability zero. Let $A_i^\star$ be the event $u = U_i$. Then $A_i^\star$ occurs if and only if A_{i1} or A_{i2} or $\cdots$ or A_{iM} occurs. Hence

$$Pr(u = U_i) = Pr(A_i^\star) = P_{i1} + \cdots + P_{iM} = P_{i\cdot} \tag{3.1}$$

where $P_{i\cdot}$ is, by definition, $\sum^{M} P_{ij}$. Similarly

$$Pr(v = V_j) = P_{\cdot j} \tag{3.2}$$

where

$$P_{\cdot j} = \sum_i^N P_{ij}$$

Theorem 4. *If u and v are random variables then*

$$E(u + v) = Eu + Ev$$

Proof. From the definition of expected value, we have

$$\begin{aligned} E(u + v) &= \sum_i^N \sum_j^M P_{ij}(U_i + V_j) \\ &= \sum_i^N \sum_j^M P_{ij} U_i + \sum_i^N \sum_j^M P_{ij} V_j \\ &= \sum_i^N P_{i\cdot} U_i + \sum_j^M P_{\cdot j} V_j \\ &= Eu + Ev, \text{ from Eq. 3.1 and 3.2 above.} \end{aligned}$$

Using Theorems 3 and 4, we proceed to prove two generalizations that are very important.

Theorem 5. *The expected value of a sum of random variables is the sum of their expected values.*

Proof. Let $u_1, \cdots, u_r$ be random variables and let $u = \sum_i^r u_i$. Then we want to show that

$$Eu = \sum_i^r Eu_i$$

We prove Theorem 5 by induction, i.e., we prove that if the theorem is true when $r = 1$, then it is true when $r = 2$; and if it is true when $r = 2$, then it is true when $r = 3$; and so on; so that the theorem is proved for all positive integral values of r. The theorem is an identity if $r = 1$. In Theorem 4 we have proved the theorem to be true for $r = 2$. Now suppose that the theorem is true for all j, $j = 1, \cdots, r - 1$. Let $v = \sum^{r-1} u_i$; then by the hypothesis of the induction we have $Ev = \sum_i^{r-1} Eu_i$.

Also, $u = v + u_r$ and, by Theorem 4, we have

$$Eu = Ev + Eu_r$$

which completes the proof.

Theorem 6. *The expected value of a linear combination of random variables is the same linear combination of the expected values of these random variables.*

Proof. Let $u_1, \cdots, u_r$ be random variables and let $c_1, \cdots, c_r$ be constants. Let

$$u = \sum_i^r c_i u_i$$

be the linear combination of $u_1, \cdots, u_r$. Then we want to show that

$$Eu = \sum_i^r c_i Eu_i$$

By Theorem 5

$$E\sum_i^r c_i u_i = \sum_i^r Ec_i u_i$$

By Theorem 3

$$Ec_i u_i = c_i Eu_i$$

Combining these results, the proof is completed. It should be noted that since $u_1, \cdots, u_r$ are *any* random variables they may be functions of other random variables, i.e., if $u_i = f_i(v_1, \cdots, v_m)$, where $v_1, \cdots, v_m$ are random variables, and if $f(v_1, \cdots, v_m) = \sum_i^r c_i f_i(v_1, \cdots, v_m)$, then $Ef(v_1, \cdots, v_m) = \sum_i^r c_i Ef_i(v_1, \cdots, v_m)$.

Illustration 3.1. (*a*) Let u_1, u_2, and u_3 be random variables with $Eu_i = i$, and $Eu_i^2 = 2i^2$, $i = 1, 2, 3$. Let $u = \sum_i^3 iu_i$. Then, by Theorem 6,

$$Eu = \sum_i^3 iEu_i = \sum_i^3 i^2 = 14$$

Also, if $v = -u_1^2 + 2u_2^2$, then $Ev = -Eu_1^2 + 2Eu_2^2 = 14$.

(*b*) Let $E \log u_1 = b_1$ and $E10^{u_2} = b_2$, and let $u = 3 \log u_1 + (1.7)10^{u_2}$. Then $Eu = 3b_1 + 1.7b_2$.

From the consideration of Theorem 6 and Illustration 3.1 it will be clear that, when we wish to calculate the expected value of a linear combination u of the random variables $u_1, \cdots, u_r$, we first use Theorem 6 to reduce the problem to one of computing the expected values of the component random variables $u_1, \cdots, u_r$. For the computation of these latter expected values, it will often be necessary to return to the definition of expected value.

Illustration 3.2. Let u_i have the possible values $U_{i1}, \cdots, U_{iN_i}$ with probabilities $P_{i1}, \cdots, P_{iN_i}$, $P_{ij} \geq 0$, $\sum_j^{N_i} P_{ij} = 1$, $i = 1, \cdots, r$, and let

$$u = c_1u_1 + \cdots + c_ru_r$$

Then, by Theorem 6,

$$Eu = c_1Eu_1 + \cdots + c_rEu_r \tag{3.3}$$

and by the definition of expected value

$$Eu_i = \sum_j^{N_i} P_{ij}U_{ij} \qquad i = 1, \cdots, r \tag{3.4}$$

Computing the values of Eu_i from Eq. 3.4, we substitute them in Eq. 3.3 to obtain Eu.

Further illustrations of the uses of Theorems 3–6 will be found in Sec. 1, 2, and 4 of Ch. 4.

Exercises

3.1. Let $u = 4u_1 + u_2$, where $Eu_1 = 1$ and $Eu_2 = -7$. Calculate Eu.

3.2. Let $\bar{u}$ be the arithmetic mean of $u_1, u_2, \cdots, u_N$, where u_i has 2 possible values, 1 and 0, with $Pr(u_i = 1) = P_i$ and $Pr(u_i = 0) = 1 - P_i$, $i = 1, \cdots, N$. Calculate $E\bar{u}$. What does $E\bar{u}$ become if $P_1 = \cdots = P_N = P$? If u is equal to $u_1 + u_2 + \cdots + u_n$, compute Eu. Of what quantity is u an unbiased estimate?

3.3. Let $u = u_1^2 + \cdots + u_N^2$, where u_i has 2 possible values 1 and 0 with $Pr(u_i = 1) = P_i$ and $Pr(u_i = 0) = 1 - P_i$. Calculate Eu. What does Eu become if $P_1 = \cdots = P_N = P$? Of what quantity is u an unbiased estimate?

3.4. Let $u = 5u_1 + 3u_2$, where u_1 has possible values 1, 2, 3 which are equally probable and u_2 has possible values -1, 0, 1 which have probabilities $\frac{1}{6}$, $\frac{2}{3}$, $\frac{1}{6}$. Compute Eu.

3.5. Let $\bar{u}$ be the arithmetic mean of $u_1, \cdots, u_n$, where u_i has possible values $U_1, \cdots, U_N$ for all i and these possible values are equally probable. Compute $E\bar{u}$. If u is the total of $u_1, \cdots, u_n$, compute Eu. Of what quantity is u an unbiased estimate?

3.6. Show that, if $\bar{u}$ is the arithmetic mean of $u_1, \cdots, u_n$ and u is their sum, then it always follows that $Eu = nE\bar{u}$.

4. Variance, covariance, mean square error, rel-variance, coefficient of variation. The variance, σ_u^2, of the random variable u is defined by the equation

$$\sigma_u^2 = E(u - Eu)^2 \tag{4.1}$$

Thus, if u is a random variable with possible values $U_1, \cdots, U_N$ and probabilities $P_1, \cdots, P_N$, then $\sigma_u^2 = \sum_i^N P_i(U_i - Eu)^2$.

It is easy to see that $\sigma_u^2 \geq 0$, and that $\sigma_u^2 = 0$ only if $U_i = Eu$ for all i except, perhaps, those having probability 0.

The covariance σ_{uv} of the two random variables u and v is defined by the equation

$$\sigma_{uv} = E(u - Eu)(v - Ev) \tag{4.2}$$

Thus, if u is a random variable with possible values $U_1, \cdots, U_N$ and v is a random variable with possible values $V_1, \cdots, V_M$, and if

$$P_{ij} = Pr(u = U_i, v = V_j)$$

then

$$\sigma_{uv} = \sum_i^N \sum_j^M P_{ij}(U_i - Eu)(V_j - Ev)$$

Hence

$$\sigma_{uu} = \sigma_u^2$$

The variance and covariance of random variables are very important in sampling theory, since measures of efficiency are largely based on them.

Also important in the interpretation of estimates are the *standard deviation*, σ_u, which is the positive square root of the variance, *the mean square error*, which is defined to be $E(u - U)^2$, where u is the estimate and U is the characteristic that is being estimated, and the *rel-variance*, which is defined to be

$$\frac{\sigma_u^2}{(Eu)^2} \tag{4.3}$$

The positive square root of the rel-variance is called the coefficient of variation. The relation between the variance and the mean square error will be found in Theorem 7 below. Inasmuch as the standard deviation, the mean square error, the rel-variance, and the coefficient of variation all depend so closely on the variance, we shall not discuss them in detail in this chapter. However, they are often used in later chapters.

Theorem 7. *If c and d are any constants and if $U = Eu$, $V = Ev$, then*

$$E(u - c)^2 = \sigma_u^2 + (U - c)^2 \tag{4.4}$$

i.e., the mean square error is the sum of the variance and the square of the bias. Hence, the minimum value of $E(u - c)^2$ occurs when $c = U$. Moreover,

$$E(u - c)(v - d) = \sigma_{uv} + (U - c)(V - d) \tag{4.4a}$$

where σ_{uv} is given by Eq. 4.2.

Proof. Since $u - c = u - U + U - c$,

$$(u - c)^2 = (u - U)^2 + 2(u - U)(U - c) + (U - c)^2 \tag{4.5}$$

Equation 4.4 follows from Theorem 6 and the fact that

$$E2(u - U)(U - c) = 2(U - c)E(u - U) = 0$$

The fact that the minimum value of $E(u - c)^2$ occurs when $c = U$ then follows from the fact that $(U - c)^2$ in Eq. 4.5 is greater than 0 whenever $U \neq c$. The proof that Eq. 4.4a holds follows the same reasoning as the proof that Eq. 4.4 holds. Note that Eq. 4.4a is Eq. 4.4 when $u = v$ and $c = d$.

Corollary. If u is a random variable, then

$$\sigma_u^2 = Eu^2 - (Eu)^2 \tag{4.6}$$

$$Eu^2 \geq (Eu)^2 \tag{4.7}$$

$$(Eu^2)^{1/2} \geq E|u| \tag{4.8}$$

where $|u|$ stands for the absolute value of u, i.e., $|u| = u$ if $u \geq 0$ and $|u| = -u$ if $u < 0$.

Proof. Equation 4.6 follows from Eq. 4.4 in Theorem 7, with $c = 0$. To prove that $Eu^2 \geq (Eu)^2$ we first note that from Theorem 1 $\sigma_u^2 = E(u - U)^2 \geq 0$. The proof thus follows from Eq. 4.6.

Finally, $(Eu^2)^{1/2} \geq E|u|$ is obtained from the following:

$$\sigma_{|u|}^2 = E|u|^2 - (E|u|)^2 \text{ from Eq. 4.6}$$

$$E|u|^2 \geq (E|u|)^2 \text{ from Inequality 4.7 above}$$

and since $|u|^2 = u^2$, we have

$$Eu^2 \geq (E|u|)^2$$

and

$$(Eu^2)^{1/2} \geq E|u|$$

Illustration 4.1. (*a*) Let u have possible values $U_1, \cdots, U_N$ all of which are equally probable. Then

$$\sigma_u^2 = \frac{1}{N}\sum_i^N (U_i - \bar{U})^2$$

We have already seen that $Eu = \bar{U}$. Also, $Pr(u = U_i) = Pr\{(u - Eu)^2 = (U_i - Eu)^2\}$. Hence

$$E(u - Eu)^2 = \sum_i^N \frac{1}{N}(U_i - \bar{U})^2$$

We could also obtain this result by using Theorem 7 with $c = 0$. From Theorem 7 with $c = 0$ we have

$$\sigma_u^2 = Eu^2 - U^2$$

Since $Pr(u^2 = U_i^2) = Pr(u = U_i)$ we have

$$Eu^2 = \frac{1}{N}\sum_i^N U_i^2$$

so that

$$\sigma_u^2 = \frac{1}{N}\sum_i^N U_i^2 - \bar{U}^2$$

Since

$$\sum_i^N (U_i - \bar{U})^2 = \sum_i^N U_i^2 - N\bar{U}^2$$

we have the desired result.

(*b*) Suppose that u has possible values $U_1, \cdots, U_N$, v has possible values $V_1, \cdots, V_N$, and that

$$Pr(u = U_i, v = V_j) = 0 \qquad \text{if } i \neq j$$

$$Pr(u = U_i, v = V_i) = P_i \qquad i = 1, \cdots, N$$

Then, since

$$Pr(u = U_i) = \sum_j^N Pr(u = U_i, v = V_j)$$

it follows that

$$Pr(u = U_i) = P_i$$

and similarly

$$Pr(v = V_i) = P_i \qquad i = 1, \cdots, N$$

Hence,

$$Eu = \sum_i^N P_i U_i$$

$$\sigma_u^2 = \sum_i^N P_i (U_i - Eu)^2$$

$$Ev = \sum_i^N P_i V_i$$

$$\sigma_v^2 = \sum_i^N P_i (V_i - Ev)^2$$

and

$$\sigma_{uv} = \sum_i^N P_i (U_i - Eu)(V_i - Ev)$$

Exercises

4.1. If u has possible values 1 and 0 with probabilities P and Q, then show that $\sigma_u^2 = PQ = 1 - P$.

4.2. If u has possible values $U_1, \cdots, U_N$ and v has possible values $V_1, \cdots, V_N$, if $Pr(u = U_i, v = V_j) = 0$ if $i \neq j$, and if the possible values of u are equally probable, then show that

$$\sigma_{uv} = \frac{1}{N} \sum_{i}^{N} (U_i - \bar{U})(V_i - \bar{V})$$

4.3. If u has possible values $\bar{U}_1, \cdots, \bar{U}_M$ where

$$\bar{U}_i = \frac{1}{N_i} \sum_{j}^{N_i} U_{ij} \qquad i = 1, \cdots, M$$

and if

$$Pr(u = \bar{U}_i) = \frac{N_i}{N} \qquad i = 1, \cdots, M$$

where

$$N = N_1 + \cdots + N_M$$

then show that

$$\sigma_u^2 = \sum_{i}^{M} \frac{N_i}{N} (\bar{U}_i - \bar{U})^2$$

where

$$\bar{U} = \frac{1}{N} \sum_{i}^{M} \sum_{j}^{N_i} U_{ij}$$

Theorem 8. *If u and v are independent, then $Euv = EuEv$. Also, if $u_1, \cdots, u_k$ are independent, then $Eu_1u_2 \cdots u_k = Eu_1Eu_2 \cdots Eu_k$.*

Proof. By definition

$$Euv = \sum_{i}^{N} \sum_{j}^{M} P_{ij} U_i V_j$$

If u and v are independent, then

$$P_{ij} = P_{i\cdot} P_{\cdot j}$$

where

$$P_{i\cdot} = Pr(u = U_i), \quad P_{\cdot j} = Pr(v = V_j)$$

Substituting, we see that

$$Euv = \sum_{i}^{N} P_{i\cdot} U_i \sum_{j}^{M} P_{\cdot j} V_j$$
$$= EuEv$$

To evaluate $Eu_1 \cdots u_k$ we note that, if $u_1, \cdots, u_k$ are independent, then the product $u_1 \cdots u_{k-1}$ is independent of u_k. Hence, by the first part of the theorem we have $Eu_1 \cdots u_k = E(u_1 \cdots u_{k-1}) \cdot Eu_k$. Continuing this procedure, we finally prove the theorem in $k - 1$ steps.

Theorem 9. *If u and v are independent, then $\sigma_{uv} = 0$.*

Proof. From Theorem 7 with $c = d = 0$, we have $\sigma_{uv} = Euv - EuEv$. Then the theorem follows from Theorem 8.

It is possible to have $\sigma_{uv} = 0$ without u and v being independent. For example, let the possible values of u be ± 1, ± 2, each having probability $\frac{1}{4}$, and let $v = u^2$. Then

$$Pr(u = +1, v = 1) = Pr(u = -1, v = 1) = \tfrac{1}{4}$$

$$Pr(u = +2, v = 4) = Pr(u = -2, v = 4) = \tfrac{1}{4}$$

and

$$Pr(u = 1, v = 4) = 0$$

whereas

$$Pr(u = 1) = \tfrac{1}{4} \quad \text{and} \quad Pr(v = 4) = \tfrac{1}{2}$$

and hence

$$Pr(u = 1, v = 4) \neq Pr(u = 1)\, Pr(v = 4)$$

Thus, u and v are dependent. Yet $Eu = 0$ and $Euv = 0$ so that $\sigma_{uv} = 0$. If $\sigma_{uv} = 0$, we shall say that u and v have zero covariance.

Let us define the correlation coefficient ρ_{uv} by the equation $\rho_{uv} = \sigma_{uv}/\sigma_u\sigma_v$. Then, if $\sigma_{uv} = 0$, we have $\rho_{uv} = 0$, and say that u and v are uncorrelated.

Theorem 10. *The correlation coefficient* ρ_{uv} *is such that* $-1 \leq \rho_{uv} \leq 1$ *and* $\rho_{uv} = \pm 1$ *if and only if* $u = c + dv$, *where* c *and* d *are constants, except for values of* u *and* v *having zero probability.*

Proof. Since

$$E\left(\frac{u - Eu}{\sigma_u} - \frac{v - Ev}{\sigma_v}\right)^2 \geq 0$$

it follows that, if we expand the square, then, from Theorem 6 and the definition of variance and covariance, we have

$$1 - 2\rho_{uv} + 1 \geq 0 \tag{4.9}$$

$$1 - \rho_{uv} \geq 0$$

i.e.,

$$1 \geq \rho_{uv}$$

Similarly, since

$$E\left(\frac{u - Eu}{\sigma_u} + \frac{v - Ev}{\sigma_v}\right)^2 \geq 0$$

it follows that

$$1 + \rho_{uv} \geq 0$$

so that $-1 \leq \rho_{uv}$.

If $\rho_{uv} = 1$, then the left member of the Inequality 4.9 is zero, and since

$$E\left(\frac{u - Eu}{\sigma_u} - \frac{v - Ev}{\sigma_v}\right)^2 = 0$$

only if the quantity in parentheses is zero, excluding possible values that have zero probability, we have

$$\frac{u - Eu}{\sigma_u} = \frac{v - Ev}{\sigma_v}$$

or

$$u = Eu - \frac{\sigma_u}{\sigma_v} Ev + \frac{\sigma_u}{\sigma_v} v$$

Hence, the third part of Theorem 10 is proved, if $\rho_{uv} = 1$, with $c = Eu - (\sigma_u/\sigma_v)Ev$ and $d = \sigma_u/\sigma_v$. A similar result is obtained if $\rho_{uv} = -1$.

Remark. Theorem 10 is a special case of the more general theorem $Euv \leq (Eu^2Ev^2)^{1/2}$ for any random variables u and v having finite second moments.

Just as the theorem on the expected value of a linear combination of random variables plays a fundamental part in sampling, so does the theorem on the variance of a linear combination of random variables.

Theorem 11. *If* $u = \sum_i^k c_i u_i$, *where* $u_1, \cdots, u_k$ *are random variables and* $c_1, \cdots, c_k$ *are constants, then*

$$\sigma_u^2 = \sum_i^k \sum_j^k c_i c_j \sigma_{u_i u_j}$$

where $\sigma_{u_i u_j}$ *is the covariance of* u_i *and* u_j.

Proof. Since $u - Eu = \sum_i^k c_i(u_i - Eu_i)$, it follows upon squaring both sides of the equation (see Illustration 2.2 of Ch. 2) that

$$E(u - Eu)^2 = E \sum_i^k \sum_j^k c_i c_j (u_i - Eu_i)(u_j - Eu_j)$$

$$= \sum_i^k \sum_j^k c_i c_j E(u_i - Eu_i)(u_j - Eu_j)$$

by Theorem 6. Then, the theorem follows from the definition of the covariance.

The following four corollaries are obvious conclusions from Theorem 11. Their proofs are left as exercises.

Corollary 1. If the u_i are uncorrelated, then

$$\sigma_u^2 = \sum_i^k c_i^2 \sigma_{u_i}^2$$

Corollary 2. If the u_i are uncorrelated and have equal variances, σ^2, then

$$\sigma_u^2 = \sigma^2 \sum_i^k c_i^2$$

Corollary 3. If the u_i are uncorrelated and have equal variances, σ^2, and

$$\bar{u} = \frac{u_1 + \cdots + u_k}{k}$$

then

$$\sigma_{\bar{u}}^2 = \frac{\sigma^2}{k}$$

Also, if $u = u_1 + \cdots + u_k$, then

$$\sigma_u^2 = k\sigma^2$$

Corollary 4. The variance of a constant times a random variable is the constant squared times the variance of the random variable, i.e.,

$$\sigma_{cu}^2 = c^2 \sigma_u^2$$

Illustration 4.2. In stratified sampling the population is first divided into L classes or, as they are usually called, strata, such that each element of the population is in one and only one stratum. Then, samples are independently selected from each stratum. If $u = c_1 u_1 + \cdots + c_L u_L$, where u_i is based on the sample from the ith stratum, then by Theorem 6 and Corollary 1 to Theorem 11, we have

$$Eu = \sum_i^L c_i E u_i$$

and

$$\sigma_u^2 = \sum_i^L c_i^2 \sigma_i^2$$

where σ_i^2 is the variance of u_i. Hence, the knowledge of the expected value and variance of each of the u_i permits us to evaluate the expected value and variance of u when stratified sampling is used. The following theorem is sometimes useful.

Theorem 12. *If* $u = \sum_i^k c_i u_i$ *and* $v = \sum_j^m d_j v_j$, *then*

$$\sigma_{uv} = \sum_i^k \sum_j^m c_i d_j \sigma_{u_i v_j}$$

Proof. The proof almost repeats that of Theorem 11. It is obtained by using Theorem 6 and the fact that, by Illustration 2.2 of Ch. 2, we have

$$(u - Eu)(v - Ev) = \sum_i^k \sum_j^m c_i d_j (u_i - Eu_i)(v_j - Ev_j)$$

Corollary. If u_i and v_j, $i \neq j$, are uncorrelated, and $k = m$, then

$$\sigma_{uv} = \sum_i^m c_i d_i \sigma_{u_i v_i}$$

and if $c_i = d_i = 1/m$, then

$$\sigma_{uv} = \frac{1}{m^2} \sum_i^m \sigma_{u_i v_i}$$

The proof is omitted.

Illustration 4.3. (*a*) If $\sigma_{u_i}^2 = \sigma^2$, $i = 1, \cdots, n$, and $\sigma_{u_i u_j} = c$, $i \neq j$, $i, j = 1, \cdots, n$, then

$$\sigma_{\bar{u}}^2 = \frac{\sigma^2}{n} + \frac{n-1}{n} c$$

where $\bar{u} = \sum_i^n u_i/n$. To see this, we turn to Theorem 11. To reduce that theorem to this special case, we first put $k = n$, $c_i = 1/n$, thus obtaining

$$\sigma_{\bar{u}}^2 = \frac{1}{n^2} \sum_i^n \sum_j^n \sigma_{u_i u_j}$$

We have already noted that $\sigma_{u_i u_i} = \sigma_{u_i}^2$. Also, it is easy to verify that the double summation above contains n^2 terms of which exactly n are of the form $\sigma_{u_i u_i}$, so that the remaining $n^2 - n = n(n-1)$ terms are of the form $\sigma_{u_i u_j}$, $i \neq j$. Hence

$$\sigma_{\bar{u}}^2 = \frac{1}{n^2} \sum_i^n \sigma_{u_i u_i} + \frac{1}{n^2} \sum_{i \neq j}^n \sigma_{u_i u_j}$$

where the first summation has n terms and the second summation has $n(n-1)$ terms. Therefore

$$\sigma_{\bar{u}}^2 = \frac{n}{n^2} \sigma^2 + \frac{n(n-1)}{n^2} c = \frac{\sigma^2}{n} + \frac{n-1}{n} c$$

(*b*) Let u_i have two possible values, 1 and 0, and let $Pr(u_i = 1) = P_i$, $Pr(u_i = 0) = Q_i = 1 - P_i$, $i = 1, \cdots, n$. Let $u = \sum_i^n c_i u_i$. Then, if the u_i are uncorrelated, it follows from Corollary 1 to Theorem 11 that

$$\sigma_u^2 = \sum_i^n c_i^2 \sigma_{u_i}^2$$

In this special case

$$\begin{aligned} \sigma_{u_i}^2 &= E(u_i - Eu_i)^2 = (1 - P_i)^2 P_i + (0 - P_i)^2 Q_i = Q_i^2 P_i + P_i^2 Q_i \\ &= P_i Q_i (P_i + Q_i) = P_i Q_i \end{aligned}$$

Hence

$$\sigma_u^2 = \sum_i^n c_i^2 P_i Q_i$$

Further illustrations of the uses of Theorems 7–12 will be found in Chapters 4 and 5.

Exercises

4.4. Consider

$$\bar{u} = \frac{u_1 + u_2 + \cdots + u_M}{M}$$

where $u_1, u_2, \cdots, u_M$ are random variables. Show that

$$\sigma_{\bar{u}}^2 = \frac{1}{M^2} \sum_{i,j}^M \sigma_{u_i u_j}$$

Now assume that u_i takes on the values $U_{i1}, U_{i2}, \cdots, U_{iN}$ with equal probability. Show that $E\bar{u} = \bar{U}$, where

$$\bar{U} = \frac{1}{MN} \sum_i^M \sum_\alpha^N U_{i\alpha}$$

Assume, also, that

$$Pr(u_i = U_{i\alpha}, u_j = U_{j\beta}) = 0 \qquad \alpha \neq \beta = 1, 2, \cdots, N$$

Then

$$\sigma_{u_i u_j} = \frac{\sum_\alpha^N (U_{i\alpha} - \bar{U}_i)(U_{j\alpha} - \bar{U}_j)}{N}$$

where

$$\bar{U}_i = \frac{\sum_\alpha^N U_{i\alpha}}{N} \quad \text{and} \quad \bar{U}_j = \frac{\sum_\alpha^N U_{j\alpha}}{N}$$

Evaluate $\sigma_{\bar{u}}^2$.

4.5. Let $u = 3u_1 + 5u_2$, where $\sigma_{u_1}^2 = 2$, $\sigma_{u_2}^2 = .25$, and $\sigma_{u_1 u_2} = .6$. Evaluate σ_u^2.

4.6. Show that

$$-\sigma_u \sigma_v \leq \sigma_{uv} \leq \sigma_u \sigma_v$$

for all random variables u and v having finite variances.

4.7. Show that

$$0 \leq E|uv| \leq (Eu^2)^{1/2} (Ev^2)^{1/2}$$

for all random variables.

5. Conditional expectation. The following definitions and theorems are important in evaluating expected values associated with multi-stage sampling.

Let $P_i(B^\star)$ be the conditional probability that the random variable u takes on the value U_i subject to the occurrence of the event $B^\star$, i.e.

$$P_i(B^\star) = Pr(u = U_i | B^\star) \qquad i = 1, \cdots, N$$

Then, by definition, *the conditional expectation of the random variable u, given $B^\star$, is*

$$E(u|B^\star) = \sum_i^N P_i(B^\star)U_i$$

In other words, to calculate the conditional expectation we apply the same procedure as that for the expected value except that we use conditional probabilities.

Illustration 5.1. Let u and v be two random variables with $Pr(u = U_i,\ v = V_j) = P_{ij},\ i = 1, \cdots, N,\ j = 1, \cdots, M$. Compute $E(u|v = V_j)$. Since $v = V_j$ if any of the elementary events $u = U_i$, $v = V_j$, $i = 1, \cdots, N$, should occur, it follows from the definition of the probability of an event that $Pr(v = V_j) = \sum_i^N P_{ij} = P_{\cdot j}$. Also, by the definition of conditional probability,

$$Pr(u = U_i|v = V_j) = \frac{Pr(u = U_i, v = V_j)}{Pr(v = V_j)}$$

$$= \frac{P_{ij}}{P_{\cdot j}} \qquad i = 1, \cdots, N$$

Hence, defining $B^\star$ to be the event $v = V_j$, we have

$$E(u|v = V_j) = \sum_i^N U_i \frac{P_{ij}}{P_{\cdot j}}$$

Illustration 5.2. (*a*) Let u be a random variable with possible values $U_1, \cdots, U_N$. Then $E(u|u = U_1) = U_1$.

Proof. Since

$$Pr(u = U_i, u = U_1) = Pr(u = U_1) \qquad \text{if } i = 1$$

it follows that

$$Pr(u = U_i|u = U_1) = 1 \qquad \text{if } i = 1$$

$$= 0 \qquad \text{otherwise}$$

When we substitute in the definition of conditional expectation, the result follows.

(*b*) Let u and v be random variables with possible values $U_1, \cdots, U_N, V_1, \cdots, V_M$. Then $E(uv|u = U_1) = U_1E(v|u = U_1)$. If $i = 1$,

$$Pr(u = U_i, v = V_j|u = U_1) = \frac{Pr(u = U_1, v = V_j)}{Pr(u = U_1)}$$

$$= Pr(v = V_j|u = U_1)$$

If $i \neq 1$,

$$Pr(u = U_i, v = V_j, u = U_1) = 0$$

and hence

$$Pr(u = U_i, v = V_j | u = U_1) = 0$$

Hence, by the definition of conditional expectation, we have

$$E(uv | u = U_1) = \sum_j^M U_1 V_j \, Pr(v = V_j | u = U_1)$$
$$= U_1 E(v | u = U_1)$$

It is clear that a theorem comparable to Theorem 6 can be stated in terms of conditional expectations as

Theorem 13. *The conditional expected value of a linear combination of random variables is the same linear combination of the conditional expected values of these random variables, the same condition being used throughout. In symbols, if $u_1, \cdots, u_r$ are random variables, and $u = c_1 u_1 + \cdots + c_r u_r$, then*

$$E(u | B^\star) = \sum_i^r c_i E(u_i | B^\star)$$

The proof directly parallels that given for Theorem 6. It is left as an exercise.

Exercise 5.1. Prove theorems comparable to Theorems 3, 4, and 5 stated in terms of conditional expectations.

Sometimes, the event $B^\star$ is one of the states of a random event $b^\star$. For example, in Illustration 5.1 the event $v = V_j$ is one of the states of the random event "v takes on one of its possible values." In such a case $E(v | B^\star)$ is one of the values of a random variable. If the random event $b^\star$ has possible states $B_1^\star, \cdots, B_M^\star$ with probabilities $P_1, \cdots, P_M$, we define $E(u | b^\star)$ to be the random variable which has possible values $E(u | B_j^\star), j = 1, \cdots, M$, with probabilities $P_1, \cdots, P_M$. Then, we have the basic computing theorem:

Theorem 14. *The expected value of a random variable is equal to the expected value of the conditional expectation of that random variable, the condition being a random event. In symbols, if u is a random variable and $b^\star$ is a random event, then*

$$Eu = E[E(u | b^\star)] = \sum_j^M P_j E(u | B_j^\star)$$

where $P_j = Pr(b^\star = B_j^\star)$.

Proof. By the definition of conditional expectation

$$E(u | B_j^\star) = \sum_i^N U_i P_i(B_j^\star)$$

and by the definition of conditional probability

$$P_i(B_j^\star) = \frac{Pr(u = U_i, b^\star = B_j^\star)}{P_j}$$

so that

$$\sum_j^M P_j E(u|B_j^\star) = \sum_j^M \sum_i^N U_i\, Pr(u = U_i, b^\star = B_j^\star)$$

$$= \sum_i^N U_i \left\{ \sum_j^M Pr(u = U_i, b^\star = B_j^\star) \right\}$$

But the event $u = U_i$ will occur if and only if one of the elementary events $u = U_i$, $b^\star = B_j^\star$, $j = 1, \cdots, M$ should occur. Hence by Ch. 2, Sec. 6, Theorem 5 (p. 28), we have

$$Pr(u = U_i) = \sum_j^M Pr(u = U_i, b^\star = B_j^\star)$$

so that

$$\sum_j^M P_j E(u|B_j^\star) = \sum_i^N U_i\, Pr(u = U_i) = Eu$$

Corollary 1. Let u and v be random variables, and let $E(u|v)$ denote the random variable having possible values $E(u|v = V_j)$ with probabilities $Pr(v = V_j), j = 1, \cdots, M$, where $V_1, \cdots, V_M$ are the possible values of v. Then

$$Eu = E\{E(u|v)\}$$

The proof consists in using Theorem 14, the possible state $B_j^\star$ being $v = V_j, j = 1, \cdots, M$.

Corollary 2. With the same conditions as those given in Corollary 1,

$$Euv = E\{vE(u|v)\}$$

Proof. Since, as is shown in detail in Illustration 5.2(*b*),

$$E(uv|v = V_j) = V_j E(u|v = V_j) \qquad j = 1, \cdots, M$$

we have

$$E(uv|v) = vE(u|v)$$

Then, by Theorem 14

$$Euv = E\{E(uv|v)\} = E\{vE(u|v)\}$$

Illustration 5.3. Suppose that a population consists of 3 blocks on which 5, 6, and 9 families live. A block is selected with probability proportionate to size, and from the selected block 1 family is selected by an epsem. Let u be the number of persons in the selected family. Compute Eu. Let $b^\star$ be the random event having 3 states $B_1^\star$, $B_2^\star$, and $B_3^\star$, where, if $B_i^\star$ occurs, the ith block is selected. Then

$$Pr(b^\star = B_1^\star) = \tfrac{5}{20}, \quad Pr(b^\star = B_2^\star) = \tfrac{6}{20}, \quad \text{and} \quad Pr(b^\star = B_3^\star) = \tfrac{9}{20}$$

since the block is being selected with probability proportionate to size. Now,

$$E(u|B_i^\star) = \sum_j^{N_i} U_{ij} \frac{1}{M_i} = \frac{U_i}{M_i}$$

where U_{ij} is the number of persons in the jth family of the ith block, M_i is the number of families in the ith block, and U_i is the number of persons in the ith block. Therefore, by Theorem 14, we have

$$Eu = \sum_i^3 Pr(b^\star = B_i^\star)E(u|B_i^\star)$$

$$= \sum^3 \frac{M_i}{M}\frac{U_i}{M_i} = \frac{\sum_i^3 U_i}{M}$$

where

$$\frac{M_i}{M} = Pr(b^\star = B_i^\star)$$

and Eu is the average number of persons per family.

Further illustrations of the uses of Theorems 13 and 14 may be found in Chapters 4, 6, 7, and 9.

Exercises

5.2. Let v have possible values 0, 1, 2 with probabilities $\frac{1}{6}$, $\frac{1}{3}$, $\frac{1}{2}$, and let $E(u|v = i) = c_i$, where $c_0 = 3$, $c_1 = -1$, and $c_2 = 7$. Then evaluate Eu.

5.3. Let u have possible values $U_1, \cdots, U_N$, and v have possible values $V_1, \cdots, V_M$, and let $P_{ij} = Pr(u = U_i, v = V_j)$. If $U_i = i$, $i = 1, \cdots, N$, and $V_j = j$, $j = 1, \cdots, M$, evaluate:

$$E(u|v < 3), \quad E(uv|u + v < 4), \quad E(u^2 - v^2|u = 1)$$

6. Conditional variance and covariance. Just as conditional expectation is useful in multi-stage and other sampling problems, so are the conditional variance and covariance.

Let u be a random variable and let $B^\star$ be an event. Then the conditional variance of u, given $B^\star$, $\sigma^2_{u|B^\star}$, is defined by the equation

$$\sigma^2_{u|B^\star} = E[\{u - E(u|B^\star)\}^2|B^\star] \tag{6.1}$$

Thus to compute the conditional variance we first compute the conditional expectation and then compute the conditional expected value of the square of the deviation between the random variable and its conditional expectation.

The conditional covariance of u and v, given $B^\star$, is similarly defined:

$$\sigma_{uv|B^\star} = E[\{u - E(u|B^\star)\}\{v - E(v|B^\star)\}|B^\star] \tag{6.2}$$

Illustration 6.1. (*a*) Let u and v be random variables with possible values, $U_1, \cdots, U_N$ and $V_1, \cdots, V_M$, and let $P_{ij} = Pr(u = U_i, v = V_j)$. Then, if $Pr(v = V_j) > 0$, it follows that

$$Pr(u = U_i | v = V_j) = \frac{P_{ij}}{P_{\cdot j}}$$

where

$$P_{\cdot j} = \sum_i^N P_{ij}$$

and

$$E(u | v = V_j) = \sum_i^N U_i \frac{P_{ij}}{P_{\cdot j}}$$

Also, using Eq. 6.1 with $B^\star$ being the event $v = V_j$, we have

$$\sigma^2_{u|(v=V_j)} = \sum_i^N \frac{P_{ij}}{P_{\cdot j}} [U_i - E(u | v = V_j)]^2$$

(*b*) If $B^\star$ is the event $v = V_1$ or V_2 or $\cdots$ or V_m $(m \leq M)$, then

$$Pr(u = U_i, B^\star) = \sum_j^m P_{ij}$$

and

$$Pr(B^\star) = \sum_i^N \sum_j^m P_{ij}$$

so that

$$Pr(u = U_i | B^\star) = \frac{\sum_j^m P_{ij}}{\sum_i^N \sum_j^m P_{ij}}$$

Then

$$E(u | B^\star) = \sum_i^N U_i \, Pr(u = U_i | B^\star)$$

and

$$\sigma^2_{u|B^\star} = \sum_i^N \frac{\sum_j^m P_{ij}}{\sum_i^N \sum_j^m P_{ij}} [U_i - E(u | B^\star)]^2$$

(*c*) Let $B^\star$ be defined as in Illustration 6.1(*b*) above. Then

$$Pr(u = U_i, v = V_j, B^\star) = P_{ij} \qquad \begin{cases} i = 1, \cdots, N \\ j = 1, \cdots, m \end{cases}$$

$$= 0 \qquad \begin{cases} i = 1, \cdots, N \\ j = m + 1, \cdots, M \end{cases}$$

so that

$$Pr(u = U_i, v = V_j | B^\star) = \frac{P_{ij}}{Pr(B^\star)} \qquad \begin{cases} i = 1, \cdots, N \\ j = 1, \cdots, m \end{cases}$$

$$= 0 \qquad \text{otherwise}$$

and

$$Pr(v = V_j | B^\star) = \sum_i^N \frac{P_{ij}}{Pr(B^\star)} \qquad j = 1, \cdots, m$$

$$= 0 \qquad \text{otherwise}$$

Hence

$$E(v | B^\star) = \sum_j^m V_j \, Pr(v = V_j | B^\star)$$

and

$$\sigma_{uv|B^\star} = \sum_i^N \sum_j^m \frac{P_{ij}}{Pr(B^\star)} [U_i - E(u|B^\star)][V_j - E(v|B^\star)]$$

If $b^\star$ is a random event having states $B_1^\star, \cdots, B_M^\star$ with probabilities $P_1, \cdots, P_M$, then by $\sigma^2_{u|b^\star}$ we mean the random variable having possible values $\sigma^2_{u|B_1^\star}, \cdots, \sigma^2_{u|B_M^\star}$ with probabilities $P_1, \cdots, P_M$. Similarly $\sigma_{uv|b^\star}$ is a random variable having possible values $\sigma_{uv|B_1^\star}, \cdots, \sigma_{uv|B_M^\star}$ with probabilities $P_1, \cdots, P_M$. Then, from the definition of expected value, it follows that

$$E\sigma^2_{u|b^\star} = \sum_j^M P_j \sigma^2_{u|B_j^\star}$$

$$E\sigma_{uv|b^\star} = \sum_j^M P_j \sigma_{uv|B_j^\star}$$

Illustration 6.2. In Illustration 6.1(u) let $b^\star$ be the random event having possible states $B_1^\star, \cdots, B_M^\star$, where $B_j^\star$ is the event $v = V_j$, and let

$$Pr(b^\star = B_j^\star) = Pr(v = V_j) = \sum_i^N P_{ij} = P_{.j}$$

Then $\sigma^2_{u|b^\star}$ is the random variable having possible values $\sigma^2_{u|B_j^\star} = \sigma^2_{u|(v=V_j)}$ and

$$Pr(\sigma^2_{u|b^\star} = \sigma^2_{u|B_j^\star}) = P_{.j} \qquad j = 1, \cdots, M$$

Also

$$E\sigma^2_{u|b^\star} = \sum_j^M P_{.j} \sigma^2_{u|B_j^\star}$$

where $\sigma^2_{u|B_j^\star}$ was evaluated in Illustration 6.1(a).

Theorem 15. *If u and v are random variables, and $b^\star$ is a random event, then*

$$\sigma^2_u = E\sigma^2_{u|b^\star} + \sigma^2_{E(u|b^\star)} \tag{6.3}$$

and

$$\sigma_{uv} = E\sigma_{uv|b^\star} + \sigma_{E(u|b^\star)E(v|b^\star)} \tag{6.4}$$

Proof. Since Eq. 6.3 is a special case of Eq. 6.4, we shall prove only Eq. 6.4. Since $u - Eu = u - E(u|b^\star) + E(u|b^\star) - Eu$ and $v - Ev = v - E(v|b^\star) + E(v|b^\star) - Ev$, we have

$$\sigma_{uv} = E[u - E(u|b^\star)][v - E(v|b^\star)] + E[u - E(u|b^\star)][E(v|b^\star) - Ev]$$
$$+ E[E(u|b^\star) - Eu][v - E(v|b^\star)] + E[E(u|b^\star) - Eu][E(v|b^\star) - Ev]$$

By Theorem 14 we have

$$E[E(u|b^\star)] = Eu$$
$$E[uE(v|b^\star)] = E[E(u|b^\star)E(v|b^\star)]$$

so that

$$E[u - E(u|b^\star)][E(v|b^\star) - Ev]$$
$$= E[uE(v|b^\star) - uEv - E(u|b^\star)E(v|b^\star) + E(u|b^\star)Ev]$$
$$= 0$$

Similarly

$$E[E(u|b^\star) - Eu][v - E(v|b^\star)] = 0$$

Now

$$E[u - E(u|b^\star)][v - E(v|b^\star)] = E\sigma_{uv|b^\star}$$

Finally

$$E[E(u|b^\star) - Eu][E(v|b^\star) - Ev] = \sigma_{E(u|b^\star)E(v|b^\star)}$$

by the definition of covariance. Hence

$$\sigma_{uv} = E\sigma_{uv|b^\star} + \sigma_{E(u|b^\star)E(v|b^\star)}$$

as was to be proved.

Exercises

6.1. Suppose that a random variable u_1 has possible values $U_1, U_2, \ldots, U_N$, with equal probability, and that if u_1 assumes the value U_i, then the random variable u_2 takes on the possible values $U_1, U_2, \cdots, U_N$, except U_i, with equal probability. Show that $\sigma^2_{u_2} = \sigma^2$, where $\sigma^2 = \sum^N (U_i - \bar{U})^2/N$.

Hint:

$$\sigma^2_{u_2} = E\sigma^2_{u_2|u_1} + \sigma^2_{E(u_2|u_1)}$$

$$Eu_i = \frac{\sum_i^N U_i}{N} = \bar{U}$$

$$E(u_2 | u_1 = U_i) = \frac{\sum_j^N U_j - U_i}{N-1} = \frac{N\bar{U} - U_i}{N-1}$$

$$\sigma_{u_1}^2 = \frac{1}{N} \sum_j^N (U_j - \bar{U})^2 = \sigma^2$$

$$\sigma_{E(u_2|u_1)}^2 = \sigma_{u'_1}^2 = \frac{\sigma^2}{(N-1)^2}$$

where

$$u'_1 = \frac{N\bar{U} - u_1}{N-1}$$

and, similarly,

$$E\sigma_{u_2|u_1}^2 = \frac{N(N-2)}{(N-1)^2} \sigma^2$$

6.2. Give the details of the proof that

$$\sigma_{u_1|u_1}^2 = 0$$

$$\sigma_{uv|v} = 0$$

For some illustrations of the application of the above theorems in the evaluation of expected values and variances of estimates for some standard survey designs, see Sec. 2, Ch. 4; Sec. 1, Ch. 5; and Sec. 1, Ch. 6.

Often, as in sampling designs involving three or more stages, we need to consider the following extension of Theorem 15.

Let $b_1^\star, b_2^\star, \cdots, b_K^\star$ be random events, and denote by

$$E(u | j!)$$

the expected value of u, given the results of the first j random events, $j = 1, \cdots, K$.

Define

$$z_j = E(u | j!) \qquad j = 1, \cdots, K \tag{6.5}$$

so that

$$z_K = E(u | K!) = u$$

By

$$\sigma_{z_j|(j-1)!}^2 \qquad j = 1, \cdots, K$$

is meant the variance of z_j holding only the first $j - 1$ random events fixed, so that

$$\sigma_{z_1|0!}^2 = \sigma_{E(u|1!)}^2 = \sigma_{z_1}^2$$

Finally, by

$$E\sigma_{z_j|(j-1)!}^2$$

is meant the expected value of $\sigma_{z_j|(j-1)!}^2$ over all possible states of the first $j - 1$ random events. Then we have the following theorem:

Theorem 16. *If u is a random variable, then*

$$\sigma_u^2 = E\sigma^2_{z_K|(K-1)!} + E\sigma^2_{z_{K-1}|(K-2)!} + \cdots + E\sigma^2_{z_2|1!} + \sigma^2_{z_1}$$

where z_j is defined by Eq. 6.5.

The proof is an immediate extension of Theorem 15.

Corollary. In applying Theorem 16 to a K-stage sampling design, u is the sample estimate and the jth random event is "selection of jth-stage sampling units," $j = 1, 2, \cdots, K$. Thus, z_j is the conditional expected value of u, holding the results of the first j stages of sampling fixed; $\sigma^2_{z_j|(j-1)!}$ is the conditional variance of z_j, holding the results of the first $j-1$ stages fixed; and $E\sigma^2_{z_j|(j-1)!}$ is the expected value of this conditional variance over all possible results of the first $j-1$ stages of sampling, $j = 1, 2, \cdots, K$. Note that $z_K = u$, since the conditional expectation of u for all K stages fixed is u itself.

The quantity

$$E\sigma^2_{z_j|(j-1)!} \qquad \text{for } j = 1, 2, \cdots, K$$

is referred to as the contribution to the variance of the jth stage of sampling.

In multi-stage sampling we use the following notation:

$$z_j = E(u|[1, 2, \cdots, j]) \qquad j = 1, 2, \cdots, K$$

$$\sigma^2_{z_j|(j-1)!} = \sigma^2_{E(u|[1, 2, \cdots, j])|[1, 2, \cdots, j-1]} \qquad j = 1, 2, \cdots, K$$

Thus, for a three-stage design we write

$$\sigma_u^2 = E\sigma^2_{u|[1, 2]} + E\sigma^2_{E(u|[1, 2])|[1]} + \sigma^2_{E(u|[1])}$$

A result similar to Theorem 16 can be obtained for σ_{uv}, using the following notation.

The quantities $E(v|j!)$, $w_j = E(v|j!)$, $\sigma^2_{w_j|(j-1)!}$, $\sigma_{z_jw_j|(j-1)!}$, $E\sigma^2_{w_j|(j-1)!}$, and $E\sigma_{z_jw_j|(j-1)!}$ are defined as were similar expressions in u and the z's; for example, $\sigma_{z_jw_j|(j-1)!}$ is the covariance of z_j and w_j, holding the first $j-1$ random events fixed.

Theorem 17. *If u and v are random variables, then*

$$\sigma_{uv} = E\sigma_{z_Kw_K|(K-1)!} + E\sigma_{z_{K-1}w_{K-1}|(K-2)!} + \cdots + E\sigma_{z_2w_2|1!} + \sigma_{z_1w_1}$$

The proof is omitted.

Corollary. A result similar to the corollary to Theorem 16 can be obtained for σ_{uv} by applying Theorem 17 to a K-stage sampling design.

In the application of the theorems above, if one wishes to determine the contributions from all stages of sampling it may sometimes be convenient to apply Theorem 15 in successive applications. When one wants to determine the contribution to the variance or covariance from a particular stage of sampling, then the corollary to Theorem 16 or 17 is appropriate. These corollaries were used in developing the variance for a three-stage sampling design in Sec. 4 of Ch. 7. The derivation of the variance for a multi-stage sampling design by a successive application of Theorem 15 is given in Chapter 9. The latter use makes it possible to indicate, for example, the contribution to the variance from the first stage of sampling and the combined contribution from all subsequent stages.

7. The Tchebycheff inequality. Convergence in probability, Consistency. In the preceding sections, we have developed the tools by which we compute expected values and variances of survey designs. In order to utilize these quantities, however, it must be possible to evaluate at least approximately the probabilities of specified differences between the estimates and the quantities we wish to estimate. It is to questions arising from this need that we turn in the present section.

a. The Tchebycheff and Markov inequalities. One of the more remarkable theorems of the theory of probability is that due to Tchebycheff by which bounds are derived for the probabilities of the difference between any random variable and a preassigned value, the latter usually being taken to be the expected value of the random variable. The Markov inequality is a generalization of the Tchebycheff inequality.

Theorem 18 (the Tchebycheff inequality). *Let u be a random variable and let c be a constant. Then, for any $\varepsilon > 0$,*

$$\left.\begin{aligned} Pr(|u - c| > \varepsilon) &< \frac{\sigma_u^2 + (Eu - c)^2}{\varepsilon^2} \\ Pr(|u - c| \leq \varepsilon) &\geq 1 - \frac{\sigma_u^2 + (Eu - c)^2}{\varepsilon^2} \end{aligned}\right\} \tag{7.1}$$

so that if $c = Eu$ the inequalities become

$$Pr(|u - Eu| > \varepsilon) < \frac{\sigma_u^2}{\varepsilon^2}$$

$$Pr(|u - Eu| \leq \varepsilon) \geq 1 - \frac{\sigma_u^2}{\varepsilon^2}$$

Proof. Suppose that u has possible values $U_1, \cdots, U_N$ with probabilities $P_1, \cdots, P_N$. Suppose that of the N differences $U_1 - c$, $U_2 - c$, $\cdots$, $U_N - c$, exactly m of them, say $U_1 - c, \cdots, U_m - c$, have their absolute values $> \varepsilon$, while the other $N - m$ have absolute values $\leq \varepsilon$. Then

$$Pr(|u - c| > \varepsilon) = P_1 + \cdots + P_m$$

By definition

$$E(u-c)^2 = \sum_i^N P_i(U_i - c)^2$$

$$\geq \sum_i^m P_i(U_i - c)^2$$

$$> \sum_i^m P_i \varepsilon^2$$

since, if $|U_i - c| > \varepsilon$, then $(U_i - c)^2 > \varepsilon^2$. Hence

$$\frac{E(u-c)^2}{\varepsilon^2} > \sum_i^m P_i$$

Now, by Theorem 7 of Sec. 4 (p. 51) of this chapter,

$$E(u-c)^2 = \sigma_u^2 + (Eu - c)^2$$

which completes the proof of the first part of the theorem. The second part follows from the fact that

$$Pr(|u-c| > \varepsilon) + Pr(|u-c| \leq \varepsilon) = 1$$

The importance of the Tchebycheff inequality arises from its generality and its usefulness in proving limiting results. The bound set by the inequality is valid for any random variable. However, if something more than the theorem requires is known about the random variable, then closer bounds can usually be set.

Some general conclusions may be given at once.

Corollary 1 (the Markov inequality). Let u be a random variable. Then, for any $\varepsilon > 0$,

$$Pr(|u - Eu| > \varepsilon) < \frac{E|u - Eu|^k}{\varepsilon^k}$$

and

$$Pr(|u - Eu| \leq \varepsilon) \geq 1 - \frac{E|u - Eu|^k}{\varepsilon^k}$$

if $k \geq 1$.

Proof. Since, by the definition of expected value,

$$E|u - Eu|^k = \sum_i^N P_i |U_i - Eu|^k$$

the steps in proving Corollary 1 follow exactly those of Theorem 18 and are left as an exercise.

Corollary 2. Let $u_1, \cdots, u_n$ be uncorrelated and have the same expected value $\bar{U}$ and same variance σ^2. Then, for any $\varepsilon > 0$,

$$\lim_{n\to\infty} Pr(|\bar{u} - \bar{U}| > \varepsilon) = 0$$

Proof. From Corollary 3 to Theorem 11, we have

$$\sigma_{\bar{u}}^2 = \frac{\sigma^2}{n}$$

Using the Tchebycheff inequality and letting $n \to \infty$, the result is immediate.

We can easily generalize Corollary 2 to the case where the random variables $u_1, \cdots, u_n$ are correlated.

Corollary 3. Let $Eu_i = U_i$ and $\sigma_{u_i u_j} = \sigma_{ij}$, $i, j = 1, \cdots, n$, and let $\bar{U}$ be the arithmetic mean of $U_1, U_2, \cdots, U_n$. Then, for any $\varepsilon > 0$, we have

$$Pr(|\bar{u} - \bar{U}| > \varepsilon) < \frac{\sum_{i,j}^{n} \sigma_{ij}}{n^2 \varepsilon^2}$$

and hence if

$$\lim_{n\to\infty} \frac{\sum_{i,j}^{n} \sigma_{ij}}{n^2} = 0$$

it follows that

$$\lim_{n\to\infty} Pr(|\bar{u} - \bar{U}| > \varepsilon) = 0$$

Exercises

7.1. Let u have possible values 1, 2, 3, 4, 5, each being equally probable. If $\varepsilon = .1$, 1, and 1.5, compare the probabilities of the inequality $|u - Eu| > \varepsilon$ obtained from the Tchebycheff inequality with the true probabilities.

7.2. By setting $\varepsilon = 3\sigma$ in Eq. 7.1, show that no more than one-ninth of the possible results will differ from the average of all results by more than three times the standard deviation, no matter what the population.

Remark 1. With the aid of the Tchebycheff inequality it is possible to show in what sense the relative frequencies of occurrence of the possible values of a random variable tend to the probabilities of those possible values.

Let $u_1, \cdots, u_n$ be n independent random variables each having possible values $U_1, \cdots, U_N$ and probabilities $P_1, \cdots, P_N$. Then, we say that $u_1, \cdots, u_n$ are n independent random variables having the same distributions, i.e., same possible values and probabilities. Let v_{ij} be a random variable having two possible values 1 and 0. Let $v_{ij} = 1$ if and only if $u_j = U_i$ and let $v_{ij} = 0$ otherwise. Then, $v_{i1}, \ldots, v_{in}$ are independent and

$$\begin{aligned} Pr(v_{ij} = 1) &= Pr(u_j = U_i) = P_i \\ Pr(v_{ij} = 0) &= Pr(u_j \neq U_i) = 1 - P_i \end{aligned} \quad \begin{cases} i = 1, \cdots, N \\ j = 1, \cdots, n \end{cases}$$

Also, if

$$\bar{v}_{in} = \frac{\sum_{j}^{n} v_{ij}}{n}$$

then the possible values of $\bar{v}_{in}$ are the relative frequencies with which U_i may occur in performing the n operations. Since

$$Ev_{ij} = 1P_i + 0(1 - P_i) = P_i$$

it follows from Theorem 6 that $E\bar{v}_{in} = P_i$. Since $\sigma^2_{v_{ij}} = P_i(1 - P_i)$ for all j, and since $v_{i1}, \cdots, v_{in}$ are independent, it follows from Corollary 3 to Theorem 11 that

$$\sigma^2_{\bar{v}_{in}} = \frac{P_i(1 - P_i)}{n}$$

Then, from Theorem 18, we have

$$Pr(|\bar{v}_{in} - P_i| > \varepsilon) < \frac{P_i(1 - P_i)}{n\varepsilon^2}$$

so that

$$\lim_{n\to\infty} Pr(|\bar{v}_{in} - P_i| > \varepsilon) = 0$$

i.e., the probability that the relative frequency of U_i differs from its probability by more than any positive quantity, however small, will tend to zero as n increases.

b. Convergence in probability. Consistency. We have already defined biased and unbiased estimates in Sec. 2 of this chapter. We shall now define consistency, which is often a more useful property than unbiasedness. To do this we shall first define convergence in probability, from which the definition of consistency follows almost immediately.

Definition of convergence in probability. We call a set of random variables $v_1, v_2, \cdots$ that are identified by the positive integers a *sequence* of random variables; i.e., given the positive integer i, there is exactly one random variable v_i that is identified, $i = 1, 2, \cdots$.

A sequence of random variables $v_1, v_2, \cdots$ is said to *converge in probability* to a random variable or constant v if, for every $\varepsilon > 0$,

$$\lim_{n\to\infty} Pr(|v_n - v| > \varepsilon) = 0$$

or, equivalently,

$$\lim_{n\to\infty} Pr(|v_n - v| \leq \varepsilon) = 1$$

Definition of consistency. A sequence of estimates $u_1, u_2, \cdots$ is said to be a *consistent* estimate of U if the sequence $u_1, u_2, \cdots$ converges in probability to U. Thus, if $u_1, u_2, \cdots$ are uncorrelated random variables having a common mean $\bar{U}$ and common variance σ^2, and if $\bar{u}_1, \bar{u}_2, \cdots$ is the sequence of arithmetic means, where

$$\bar{u}_n = \frac{u_1 + \cdots + u_n}{n} \qquad n = 1, 2, \cdots$$

then we have proved in Corollary 2 to Theorem 18 that the sequence of arithmetic means converges in probability to the expected value $\bar{U}$. Therefore $\bar{u}_n$ is a consistent estimate of $\bar{U}$. Also, we have shown in Remark 1 (p. 71) that the relative frequencies with which each of the possible results occurs when the same operation is repeated will converge in probability to the probability of the occurrence of that possible result as the number of repetitions increases. Thus, the sequence of relative frequencies $\bar{v}_{i1}, \bar{v}_{i2}, \cdots$ defined in the Remark is a consistent estimate of P_i, $i = 1, 2, \cdots, N$.

In general, the expected value of the limiting value of a sequence of random variables is equal to the limiting value of the sequence of the expected values of the random variables. There are some minor exceptions, such as in Illustration 7.1 below, but these exceptions cannot occur if all the variables are constrained to be less than some common upper bound.

Illustration 7.1. Convergence in probability may occur without the expected values of the sequence of random variables tending to the expected value of the limiting random variable. For example, let u_n have two possible values 0 and n, and let

$$Pr(u_n = 0) = \frac{n-1}{n}$$

$$Pr(u_n = n) = \frac{1}{n}$$

Then the sequence of random variables $u_1, u_2, \cdots$ converges in probability to 0, since, for any $\varepsilon > 0$,

$$\lim_{n\to\infty} Pr(|u_n - 0| > \varepsilon) = 0$$

However,

$$Eu_n = 0\left(\frac{n-1}{n}\right) + n\left(\frac{1}{n}\right) = 1$$

for all n so that

$$\lim_{n\to\infty} Eu_n \neq E\,(\operatorname*{plim}_{n\to\infty} u_n)$$

where "plim" stands for "limit in the sense of probability."

The definition of consistency given above is useful when the sampling is done with replacement or the population is infinite. However, a great deal of its usefulness is lost when sampling from a finite population. For example, suppose that a population consists of 2000 elementary units, classified into two groups each consisting of 1000 elementary units, all units in the first group being 1's and all units in the second group being 0's. Suppose, also, that the sample is obtained by selecting elements from group 1 until the number of elementary units in the first group is exhausted and then selecting the remaining elements for the sample from the second group. Now suppose that the estimate u_n of the proportion of 1's in the population is defined to be the proportion of 1's in the sample of size n. Then $u_n = 1$, $n = 1, \cdots, 1000$, and $u_n = 1000/n$, $n = 1001, \cdots, 2000$. If the sample were selected with replacement, u_n would not be a consistent estimate of the proportion of 1's in the population, since the estimate would always be 1.

To avoid this type of contradiction, we shall make the following assumptions for any finite population. We shall require that to meet the definition of consistency the following two conditions be satisfied:

(1) As the size of sample n increases, the size of population N will also increase, and for all n and N we will have $n < tN$, where $0 < t < 1$.
(2) As the size of population increases, the quantity U that we want to estimate will remain constant.

If these two assumptions are made, then the sequence of estimates discussed in the example above would be inconsistent.

When "lim" is written, we mean the limit as n becomes infinite, subject to the above two conditions if we are sampling from a finite population.

We give two simple theorems that enable us to prove consistency in many cases.

Theorem 19. *Let* $Eu_n = b_n$ *and let* $\sigma^2_{u_n} = \sigma^2_n$. *Then a sufficient condition that the sequence* $u_1, u_2, \cdots$ *be a consistent estimate of* U *is that both of the following conditions are satisfied*:

(1) $\lim |b_n - U| = 0$
(2) $\lim \sigma^2_n = 0$

Proof. By Tchebycheff's inequality (Theorem 18), we show that

$$Pr(|u_n - U| > \varepsilon) < \frac{\sigma^2_n + (b_n - U)^2}{\varepsilon^2}$$

It follows that if (1) and (2) hold, then the sequence $u_1, u_2, \cdots$ is a consistent estimate of the sequence U.

Corollary. If u_n is an unbiased estimate of U, $n = 1, 2, \cdots$, and if $\lim \sigma_n^2 = 0$, then $u_1, u_2, \cdots$ is a consistent estimate of U.

Proof. The proof consists of noting that $b_n = U$ in Theorem 19.

The following theorem states that any continuous function of consistent estimates is itself consistent and the value to which it tends is the same function of the limiting values of the consistent estimates.

Theorem 20. *Let the sequence u_{in} be a consistent estimate of U_i, $i = 1, \cdots, k$, and let $f(t_1, \cdots, t_k)$ be a continuous function of $t_1, \cdots, t_k$. Then the sequence $f(u_{1n}, \cdots, u_{kn})$ is a consistent estimate of $f(U_1, \cdots, U_k)$.*

Proof. Since $f(t_1, \cdots, t_k)$ is continuous, it follows that, given any $\varepsilon > 0$, we can find $\delta > 0$ such that, if $|u_{1n} - U_1| < \delta, \cdots, |u_{kn} - U_k| < \delta$, then $|f(u_{1n}, \cdots, u_{kn}) - f(U_1, \cdots, U_k)| < \varepsilon$. Hence

$$Pr[|f(u_{1n}, \cdots, u_{kn}) - f(U_1, \cdots, U_k)| < \varepsilon] \geq Pr(|u_{1n} - U_1| < \delta, \cdots, |u_{kn} - U_k| < \delta)$$

Now

$$Pr(|u_{1n} - U_1| < \delta, \cdots, |u_{kn} - U_k| < \delta) + Pr(\text{at least one of the inequalities } |u_{in} - U_i| \geq \delta) = 1$$

since one of the two events in parentheses is certain to occur. Also

$$Pr(\text{at least one of the inequalities } |u_{in} - U_i| \geq \delta) \leq \sum_i^k Pr(|u_{in} - U_i| \geq \delta)$$

since if the event in the parenthesis on the left occurs, one or more of the events on the right will occur. But, since the sequence $u_{in} - U_i$ converges in probability to zero, $i = 1, \cdots, k$, it then follows that the right side tends to zero, and hence

$$\lim Pr(\text{at least one of the } k \text{ inequalities } |u_{in} - U_i| \geq \delta) = 0$$

so that for any $\varepsilon > 0$

$$\lim Pr[|f(u_{1n}, \cdots, u_{kn}) - f(U_1, \cdots, U_k)| < \varepsilon] = 1$$

which completes the proof.

Corollary 1. If $f(t_1, \cdots, t_k)$ is a polynomial in $t_1, \cdots, t_k$, then $f(u_{1n}, \cdots, u_{kn})$ is a consistent estimate of $f(U_1, \cdots, U_k)$ if u_{in} is a consistent estimate of U_i, $i = 1, \cdots, k$.

Corollary 2. A rational function of consistent estimates is a consistent estimate of the same rational function of the quantities being estimated provided the denominator does not vanish. More formally, if $f(t_1, \cdots, t_k)$ is a rational function of $t_1, \cdots, t_k$, then $f(u_{1n}, \cdots, u_{kn})$ is a consistent estimate of $f(U_1, \cdots, U_k)$ for all possible values of

$U_1, \cdots, U_k$ except those for which the denominator of $f(t_1, \cdots, t_k)$ vanishes provided that u_{in} is a consistent estimate of U_i, $i = 1, \cdots, k$.

Proof. A rational function is continuous for all finite values of its variables except those for which its denominator vanishes. Hence, if we exclude the values for which the denominator vanishes, we will be limiting ourselves to values for which $f(t_1, \cdots, t_k)$ is continuous, and Theorem 20 applies.

It is easy to show that an estimate may be consistent and yet have infinite variance as n increases. If we consider the sequence of random variables discussed in Remark 1 of this section, then we have already shown that $u_1, u_2, \cdots$ is a consistent estimate of 0, since the sequence converges in probability to 0. We have also shown that $Eu_n = 1$. Now

$$Eu_n^2 = 0\left(\frac{n-1}{n}\right) + n^2\frac{1}{n} = n$$

so that $\sigma_{u_n}^2 = n - 1$ and hence $\sigma_{u_n}^2$ becomes larger and larger as n increases, even though $u_1, u_2, \cdots$ converges in probability to zero.

Remark 2. Satisfactory discussions of the normal limiting distribution may be found in almost any standard text on probability theory. It has been shown [see, for example, W. G. Madow, "On the Limiting Distributions of Estimates Based on Samples from Finite Universes," *Annals Math. Stat.*, **19** (1948), 535–545] that these limiting distributions also are valid for samples from finite populations.

APPENDIX A

SUMS OF POWERS

By the Kth sum of powers about a is meant

$$\sum_{i=1}^{M} (x_i - a)^K \qquad K = 1, 2, \cdots$$

Ordinarily we shall be concerned with sums of powers about 0 ($a = 0$) and sums of powers about $\bar{x}$ ($a = \bar{x}$, the arithmetic mean of $x_1, \cdots, x_M$). Sums of powers about 0 will be denoted by $S_1^\star, S_2^\star, \cdots$, and sums of powers about $\bar{x}$ will be denoted by $S_1, S_2, \cdots$.

In order to evaluate the S's in terms of the $S^\star$'s let us recall the binomial theorem for positive integral exponents.

Binomial theorem. *If K is any positive integer, then*

$$\left.\begin{aligned}(a+b)^K &= a^K + C_1^K a^{K-1}b + C_2^K a^{K-2}b^2 + \cdots + b^K \\ &= \sum_{i=0}^{K} C_i^K a^{K-i} b^i\end{aligned}\right\} \tag{A.1}$$

where

$$C_i^K = \frac{K!}{i!(K-i)!} \quad \textit{and} \quad i! = i(i-1)\cdots(2)(1)$$

and we define $C_0^K = 1$.

Proof. If $K = 1$ or $K = 2$, the theorem may be proved by expanding the parenthesis. Suppose now that the theorem is true for K. Then we will prove it to be true for $K + 1$. Since it is true for $K = 1$, this will then complete the proof for all positive integers. Since $(a + b)^{K+1} = (a + b)(a + b)^K$, and since Eq. A.1 is assumed to hold, we have

$$\begin{aligned}(a+b)^{K+1} &= (a+b)\sum_{i=0}^{K} C_i^K a^{K-i}b^i \\ &= \sum_{i=0}^{K} C_i^K a^{K-i+1}b^i + \sum_{i=0}^{K} C_i^K a^{K-i}b^{i+1} \\ &= a^{K+1} + (C_1^K + C_0^K)a^K b + (C_2^K + C_1^K)a^{K-1}b^2 + \\ &\qquad \cdots + (C_K^K + C_{K-1}^K)ab^K + b^{K+1}\end{aligned} \tag{A.2}$$

Hence we want to evaluate $C_{i+1}^K + C_i^K$. Now by the definition we have

$$C_{i+1}^K + C_i^K = \frac{K!}{(i+1)!\,(K-i-1)!} + \frac{K!}{i!\,(K-i)!}$$

$$= \frac{K!}{i!\,(K-i-1)!}\left(\frac{1}{i+1} + \frac{1}{K-i}\right)$$

$$= \frac{(K+1)!}{(i+1)!\,(K-i)!} = C_{i+1}^{K+1}$$

Hence we have proved the important equation

$$C_{i+1}^{K+1} = C_{i+1}^K + C_i^K \qquad i = 0,\cdot\cdot\cdot, K;\ K > 0 \tag{A.3}$$

Substituting in Eq. A.2, we have

$$(a+b)^{K+1} = \sum_{i=0}^{K+1} C_i^{K+1} a^{K+1-i} b^i \tag{A.4}$$

which is the same formula as Eq. A.1 but with K replaced by $K+1$. Hence, the theorem is true for $K+1$ if it is true for K. (This is a proof by induction on K.)

Let us now apply the binomial theorem and some previous results to obtain some relations between S and $S^\star$.

Theorem A.1. *If we define* $S_0 = S_0^\star = M$, *then*

$$S_K = \sum_{j=0}^{K} C_j^K(-1)^j \bar{x}^j S_{K-j}^\star \tag{A.5}$$

$$S_K^\star = \sum_{j=0}^{K} C_j^K \bar{x}^j S_{K-j} \tag{A.6}$$

Proof. By definition $S_K = \sum_{i=1}^{M} (x_i - \bar{x})^K$, and by the binomial theorem

$$(x_i - \bar{x})^K = \sum_{j=0}^{K} C_j^K(-1)^j \bar{x}^j x_i^{K-j}$$

Hence

$$S_K = \sum_{i=1}^{M} \sum_{j=0}^{K} C_j^K(-1)^j \bar{x}^j x_i^{K-j}$$

$$= \sum_{j=0}^{K} \sum_{i=1}^{M} C_j^K(-1)^j \bar{x}^j x_i^{K-j} \quad \text{(by Ex. 2.10, Ch. 2)}$$

$$= \sum_{j=0}^{K} C_j^K(-1)^j \bar{x}^j \sum_{i=1}^{M} x_i^{K-j} \quad \text{(by Eq. 2.1, Ch. 2)}$$

Hence Eq. A.5 is true. To prove Eq. A.6 replace x_i by $x_i - \bar{x} + \bar{x}$ in the definition of $S_K^\star$ and apply the binomial theorem with $x_i - \bar{x} = a$, $\bar{x} = b$. This is left as an exercise.

Let us now write out Eq. A.5 for $K = 1, 2, 3, 4$.

$$S_1 = S_1^\star - M\bar{x} = 0$$

$$S_2 = S_2^\star - 2\bar{x}S_1^\star + M\bar{x}^2 = S_2^\star - M\bar{x}^2$$

$$S_3 = S_3^\star - 3\bar{x}S_2^\star + 3\bar{x}^2S_1^\star - M\bar{x}^3 = S_3^\star - 3\bar{x}S_2^\star + 2M\bar{x}^3$$

$$\begin{aligned} S_4 &= S_4^\star - 4\bar{x}S_3^\star + 6\bar{x}^2S_2^\star - 4\bar{x}^3S_1^\star + M\bar{x}^4 \\ &= S_4^\star - 4\bar{x}S_3^\star + 6\bar{x}^2S_2^\star - 3M\bar{x}^4 \end{aligned}$$

It is easy to verify that the last two terms of S_K will always combine, since they are

$$\begin{aligned} &C_{K-1}^K(-1)^{K-1}\bar{x}^{K-1}S_1^\star + (-1)^K M\bar{x}^K \\ &\qquad = K(-1)^{K-1}M\bar{x}^K + (-1)^K M\bar{x}^K = (-1)^{K-1}(K-1)M\bar{x}^K \end{aligned}$$

Also, from Eq. A.6 we have

$$S_1^\star = S_1 + M\bar{x} = M\bar{x}$$

$$S_2^\star = S_2 + 2\bar{x}S_1 + M\bar{x}^2 = S_2 + M\bar{x}^2$$

$$S_3^\star = S_3 + 3\bar{x}S_2 + 3\bar{x}^2S_1 + M\bar{x}^3 = S_3 + 3\bar{x}S_2 + M\bar{x}^3$$

$$\begin{aligned} S_4^\star &= S_4 + 4\bar{x}S_3 + 6\bar{x}^2S_2 + 4\bar{x}^3S_1 + M\bar{x}^4 \\ &= S_4 + 4\bar{x}S_3 + 6\bar{x}^2S_2 + M\bar{x}^4 \end{aligned}$$

since $S_1 = 0$ as previously shown.

Let us now evaluate $\sum_{i=1}^N i$, $\sum_{i=1}^N i^2$, and $\sum_{i=1}^N i^3$. Let

$$S_K^\star(N) = \sum_{i=1}^N i^K$$

then

$$S_K^\star(N+1) - S_K^\star(N) = (N+1)^K \tag{A.7}$$

But also

$$S_K^\star(N+1) = 1 + \sum_{i=1}^N (i+1)^K$$

so that

$$\left.\begin{aligned} S_K^\star(N+1) - S_K^\star(N) &= 1 + \sum_{i=1}^N [(i+1)^K - i^K] \\ &= 1 + C_1^K S_{K-1}^\star(N) + C_2^K S_{K-2}^\star(N) \\ &\qquad + \cdots + C_{K-1}^K S_1^\star(N) + N \end{aligned}\right\} \tag{A.8}$$

Combining Eq. A.7 and A.8, we have

$$\left.\begin{aligned} &C_1^K S_{K-1}^\star(N) + C_2^K S_{K-2}^\star(N) + \cdots + C_{K-1}^K S_1^\star(N) \\ &\qquad = (N+1)^K - (N+1) \\ &\qquad = (N+1)[(N+1)^{K-1} - 1] \end{aligned}\right\} \tag{A.9}$$

Putting $K = 2, 3, 4$ in A.9, we obtain

$$2S_1^\star(N) = (N+1)N$$

$$3S_2^\star(N) + 3S_1^\star(N) = (N+1)(N^2+2N)$$

$$4S_3^\star(N) + 6S_2^\star(N) + 4S_1^\star(N) = (N+1)(N^3+3N^2+3N)$$

so that, as may be verified,

$$\left.\begin{aligned} S_1^\star(N) &= \frac{N(N+1)}{2} \\ S_2^\star(N) &= \frac{N(N+1)(2N+1)}{6} \\ S_3^\star(N) &= \frac{N^2(N+1)^2}{4} \end{aligned}\right\} \tag{A.10}$$

APPENDIX B

MOMENTS

The rth moment of the random variable u about the quantity c is defined to be

$$E(u - c)^r$$

If u has possible values $U_1, \cdots, U_N$ with probabilities $P_1, \cdots, P_N$, then, by the definition of expected value

$$E(u - c)^r = \sum_i^N P_i(U_i - c)^r$$

If u and v are two random variables, then the product moment of order r, s about c, d is defined to be

$$E(u - c)^r(v - d)^s$$

If the possible values of u are $U_1, \cdots, U_N$ and those of v are $V_1, \cdots, V_M$, and if

$$P_{ij} = Pr(u = U_i, v = V_j)$$

then

$$E(u - c)^r(v - d)^s = \sum_i^N \sum_j^M P_{ij}(U_i - c)^r(V_j - d)^s$$

If $P_{ij} = 0$ when $i \neq j$, i.e., the elementary event U_i, V_j is impossible if $i \neq j$, then we denote P_{ii} by P_i and

$$E(u - c)^r(v - d)^s = \sum_{i=1}^N P_i(U_i - c)^r(V_i - d)^s$$

where N is the minimum of N and M.

Moments about 0 and about the expected values of the random variables are the two most important special cases. We define

$$\mu_r^\star = Eu^r$$

and

$$\mu_r = E(u - Eu)^r$$

Also, by definition,

$$\mu_{rs}^\star = Eu^r v^s$$

$$\mu_{rs} = E(u - Eu)^r(v - Ev)^s$$

Remark. If r is an integer, then by the binomial theorem

$$(u - c)^r = \sum_{\alpha=0}^{r} (-1)^\alpha C_\alpha^r c^\alpha u^{r-\alpha}$$

and

$$(v - d)^s = \sum_{\beta=0}^{s} (-1)^\beta C_\beta^s d^\beta v^{s-\beta}$$

Also

$$u^r = [(u - c) + c]^r$$

$$= \sum_{\alpha=0}^{r} C_\alpha^r c^\alpha (u - c)^{r-\alpha}$$

and similarly

$$v^s = \sum_{\beta=0}^{s} C_\beta^s d^\beta (v - d)^{s-\beta}$$

Hence, if we let $\mu_1^\star = \mu$, we have

$$\mu_r = \sum_{\alpha=0}^{r} (-1)^\alpha C_\alpha^r \mu^\alpha \mu_{r-\alpha}^\star$$

$$= \sum_{\alpha=0}^{r-2} (-1)^\alpha C_\alpha^r \mu^\alpha \mu_{r-\alpha}^\star + (-1)^{r-1}(r-1)\mu^r$$

Also

$$\mu_r^\star = \sum_{\alpha=0}^{r} C_\alpha^r \mu^\alpha \mu_{r-\alpha}$$

$$= \sum_{\alpha=0}^{r-2} C_\alpha^r \mu^\alpha \mu_{r-\alpha} + \mu^r$$

since

$$\mu_1 = 0$$

It is left as an exercise to express the μ_{rs}'s as a linear combination of the $\mu_{rs}^\star$'s and conversely.

Illustration B.1. (*a*) If u is a random variable having two possible values 1 and 0 with probabilities P and $Q = 1 - P$, then compute $\mu_j^\star$ and μ_j, $j = 1, 2, 3, 4$. By definition

$$\mu_j^\star = Eu^j = \sum_{i=1}^{2} P_i U_i^j$$

$$= P1^j + Q0^j = P \qquad i = 1, \cdots$$

Similarly

$$\mu_j = P(1 - P)^j + Q(-P)^j \qquad j = 1, \cdots$$

$$= PQ^j + (-1)^j QP^j$$

$$= PQ\{Q^{j-1} + (-1)^j P^{j-1}\}$$

Hence

$$\mu_1 = 0$$

$$\mu_2 = PQ$$

$$\mu_3 = PQ(Q^2 - P^2) = PQ(Q - P)$$

$$\mu_4 = PQ(Q^3 + P^3)$$

(*b*) If u and v are random variables with possible values 0, 1 and if $Pr(u = i, v = j) = P_{ij}$, $i, j = 0, 1$, then

$$\mu_{11}^{\star} = (1)(1)P_{11} + (1)(0)P_{10} + (0)(1)P_{01} + (0)(0)P_{00} = P_{11}$$

$$\begin{aligned}\mu_{11} &= P_{11}(1 - P_{1\cdot})(1 - P_{\cdot 1}) + P_{01}(- P_{1\cdot})(1 - P_{\cdot 1}) \\ &\quad + P_{10}(1 - P_{1\cdot})(- P_{\cdot 1}) + P_{00}(- P_{1\cdot})(- P_{\cdot 1}) \\ &= P_{11}Q_{1\cdot}Q_{\cdot 1} - P_{01}P_{1\cdot}Q_{\cdot 1} - P_{10}P_{\cdot 1}Q_{1\cdot} + P_{00}P_{1\cdot}P_{\cdot 1}\end{aligned}$$

where

$$P_{1\cdot} = P_{10} + P_{11} = Pr(u = 1) = 1 - Q_{1\cdot}$$

$$P_{\cdot 1} = P_{01} + P_{11} = Pr(v = 1) = 1 - Q_{\cdot 1}$$

(*c*) Let $\bar{u}$ be the arithmetic mean of $u_1, \cdots, u_n$. Then, as may be verified by multiplication,

$$\mu_1^{\star}(\bar{u}) = \frac{1}{n}(Eu_1 + \cdots + Eu_n)$$

$$\mu_2^{\star}(\bar{u}) = \frac{1}{n^2}\left\{\sum_{i=1}^{n} Eu_i^2 + \sum_{i \neq j}^{n} Eu_iu_j\right\}$$

$$\mu_3^{\star}(\bar{u}) = \frac{1}{n^3}\left\{\sum_{i=1}^{n} Eu_i^3 + 3\sum_{i \neq j}^{n} u_i^2u_j + \sum_{i \neq j \neq k}^{n} u_iu_ju_k\right\}$$

$$\begin{aligned}\mu_4^{\star}(\bar{u}) = \frac{1}{n^4}\Bigg\{&\sum_{i=1}^{n} Eu_i^4 + 4\sum_{i \neq j}^{n} Eu_i^3u_j + 3\sum_{i \neq j}^{n} Eu_i^2u_j^2 + 6\sum_{i \neq j \neq k}^{n} Eu_i^2u_ju_k \\ &+ \sum_{i \neq j \neq k \neq m}^{n} Eu_iu_ju_ku_m\Bigg\}\end{aligned}$$

Also $\mu_j(\bar{u})$ is of the same form as $\mu_j^{\star}(\bar{u})$ but with $u_i - Eu_i$ in place of u_i.

Exercises

B.1. If u has possible values U_1, U_2 with probabilities P, Q, compute $\mu_j^{\star}$, μ_j, $j = 1, 2, 3, 4$.

B.2. Verify

(*a*) $\mu_2 = \mu_2^\star - \mu^2$

$\mu_3 = \mu_3^\star - 3\mu\mu_2^\star + 2\mu^3$

$\mu_4 = \mu_4^\star - 4\mu\mu_3^\star + 6\mu^2\mu_2^\star - 3\mu^4$

(*b*) $\mu_2^\star = \mu_2 + \mu^2$

$\mu_3^\star = \mu_3 + 3\mu\mu_2 + \mu^3$

$\mu_4^\star = \mu_4 + 4\mu\mu_3 + 6\mu^2\mu_2 + \mu^4$

B.3. If u has possible values $1, \cdots, N$ with equal probabilities, compute $\mu_j^\star, \mu_j$. Evaluate $\mu_j^\star, \mu_j, j = 1, 2, 3, 4$, if $N = 11$.

APPENDIX C

Rapidity of Approach to Limit

Let f_n be some function of n and suppose that

$$\lim_{n\to\infty} n^h |f_n| = K < \infty$$

where K is a constant. Then we write

$$f_n = 0(n^{-h})$$

and say that f_n tends to K with n^{-h}. If $K = 0$ we will sometimes say

$$f_n = o(n^{-h})$$

Similarly, if f_{nN} is a function of n and N and if

$$\lim_{n,\, N\to\infty} n^h |f_{nN}| = K < \infty$$

$$n < tN$$

$$0 < t < 1$$

where K is a constant, then we shall also write

$$f_{nN} = 0(n^{-h})$$

Of course h may be 0, which we indicate by replacing n^{-h} by 1. If $f_n = 0(1)$ or $f_N = 0(1)$ or $f_{nN} = 0(1)$, then f_n or f_N or f_{nN} is bounded as n, N, or both n and N become infinite. Thus if

$$f_{nN} = \frac{N-n}{N-1}\frac{\sigma^2}{n}$$

it follows that

$$\frac{N-n}{N-1}\frac{\sigma^2}{n} = 0\left(\frac{1}{n}\right)$$

Since by the Tchebycheff inequality

$$Pr(|\bar{u}_{nN} - U_N| > \varepsilon) < \frac{1}{\varepsilon^2}\frac{N-n}{N-1}\frac{\sigma_N^2}{n}$$

if $\bar{u}_{nN}$ is the mean of a simple random sample of n elements from a population of N elements having U_N as mean and σ_N^2 as variance, it follows that, if σ_N^2 is a constant or if $\sigma_N^2 = 0(1)$, then

$$Pr(|\bar{u}_{nN} - U_N| > \varepsilon) = 0\left(\frac{1}{n}\right)$$

It should be noted that even if $\lim_{N\to\infty} \sigma_N^2 = \infty$ it may still be that $\bar{u}_{nN} - U_N$ converges in probability to 0 but the rate of approach may be less. For example, suppose that $\sigma_N^2 = \sqrt{N}\,K$ and $n = tN$, $0 < t < 1$. Then

$$Pr(|\bar{u}_{nN} - U_N| > \varepsilon) < \frac{1}{\varepsilon^2}\frac{N}{N-1}\frac{(1-t)}{t}\frac{K}{\sqrt{N}} = 0\,(N^{-1/2})$$

For simplicity we shall now suppose that $u_1, \cdots, u_n, v_1, \cdots, v_n$ are obtained by sampling the pairs u_i, v_i, with replacement from a population such that $Eu_i = U$, $Ev_i = V$, $\sigma_{u_i}^2 = \sigma_u^2$, $\sigma_{v_i}^2 = \sigma_v^2$, $\sigma_{u_iv_i} = \sigma_{uv}$, $i = 1, \cdots, n$. Then, if

$$\bar{u}_n = \frac{1}{n}\sum_{i=1}^{n} u_i, \quad \text{and} \quad \bar{v}_n = \frac{1}{n}\sum_{i=1}^{n} v_i$$

it follows that

$$\sigma_{\bar{u}_n}^2 = \frac{\sigma_u^2}{n} = 0\left(\frac{1}{n}\right)$$

$$\sigma_{\bar{v}_n}^2 = \frac{\sigma_v^2}{n} = 0\left(\frac{1}{n}\right)$$

$$\sigma_{\bar{u}_n\bar{v}_n} = \frac{\sigma_{uv}}{n} = 0\left(\frac{1}{n}\right)$$

Also

$$E(\bar{u}_n - U)^3 = \frac{1}{n^3}\left\{\sum_{i=1}^{n} E(u_i - U)^3 + 3\sum_{i\neq j} E(u_i - U)^2(u_j - U)\right.$$

$$\left. + \sum_{i\neq j\neq k} E(u_i - U)(u_j - U)(u_k - U)\right\}$$

$$= \frac{\mu_{3u}}{n^2} = 0\left(\frac{1}{n^2}\right)$$

if $\mu_{3u} = E(u_i - U)^3$, $i = 1, \cdots, n$, and similarly

$$E(\bar{v}_n - V)^3 = \frac{\mu_{3v}}{n^2} = 0\left(\frac{1}{n^2}\right)$$

if $\mu_{3v} = E(v_i - V)^3$. Now

$$E(\bar{u}_n - U)^2(\bar{v}_n - V) = \frac{1}{n^3}\left\{\sum_{i=1}^{n} E(u_i - U)^2(v_i - V) + \sum_{i\neq j} E(u_i - U)^2(v_j - V) + 2\sum_{i\neq j} E(u_i - U)(u_j - U)(v_j - V) + \sum_{i\neq j\neq k} E(u_i - U)(u_j - U)(v_k - V)\right\}$$

$$= \frac{\mu_{21}}{n^2} = 0\left(\frac{1}{n^2}\right) \quad \text{if } \mu_{21} = E(u_i - U)^2(v_i - V) \quad i = 1,\cdots, n$$

Theorem C.1. *If the random variables* $u_1, \cdots, u_n$ *are independent and* $Eu_i = U$, $E(u_i - U)^j = \mu_j$, $i = 1, \cdots, n$, $j = 1, 2, \cdots$ $(\mu_1 = 0)$, *and if* $\bar{u}_n$ *is the arithmetic mean of* $u_1, \cdots, u_n$, *then*

$$\lim_{n\to\infty} n^{\frac{j}{2}} E(\bar{u}_n - U)^j = \frac{j!}{2^{\frac{j}{2}}\left(\frac{j}{2}\right)!}\mu_2^{\frac{j}{2}} \qquad \text{if } j \text{ is even}$$

$$\lim_{n\to\infty} n^{\frac{j}{2}} E(\bar{u}_n - U)^j = 0 \qquad \text{if } j \text{ is odd}$$

Proof. We can suppose without loss of generality that $U = 0$. Then

$$E\bar{u}_n^j = \frac{1}{n^j} E(\sum_i^n u_i)^j \tag{C.1}$$

is a sum of terms like

$$E(u_1^{j_1}u_2^{j_2}\cdots u_h^{j_h}) \tag{C.2}$$

where $j_1 + j_2 + \cdots + j_h = j$.

If $U = 0$ then $Eu_i = 0$, $i = 1, \cdots, n$, so that any term such as Eq. C.2 with any of the j's equal to 1 will vanish. Hence we can limit ourselves to terms in the expansion of Eq. C.1 for which each of the j's is at least as large as 2. Then

$$E(u_1^{j_1}u_2^{j_2}\cdots u_h^{j_h}) = \mu_{j_1}\mu_{j_2}\cdots\mu_{j_h} \tag{C.3}$$

and in the expansion of Eq. C.1 we will have $\mu_{j_1}\mu_{j_2}\cdots\mu_{j_h}$ with a coefficient equal to $\dfrac{n(n-1)\cdots(n-h+1)}{h!}$ times a factor that is independent of n. Now, if j is even, then the maximum value of h will be $j/2$, which will occur if $j_1 = \cdots = j_h = 2$; and if j is odd, then the maximum value of h will be $(j-1)/2$, which will occur if one of the j's is 3 and the others are all 2. Hence, if j is even

$$E\bar{u}_n^j = \frac{1}{n^j}\left\{C_{\frac{j}{2}} \frac{n(n-1)\cdots\left(n-\frac{j}{2}+1\right)}{\left(\frac{j}{2}\right)!} \mu_2^{\frac{j}{2}}\right\} + o(n^{-\frac{j}{2}})$$

so that

$$\lim_{n\to\infty} n^{\frac{j}{2}} E\bar{u}_n^j = \frac{C_{\frac{j}{2}}}{\left(\frac{j}{2}\right)!} \mu_2^{\frac{j}{2}} \qquad \text{if } n \text{ is even}$$

Furthermore $C_{\frac{j}{2}}$ will be the number of ways of distributing j elements into $j/2$ boxes each containing 2 elements or

$$C_{\frac{j}{2}} = \frac{j!}{(2!)^{\frac{j}{2}}} = \frac{j!}{2^{\frac{j}{2}}}$$

If j is odd, then

$$E\bar{u}_n^j = \frac{1}{n^j}\left\{C_{\frac{j-1}{2}} \frac{n(n-1)\cdots\left(n-\frac{j-1}{2}+1\right)}{\left(\frac{j-1}{2}\right)!} \mu_2^{\frac{j-3}{2}} \mu_3\right\} + o(n^{-\frac{j+1}{2}})$$

But

$$\frac{1}{n^j}\left\{n(n-1)\cdots\left(n-\frac{j-1}{2}+1\right)\right\} = o(n^{-\frac{j}{2}})$$

so that, if j is odd, then

$$E\bar{u}_n^j = o(n^{-\frac{j}{2}})$$

which completes the proof.

It may be noted that Theorem C.1 holds if we sample with replacement from a finite population and let the size of population become infinite in such a manner that its moments are unchanged. It is also possible to prove theorems like Theorem C.1 for sampling without replacement and to show that

$$E(\bar{u}_n - U)^j(\bar{v}_n - V)^g = 0(n^{-\frac{j+g}{2}}) \qquad \text{if } j+g \text{ is even}$$

$$= 0(n^{-\frac{j+g+1}{2}}) \qquad \text{if } j+g \text{ is odd}$$

Corollary. If j is odd, then

$$E|\bar{u}_n - U|^j = 0(n^{-\frac{j+1}{2}})$$

Proof. For any random variable w we have

$$\sigma^2_{|w|} = E|w|^2 - (E|w|)^2 \geq 0$$

so that

$$E|w|^2 \geq (E|w|)^2$$

and since

$$|w|^2 = w^2$$

we have

$$(Ew^2)^{1/2} \geq E|w|$$

Hence, putting $w = (\bar{u}_n - U)^j$, we have

$$E|\bar{u}_n - U|^j \leq [E(\bar{u}_n - U)^{2j}]^{1/2}$$

and thus

$$\lim_{n\to\infty} n^{j/2} E|\bar{u}_n - U|^j \leq \lim_{n\to\infty} [n^j E(\bar{u}_n - U)^{2j}]^{1/2} = K_{2j}^{1/2} < \infty$$

But the order of $E|\bar{u}_n - U|^j$ must be integral. If it were $\frac{j-1}{2}$, then $\lim_{n\to\infty} n^{\frac{j}{2}} E|\bar{u}_n - U|^j$ would be infinite. Hence it must be $\frac{j+1}{2}$, which completes the proof.

We can similarly show that if $j + g$ is odd then

$$E|\bar{u}_n - U|^j|\bar{v}_n - V|^g = 0(n^{-\frac{j+g+1}{2}})$$

REFERENCES

(1) H. Cramér, *Mathematical Methods of Statistics*, Princeton University Press, 1946.

(2) W. Feller, *An Introduction to Probability Theory and Its Applications*, Vol. I, John Wiley & Sons, New York, 1950.

CHAPTER 4

Simple Random Sampling

DERIVATIONS, PROOFS, AND SOME EXTENSIONS OF THEORY FOR CH. 4 OF VOL. I*

NOTE. In this chapter we shall present a number of derivations and proofs for simple random sampling. As indicated earlier, a simple random sample of n elements (sampling units) is a sample so drawn that every combination of n elements has the same chance of being selected. Simple random sampling with replacement is introduced in some cases in order to simplify results and at the same time to provide approximate theory for sampling without replacement.

Many of the theorems and statements proved in this chapter are applicable to other types of selection methods. For example, the derivation of the variance of a ratio of random variables, and the derivation of the variance of estimates of precision are applicable to more complex systems of sampling.

1. The mathematical expectation of the arithmetic mean of a simple random sample (Vol. I, Ch. 4, Sec. 7). *To prove:* A sample mean for a simple random sample of n units drawn from a population of N units is an unbiased estimate of the population mean, i.e.,

$$E\bar{x} = \bar{X}$$

where E stands for the mathematical expectation of the expression following it,

$$\bar{x} = \frac{\sum_i^n x_i}{n}, \quad \bar{X} = \frac{\sum_i^N X_i}{N} \tag{1.1}$$

x_i is the value of the characteristic obtained on the ith drawing, and X_i is the value of the characteristic for the ith unit in the population. An estimate is said to be an unbiased estimate of a specified population

* Appropriate references to Vol. I are shown in parentheses after section or subsection headings. The number following I– after some equations gives the chapter, section, and number of that particular equation in Vol. I.

characteristic if and only if the expected value of the estimate is the population characteristic.

Proof. Since the expected value of a constant times a variable is the constant times the expected value of the variable (see Ch. 3, Sec. 3, Theorem 3, p. 47), we may write

$$E\bar{x} = E\frac{1}{n}\sum_{i}^{n} x_i = \frac{1}{n}E\sum_{i}^{n} x_i$$

Further, since the expected value of a sum of random variables is equal to the sum of their expected values (Ch. 3, Sec. 3, Theorem 5, p. 48), we may write

$$E\bar{x} = \frac{1}{n}\sum_{i}^{n} Ex_i \tag{1.2}$$

We now need to evaluate Ex_i. By definition, the expected value of a variate, u, is the sum of the products obtained by multiplying each possible value of u by its probability, summed over all the possible values u takes on. Applying this definition, we have

$$\left.\begin{aligned} Ex_i &= X_1P_1 + X_2P_2 + \cdots + X_NP_N \\ &= \sum_{j}^{N} X_jP_j \end{aligned}\right\} \tag{1.3}$$

where $X_1, \cdots, X_N$ are the possible values of x_i and $P_1, \cdots, P_N$ their respective probabilities.

We now must evaluate P_j, the probability that X_j, the jth element of the population, is selected at the ith drawing. This probability can be written

$$P_j = \frac{N-1}{N}\,\frac{N-2}{N-1}\cdots\frac{N-i+1}{N-i+2}\,\frac{1}{N-i+1} = \frac{1}{N}$$

(see Ch. 2, Sec. 6, Theorem 3, p. 28, and Sec. 7*a* of Ch. 2), where the first factor is the probability that X_j *is not* selected from the whole population of N at the first drawing, the second factor is the probability that X_j *is not* selected from the remaining $N - 1$ elements at the second drawing, and so forth, the last factor being the probability that X_j *is* selected from the remaining $N - i + 1$ elements at the ith drawing. Evidently P_j reduces to $1/N$ for all values of i, $i = 1, \cdots, n$. Therefore the probability of selecting any element at any drawing is equal to $1/N$. Substituting $P_j = 1/N$ in Eq. 1.3, we have

$$Ex_i = \frac{1}{N}\sum_{j}^{N} X_j = \bar{X} \tag{1.4}$$

which shows that the expected value of each observation is the arithmetic mean of the population if simple random sampling is used. Substituting $\bar{X}$ for Ex_i in Eq. 1.2, we have

$$E\bar{x} = \frac{1}{n}\sum_i^n \bar{X} = \frac{1}{n}(n\bar{X})$$
$$= \bar{X}$$

and hence $\bar{x}$ is an unbiased estimate of $\bar{X}$.

Exercise. Show that, if the sample of n elements is selected from a population of N elements at random with replacement, then $E\bar{x} = \bar{X}$. *Hint:* The only change in the proof occurs in the method of evaluation of P_j (which still has the value $1/N$).

2. The variance of the arithmetic mean of a simple random sample (Vol. I, Ch. 4, Sec. 9). *To prove:* The variance, $\sigma_{\bar{x}}^2$, of the sample mean of a simple random sample of n units selected from a population of N units without replacement is given by $(1-f)\dfrac{S^2}{n}$, and if the sample is selected with replacement, the variance is given by σ^2/n, where

$$f = \frac{n}{N} \text{ is the sampling fraction}$$

$$S^2 = \frac{\sum_i^N (X_i - \bar{X})^2}{N-1} \qquad \text{(2.1 or I–4.3.10)}$$

$$\sigma^2 = \frac{\sum_i^N (X_i - \bar{X})^2}{N} \qquad \text{(2.2 or I–4.3.9)}$$

X_i is the value of the characteristic for the ith unit in the population, and $\bar{x}$ and $\bar{X}$ are defined by Eq. 1.1; i.e., if we sample without replacement

$$\sigma_{\bar{x}}^2 = \frac{N-n}{N}\frac{S^2}{n} = (1-f)\frac{S^2}{n} \qquad \text{(2.3 or I–4.9.2)}$$

if we sample with replacement

$$\sigma_{\bar{x}}^2 = \frac{\sigma^2}{n} \qquad (2.4)$$

*Proof.** The variance of any random variable is defined to be the expected value of the square of the deviation between that variable and

* See Remark 2 for an alternative proof that does not make use of conditional expected value theorems.

its expected value (Eq. 4.1, Ch. 3), i.e.,

$$\sigma_{\bar{x}}^2 = E(\bar{x} - E\bar{x})^2$$

and since, from Sec. 1, $E\bar{x} = \bar{X}$, we have

$$\sigma_{\bar{x}}^2 = E(\bar{x} - \bar{X})^2 \tag{2.5}$$

Since

$$\bar{x} = \frac{\sum_i^n x_i}{n}$$

then $\bar{x}$ is a linear combination of random variables; i.e., $\bar{x}$ can be written in the form $u = \sum_i^n c_i u_i$, where $c_i = 1/n$ and $u_i = x_i$. Therefore, by Theorem 11, Ch. 3 (p. 56),

$$\sigma_{\bar{x}}^2 = \sum_i^n \sum_j^n \frac{1}{n^2} \sigma_{ij} \tag{2.6}$$

where

$$\sigma_{ij} = E(x_i - \bar{X})(x_j - \bar{X}) \qquad i, j = 1, 2, \cdots, n$$

Equation 2.6 can be rewritten

$$\sigma_{\bar{x}}^2 = \frac{1}{n^2} \sum_i^n \sigma_i^2 + \frac{1}{n^2} \sum_{i \neq j}^n \sigma_{ij} \tag{2.7}$$

Thus, for a sample of 3, we have

$$\begin{aligned} \bar{x} &= \tfrac{1}{3}x_1 + \tfrac{1}{3}x_2 + \tfrac{1}{3}x_3 \\ &= \tfrac{1}{3}(x_1 + x_2 + x_3) \end{aligned}$$

and the variance of $\bar{x}$ is

$$\begin{aligned} \sigma_{\bar{x}}^2 &= \tfrac{1}{9}(\sigma_1^2 + \sigma_2^2 + \sigma_3^2 + \sigma_{12} + \sigma_{13} + \sigma_{21} + \sigma_{23} + \sigma_{31} + \sigma_{32}) \\ &= \tfrac{1}{9}(\sigma_1^2 + \sigma_2^2 + \sigma_3^2 + 2\sigma_{12} + 2\sigma_{13} + 2\sigma_{23}) \end{aligned}$$

where σ_1^2, σ_2^2, and σ_3^2 are the variances of the values obtained on the first, second, and third selections, respectively. The remaining terms, $\sigma_{12} = \sigma_{21}$, $\sigma_{13} = \sigma_{31}$, and $\sigma_{23} = \sigma_{32}$, are the covariances of the values taken on by *i*th and *j*th selections with *i* not equal to *j*. Thus, σ_{12} is the covariance between the values taken by the first and second selections. Note that there are 3 variance terms and 6 covariance terms. In general, for a sample of size *n*, the first term of Eq. 2.7 will consist of *n* variance terms and the second will consist of $n(n - 1)$ covariance terms. The $\sum_{i \neq j}^n$ in Eq. 2.7 denotes the sum over these $n(n - 1)$ covariance terms. Now, since $(x_i - \bar{X})^2$ takes on the possible values $(X_1 - \bar{X})^2, (X_2 - \bar{X})^2, \cdots, (X_N - \bar{X})^2$, each with probability $1/N$, we have, for σ_i^2, in Eq. 2.7,

$$\sigma_i^2 = E(x_i - \bar{X})^2 = \frac{1}{N}\sum_i^N (X_i - \bar{X})^2 = \sigma^2 \tag{2.8}$$

Hence, $\sigma_i^2 = \sigma^2$ for all i.

Now, to evaluate the second term of Eq. 2.7, we must find

$$\sigma_{ij} = E(x_i - \bar{X})(x_j - \bar{X})$$

By Corollary 2 of Theorem 14, Ch. 3 (p. 62), we can write

$$E(x_i - \bar{X})(x_j - \bar{X}) = E[(x_i - \bar{X})E\{(x_j - \bar{X})|(x_i - \bar{X})\}] \tag{2.9}$$

where $E\{(x_j - \bar{X})|(x_i - \bar{X})\}$ means the conditional expected value of $(x_j - \bar{X})$ for a fixed value of $(x_i - \bar{X})$.

To evaluate this conditional expected value, when sampling without replacement, one must in effect list the possible values of $(x_j - \bar{X})$ for the jth selection for a fixed value of the ith selection, and determine their probabilities of selection. Thus, for $N = 3$, if $(x_i - \bar{X}) = (X_2 - \bar{X})$, then $(x_j - \bar{X})$ takes on the values $(X_1 - \bar{X})$, and $(X_3 - \bar{X})$ with probabilities equal to $1/(3 - 1) = \frac{1}{2}$. In general, the probability of $(x_j - \bar{X})$ taking on a particular value on the jth selection other than that of $(x_i - \bar{X})$ is

$$Pr[(x_j - \bar{X})|(x_i - \bar{X})] = \frac{1}{N - 1} \tag{2.10}$$

since, with the ith selection fixed, the only possible values taken on by $(x_j - \bar{X})$ are those left after the value taken on by the ith selection is withdrawn from the population. Hence

$$E[(x_j - \bar{X})|(x_i - \bar{X})] = \frac{1}{N-1}\left[\sum_k^N (X_k - \bar{X}) - (x_i - \bar{X})\right] \tag{2.11}$$

Since

$$\sum_k^N (X_k - \bar{X}) = \sum_k^N X_k - N\bar{X} = 0$$

Eq. 2.11 becomes

$$E[(x_j - \bar{X})|(x_i - \bar{X})] = -\frac{1}{N-1}(x_i - \bar{X}) \tag{2.12}$$

Substituting Eq. 2.12 for $E[(x_j - \bar{X})|(x_i - \bar{X})]$ in Eq. 2.9, we have

$$E(x_i - \bar{X})(x_j - \bar{X}) = -\frac{E(x_i - \bar{X})^2}{N - 1}$$

and from Eq. 2.8

$$E(x_i - \bar{X})(x_j - \bar{X}) = -\frac{\sigma^2}{N-1} \tag{2.13}$$

for all i and j. Substituting Eq. 2.8 and 2.13 for σ_i^2 and σ_{ij} in Eq. 2.7, we have

$$\sigma_{\bar{x}}^2 = \frac{n}{n^2}\,\sigma^2 - \frac{n(n-1)}{n^2(N-1)}\,\sigma^2$$

$$= \frac{N-n}{(N-1)n}\,\sigma^2 \tag{2.14}$$

Since

$$\frac{N}{N-1}\,\sigma^2 = S^2$$

$$\sigma_{\bar{x}}^2 = \frac{N-n}{N}\,\frac{S^2}{n} = (1-f)\,\frac{S^2}{n} \tag{2.3}$$

If the sample is selected with replacement, $\sigma_{\bar{x}}^2$ is equal to the first term of Eq. 2.7 since σ_{ij} in that equation

$$= E(x_i - \bar{X})(x_j - \bar{X}) = E(x_i - \bar{X})E(x_j - \bar{X}) = 0 \tag{2.15}$$

Hence, for sampling with replacement,

$$\sigma_{\bar{x}}^2 = \frac{\sigma^2}{n} \tag{2.4}$$

Remark 1. For an estimated total, $x' = N\bar{x}$, the results are

$$\sigma_{x'}^2 = N^2(1-f)\,\frac{S^2}{n}$$

when sampling without replacement and

$$\sigma_{x'}^2 = N^2\,\frac{\sigma^2}{n}$$

when sampling with replacement. This follows from Corollary 4, Theorem 11, Ch. 3 (p. 57).

Remark 2. Following is an alternate proof which makes use only of the definition of expected value and basic expected value theorems. Returning to Eq. 2.5, we have

$$\sigma_{\bar{x}}^2 = E(\bar{x} - \bar{X})^2$$

$$= E\left\{\frac{\sum_i^n (x_i - \bar{X})}{n}\right\}^2 = \frac{1}{n^2}\,E\left\{\sum_i^n (x_i - \bar{X})\right\}^2 \tag{2.16}$$

Now, the square of an algebraic sum of n items can be written as the sum of the squares of the n individual items plus the sum of the cross products of the $n(n-1)$ possible pairs of the n items, i.e.,

$$\left\{\sum_i^n (x_i - \bar{X})\right\}^2 = \sum_i^n (x_i - \bar{X})^2 + \sum_{i \neq k}^n (x_i - \bar{X})(x_k - \bar{X})$$

(See also Eq. 2.3, Sec. 2, Ch. 2.) Using this expansion in Eq. 2.16, we may write

$$\sigma_{\bar{x}}^2 = \frac{1}{n^2} E\left\{\sum_i^n (x_i - \bar{X})^2 + \sum_{i \neq k}^n (x_i - \bar{X})(x_k - \bar{X})\right\}$$

$$= \frac{1}{n^2} \sum_i^n E(x_i - \bar{X})^2 + \frac{1}{n^2} \sum_{i \neq k}^n E(x_i - \bar{X})(x_k - \bar{X}) \tag{2.17}$$

The probability of obtaining $(X_j - \bar{X})^2$ on the ith draw is the same as the probability of obtaining X_j on the ith draw, namely $1/N$. Therefore

$$E(x_i - \bar{X})^2 = \frac{\sum_j^N (X_j - \bar{X})^2}{N}$$

$$= \sigma^2 \tag{2.18}$$

Consider now the second term in the right-hand member of Eq. 2.17. When the sampling is carried out without replacement, the probability of obtaining $(X_j - \bar{X})$ on the ith draw and $(X_m - \bar{X})$ on the kth draw is $1/N(N-1)$. This follows from the fact that the probability of obtaining X_j on the ith draw is $1/N$ and the probability of obtaining X_m on the kth draw, knowing that X_j has been drawn, is $1/(N-1)$. Hence

$$E(x_i - \bar{X})(x_k - \bar{X}) = \frac{\sum_{j \neq m}^N (X_j - \bar{X})(X_m - \bar{X})}{N(N-1)} \tag{2.19}$$

when the sampling is without replacement. Substituting Eq. 2.18 and 2.19 into Eq. 2.17 gives

$$\sigma_{\bar{x}}^2 = \frac{1}{n^2} \sum_i^n \sigma^2 + \frac{1}{n^2} \sum_{i \neq k}^n \frac{\sum_{j \neq m}^N (X_j - \bar{X})(X_m - \bar{X})}{N(N-1)}$$

but

$$\sum_{j \neq m}^N (X_j - \bar{X})(X_m - \bar{X}) = \left\{\sum_j^N (X_j - \bar{X})\right\}^2 - \sum_j^N (X_j - \bar{X})^2$$

$$= -\sum_j^N (X_j - \bar{X})^2$$

and therefore

$$\sigma_{\bar{x}}^2 = \frac{n}{n^2} \sigma^2 - \frac{n(n-1)}{n^2} \frac{\sigma^2}{N-1}$$

$$= \frac{N-n}{N-1} \frac{\sigma^2}{n} = \frac{N-n}{N} \frac{S^2}{n} \tag{2.3}$$

3. The covariance and correlation of arithmetic means for a simple random sample (Vol. I, Ch. 4, Sec. 18). *To prove:* The covariance,

$\sigma_{\bar{x}\bar{y}}$, in a simple random sample of n units selected from a population of N units without replacement is

$$\sigma_{\bar{x}\bar{y}} = (1-f)\frac{S_{XY}}{n} \tag{3.1}$$

and if the sample is selected with replacement, the covariance is given by σ_{XY}/n, where

$$S_{XY} = \frac{\sum^{N}(X_j - \bar{X})(Y_j - \bar{Y})}{N-1} = \frac{N}{N-1}\sigma_{XY} \tag{3.2}$$

X_j is the value of a characteristic for the jth unit in the population, Y_j is the value of another characteristic for the jth unit in the population, $\bar{x}$ and $\bar{X}$ are defined in Eq. 1.1, and $\bar{y}$ and $\bar{Y}$ are similarly defined.

For simple random sampling, the correlation between two sample means $\bar{x}$ and $\bar{y}$ is independent of the sample size and is equal to the correlation between individual observations, i.e.,

$$\rho_{\bar{x}\bar{y}} = \rho_{XY} = \frac{S_{XY}}{S_X S_Y} \tag{3.3}$$

Note that $\sigma_{\bar{x}}^2$, which was derived in Sec. 2, is a special case of $\sigma_{\bar{x}\bar{y}}$, since $\sigma_{\bar{x}\bar{x}} = \sigma_{\bar{x}}^2$, where $\sigma_{\bar{x}\bar{x}}$ is defined by Eq. 3.1 with $\bar{x}$ substituted for $\bar{y}$ (X substituted for Y throughout). Similarly,

$$S_{XX} = S_X^2 \quad \text{and} \quad S_{YY} = S_Y^2$$

Proof. The proof follows steps analogous to those given in deriving the variance in Sec. 2. Thus, since $\bar{x}$ and $\bar{y}$ are linear combinations of random variables, the covariance of $\bar{x}$ and $\bar{y}$ is (by Theorem 12, Ch. 3, p. 57)

$$\sigma_{\bar{x}\bar{y}} = \frac{1}{n^2}\sum_{i}^{n}\sigma_{x_i y_i} + \frac{1}{n^2}\sum_{i \neq j}^{n}\sigma_{x_i y_j} \tag{3.4}$$

and since

$$\sigma_{x_i y_i} = \frac{\sum_{i}^{N}(X_i - \bar{X})(Y_i - \bar{Y})}{N} = \sigma_{XY} \tag{3.5}$$

and

$$\sigma_{x_i y_j} = -\frac{\sigma_{XY}}{N-1} \qquad \text{for } i \neq j \tag{3.6}$$

therefore, by substitution in Eq. 3.4, we have

$$\sigma_{\bar{x}\bar{y}} = \frac{1}{n^2}\sum_{i}^{n}\sigma_{XY} - \frac{1}{n^2}\sum_{i \neq j}^{n}\frac{\sigma_{XY}}{N-1}$$

$$= \frac{N-n}{Nn}\frac{\sum^{N}(X_i - \bar{X})(Y_i - \bar{Y})}{N-1} = \frac{N-n}{Nn}S_{XY} = (1-f)\frac{S_{XY}}{n} \tag{3.1}$$

where S_{XY} is given by Eq. 3.2. Now, by definition,

$$\rho_{\bar{x}\bar{y}} = \frac{\sigma_{\bar{x}\bar{y}}}{\sigma_{\bar{x}}\sigma_{\bar{y}}} \tag{3.7}$$

and from Eq. 2.3

$$\sigma_{\bar{x}}\sigma_{\bar{y}} = (1-f)\frac{S_X S_Y}{n} \tag{3.8}$$

and therefore, substituting Eq. 3.8 and 3.1 in Eq. 3.7, we have

$$\rho_{\bar{x}\bar{y}} = \rho_{XY} = \frac{S_{XY}}{S_X S_Y} = \frac{\sigma_{XY}}{\sigma_X \sigma_Y} \tag{3.3}$$

4. The mathematical expectation of the sample variance for a simple random sample (Vol. I, Ch. 4, Sec. 12). *To prove:* In a simple random sample of n elements s^2 is an unbiased estimate of S^2 if the sample is drawn without replacement and s^2 is an unbiased estimate of σ^2 if the sample is drawn with replacement, i.e., to prove:

(*a*) $Es^2 = S^2$ if the sample is selected without replacement.
(*b*) $Es^2 = \sigma^2$ if the sample is selected with replacement,

where

$$s^2 = \frac{\sum^n (x_i - \bar{x})^2}{n-1} \tag{4.1 or I–4.3.12}$$

and S^2 and σ^2 are defined by Eq. 2.1 and 2.2, respectively.

Proof. We note that

$$Es^2 = E\frac{\sum^n (x_i - \bar{x})^2}{n-1} = E\frac{\sum^n x_i^2}{n-1} - E\frac{n\bar{x}^2}{n-1} \tag{4.2}$$

and that

$$E\frac{\sum^n x_i^2}{n-1} = \frac{n}{n-1}\frac{\sum_i^N X_i^2}{N} \tag{4.3}$$

Now

$$E\bar{x}^2 = \sigma_{\bar{x}}^2 + \bar{X}^2 \tag{4.4}$$

and, when the sampling is without replacement,

$$\sigma_{\bar{x}}^2 = \frac{N-n}{(N-1)n}\sigma^2 \qquad \text{from Eq. 2.14}$$

Hence

$$E\frac{n\bar{x}^2}{n-1} = \frac{n}{n-1}\left[\frac{N-n}{(N-1)n}\sigma^2 + \bar{X}^2\right] \tag{4.5}$$

and, substituting Eq. 4.3 and Eq. 4.5 in Eq. 4.2, and recognizing that $\frac{N}{N-1}\sigma^2 = S^2$, we have

$$Es^2 = S^2$$

As an exercise, show that when the sampling is with replacement $Es^2 = \sigma^2$.

5. The rel-variance of the estimated variance for a simple random sample* (Vol. I, Ch. 4, Sec. 12). *To prove:* The rel-variance, $V_{s^2}^2$, of the estimated variance, s^2, of a simple random sample of n units selected with replacement is given by

$$V_{s^2}^2 = \frac{1}{n}\left(\beta - \frac{n-3}{n-1}\right) \doteq \frac{1}{n}(\beta - 1) \tag{5.1}$$

where

$$s^2 = \frac{\sum^n (x_i - \bar{x})^2}{n-1}$$

$$\beta = \frac{\mu_4}{\sigma^4} \text{ with } \mu_4 = \frac{\sum^N (X_i - \bar{X})^4}{N} \qquad (5.2 \text{ or I–4.12.7})$$

and

$$\sigma^2 = \frac{\sum^N (X_i - \bar{X})^2}{N}$$

Proof. By definition

$$V_{s^2}^2 = \frac{E(s^2 - Es^2)^2}{(Es^2)^2} = \frac{E(s^2 - \sigma^2)^2}{\sigma^4} = \frac{Es^4 - \sigma^4}{\sigma^4} \tag{5.3}$$

since, by Sec. 4, $Es^2 = \sigma^2$.

We now need to evaluate Es^4. Apply the following transformation:

$$x_i - \bar{X} = z_i$$
$$\bar{x} - \bar{X} = \bar{z}$$

Then

$$Es^4 = E\left[\frac{\sum^n (z_i - \bar{z})^2}{n-1}\right]^2$$

$$= \frac{1}{(n-1)^2} E\left[(\sum^n z_i^2)^2 - 2n\bar{z}^2 \sum^n z_i^2 + n^2\bar{z}^4\right] \tag{5.4}$$

Now

$$(\sum^n z_i^2)^2 = \sum^n z_i^4 + \sum_{i \neq j}^n z_i^2 z_j^2 \tag{5.5}$$

---

* May be taken up with Chapter 10 instead of Chapter 4.

where there are n^2 terms on the left and n terms in the first sum, and $n^2 - n$ terms in the second sum on the right. Also

$$\bar{z}^2\sum^n z_i^2 = \frac{1}{n^2}\left[\sum_i^n z_i^4 + 2\sum_{i\neq j}^n z_i^3 z_j + \sum_{i\neq j}^n z_i^2 z_j^2 + \sum_{i\neq j\neq k}^n z_i^2 z_j z_k\right] \qquad (5.6)$$

where there are n^3 terms on the left and n terms in the first sum, $n(n-1)$ terms in each of the next two sums, and $n(n-1)(n-2)$ terms in the fourth sum on the right. Finally

$$\bar{z}^4 = \frac{1}{n^4}\left[\sum_i^n z_i^4 + 4\sum_{i\neq j}^n z_i^3 z_j + 3\sum_{i\neq j}^n z_i^2 z_j^2 + 6\sum_{i\neq j\neq k}^n z_i^2 z_j z_k + \sum_{i\neq j\neq k\neq m}^n z_i z_j z_k z_m\right] \qquad (5.7)$$

where there are n^4 terms on the left and n, $n(n-1)$, $n(n-1)$, $n(n-1)(n-2)$, and $n(n-1)(n-2)(n-3)$ terms in the five summations on the right, respectively.

Since we are sampling with replacement, $z_1, \cdots, z_n$ are independent, and hence, by Theorem 8 of Ch. 3 (p. 54), we have

$$\left.\begin{aligned}
Ez_i^2 z_j^2 &= Ez_i^2 Ez_j^2 && \text{if } i \neq j \\
Ez_i^3 z_j &= Ez_i^3 Ez_j = 0 && \text{if } i \neq j \text{ since } Ez_j = 0 \\
Ez_i^2 z_j z_k &= Ez_i^2 Ez_j Ez_k = 0 && \text{if } i \neq j \neq k \text{ since } Ez_j = 0 \\
Ez_i z_j z_k z_m &= Ez_i Ez_j Ez_k Ez_m = 0 && \text{if } i \neq j \neq k \neq m \text{ since } Ez_j = 0
\end{aligned}\right\} \qquad (5.8)$$

Hence, from Eq. 5.5 and Eq. 5.8,

$$E(\sum^n z_i^2)^2 = n\mu_4 + n(n-1)\sigma^4$$

where $\mu_4 = Ez_i^4$, and $\sigma^2 = Ez_i^2 = Ez_j^2$. From Eq. 5.6 and Eq. 5.8,

$$E(\bar{z}^2\sum^n z_i^2) = \frac{1}{n^2}[n\mu_4 + n(n-1)\sigma^4] = \frac{\mu_4}{n} + \left(\frac{n-1}{n}\right)\sigma^4$$

From Eq. 5.7 and Eq. 5.8,

$$E\bar{z}^4 = \frac{1}{n^4}[n\mu_4 + 3n(n-1)\sigma^4] = \frac{\mu_4}{n^3} + \frac{3(n-1)}{n^3}\sigma^4$$

Substituting these quantities in Eq. 5.4 and subtracting σ^4, we have

$$Es^4 - \sigma^4 = \frac{1}{(n-1)^2}\left[n\mu_4 + n(n-1)\sigma^4 - 2\mu_4 - 2(n-1)\sigma^4 + \frac{\mu_4}{n} + 3\left(\frac{n-1}{n}\right)\sigma^4 - (n-1)^2\sigma^4\right]$$

$$= \frac{\mu_4}{n} + \frac{\sigma^4}{n-1}\left(n - 2 + \frac{3}{n} - n + 1\right)$$

$$= \frac{\mu_4}{n} - \frac{\sigma^4}{n-1}\frac{(n-3)}{n} \tag{5.9}$$

Finally

$$V_{s^2}^2 = \frac{Es^4 - \sigma^4}{\sigma^4} = \frac{1}{n}\left(\beta - \frac{n-3}{n-1}\right) \tag{5.10}$$

For reasonably large n we may assume $(n-3)/(n-1) = 1$, in which case $V_{s^2}^2$ is given approximately by

$$V_{s^2}^2 \doteq \frac{1}{n}(\beta - 1) \tag{5.1}$$

Remark 1. If the population mean were known and were used in estimating s^2, then

$$s^2 = \frac{\sum_i^n (x_i - \bar{X})^2}{n} \tag{5.11}$$

is an unbiased estimate of σ^2 and

$$V_{s^2}^2 = \frac{1}{n}(\beta - 1) \tag{5.12}$$

The proof, which follows the steps above, is left to the reader.

Remark 2. It is tedious but not very difficult to obtain a similar result for sampling without replacement. The result is:

$$\begin{aligned} V_{s^2}^2 = \frac{(N-1)^2}{N^2(n-1)^2}\Bigg\{&\frac{(n-1)^2}{n} - \frac{n-1}{n(N-1)}[(n-2)(n-3) - (n-1)] \\ &- \frac{4(n-1)(n-2)(n-3)}{n(N-1)(N-2)} - \frac{6(n-1)(n-2)(n-3)}{n(N-1)(N-2)(N-3)}\Bigg\}\beta \\ &+ \frac{(N-1)^2}{N^2(n-1)^2}\Bigg\{\frac{(n-1)N}{n(N-1)}[(n-1)^2 + 2] \\ &+ \frac{2(n-1)(n-2)(n-3)N}{n(N-1)(N-2)} + \frac{3(n-1)(n-2)(n-3)N}{n(N-1)(N-2)(N-3)} \\ &- \frac{N^2(n-1)^2}{(N-1)^2}\Bigg\} \end{aligned} \tag{5.13}$$

From a comparison of Eq. 5.10 and 5.13 it will be apparent why we use sampling with replacement in approximating the variance of s^2 whenever possible.

Remark 3. When sampling with replacement, the expected value of any term in which z_i appears only to the first power is zero. Hence, an easier development of $V^2_{s^2}$ can be made by recognizing that

$$s^2 = \frac{\sum_i^n z_i^2}{n} - \frac{\sum_{i \neq j}^n z_i z_j}{n(n-1)}$$

and

$$Es^4 = E\frac{\left(\sum_i^n z_i^2\right)^2}{n^2} + E\frac{\left(\sum_{i \neq j}^n z_i z_j\right)^2}{n^2(n-1)^2}$$

6. The rel-variance of the estimated standard deviation* (Vol. I, Ch. 4, Sec. 12, Eq. 12.5). *To prove:* The rel-variance, V^2_s, of the estimated standard deviation, s, is given approximately by

$$V^2_s \doteq \tfrac{1}{4}V^2_{s^2} \tag{6.1}$$

and it is desired to investigate the order of approximation.

Proof. Although the proof is expressed in terms of s, the results are valid for any positive random variable s.

Let $Es^2 = \sigma^2$. Then, since

$$\begin{aligned} E(s-\sigma)^2 &= Es^2 - 2\sigma Es + \sigma^2 \\ &= 2\sigma(\sigma - Es) \end{aligned}$$

it follows that

$$V^2_s = 2\left(\frac{\sigma - Es}{\sigma}\right) \tag{6.2}$$

and hence the investigations of approximations to the bias of s and to the rel-variance of s may be carried on simultaneously. Furthermore Eq. 6.2 shows that for *any random variable* s such that $\sigma^2 = Es^2 \neq 0$ it follows that

$$\frac{\sigma - Es}{\sigma} \geq 0 \tag{6.3}$$

so that the bias, $Es - \sigma$, is negative, i.e., the expected value of the estimated standard deviation is always less than the standard deviation.

It is not difficult to verify, by using the fact that $s^2 - \sigma^2 = (s+\sigma)(s-\sigma)$, that the following is true for all values of s and σ:

$$\begin{aligned} \frac{s-\sigma}{\sigma} = \frac{(s^2-\sigma^2)}{2\sigma^2} - \frac{(s^2-\sigma^2)^2}{8\sigma^4} + \frac{(s^2-\sigma^2)^3}{16\sigma^6} \\ - \frac{(s^2-\sigma^2)^4}{16\sigma^6(s+\sigma)^4}(s^2 + 4s\sigma + 5\sigma^2) \end{aligned} \tag{6.4}$$

* May be taken up with Chapter 10 instead of Chapter 4.

Since $Es^2 = \sigma^2$ it follows from Eq. 6.4 that

$$\frac{\sigma - Es}{\sigma} = \frac{1}{8\sigma^4} E(s^2 - \sigma^2)^2 - \frac{1}{16\sigma^6} E(s^2 - \sigma^2)^3 + \frac{1}{16\sigma^6} E \frac{(s^2 - \sigma^2)^4(s^2 + 4s\sigma + 5\sigma^2)}{(s + \sigma)^4} \tag{6.5}$$

Therefore, from Eq. 6.5 we have

$$\frac{\sigma - Es}{\sigma} - \frac{1}{8\sigma^4} E(s^2 - \sigma^2)^2 + \frac{1}{16\sigma^6} E(s^2 - \sigma^2)^3 = \frac{1}{16\sigma^6} E \frac{(s^2 - \sigma^2)^4(s^2 + 4s\sigma + 5\sigma^2)}{(s + \sigma)^4}$$

Since

$$\frac{V_s^2}{2} = \frac{\sigma - Es}{\sigma}$$

from Eq. 6.2, we have

$$V_s^2 - \frac{V_{s^2}^2}{4} + \frac{1}{8\sigma^6} E(s^2 - \sigma^2)^3 = \frac{1}{8\sigma^6} E \frac{(s^2 - \sigma^2)^4(s^2 + 4s\sigma + 5\sigma^2)}{(s + \sigma)^4} \geq 0 \tag{6.6}$$

Now

$$1 < \frac{s^2 + 4s\sigma + 5\sigma^2}{(s + \sigma)^2} < 5$$

so that

$$\frac{V_{s^2}^2}{4} - \frac{1}{8\sigma^6} E(s^2 - \sigma^2)^3 \leq V_s^2 \leq \frac{V_{s^2}^2}{4} - \frac{1}{8\sigma^6} E(s^2 - \sigma^2)^3 + \frac{5}{8\sigma^8} E(s^2 - \sigma^2)^4$$

For sufficiently large n, $\frac{1}{8\sigma^6} E(s^2 - \sigma^2)^3$ and $\frac{5}{8\sigma^8} E(s^2 - \sigma^2)^4$ are small relative to $(V_{s^2}^2)/4$, and we can say that V_s^2 is given approximately by $\frac{1}{4} V_{s^2}^2$. This is the approximation to V_s^2 for any sample design. For a simple random sample of n units we can substitute Eq. 5.10 for $V_{s^2}^2$ to obtain

$$V_s^2 \doteq \frac{\beta - \dfrac{n-3}{n-1}}{4n} \doteq \frac{\beta - 1}{4n} \tag{6.7}$$

or

$$V_s \doteq \sqrt{\frac{\beta - 1}{4n}} \qquad \text{(6.8 or I–4.12.5)}$$

7. A condition under which the approximation to the coefficient of variation of the estimated standard deviation is reasonably satisfactory* (Vol. I, Ch. 4, Sec. 12). The approximation to the coefficient of variation of the estimated standard deviation, $V_s \doteq V_{s^2}/2$, as obtained in the preceding section, will be a reasonably good approximation for sizes of sample such that V_{s^2} is less than .3. This is indicated by the following line of reasoning.

Note that, if we use the approximation to V_s given in Sec. 6 of this chapter, an approximation to the confidence limits (limits of sampling variation) on s with a reasonably large sample would be

$$\sigma\left(1 \pm \frac{tV_{s^2}}{2}\right)$$

Another formula for obtaining confidence bounds for s is

$$s = \sigma\sqrt{1 \pm tV_{s^2}} \qquad (7.1)$$

This follows from the confidence limits for s^2, which are $\sigma^2\,(1 \pm tV_{s^2})$, and from the fact that s and s^2 are mathematically dependent.

Values of V_{s^2} for several levels of error in the approximation to the confidence bounds are given in the accompanying table for $t = 1, 2,$ and 3. In the table r is the ratio of the approximation to the confidence limits to the limits given by Eq. 7.1, i.e.,

$$r = \frac{1 \pm \dfrac{tV_{s^2}}{2}}{\sqrt{1 \pm tV_{s^2}}}$$

and any value of V_{s^2} less than that shown in the table will yield a smaller value of r than is indicated in the table for the corresponding values of t and r. Hence, except when a very high probability is demanded for the confidence bounds covering the true value, such as that associated with $t = 3$, $V_{s^2}/2$ will be a reasonably good approximation for sizes of sample such that V_{s^2} is less than .3.

t	V_{s^2} for given values of r and t				
	$r = 1.10$	$r = 1.12$	$r = 1.14$	$r = 1.16$	$r = 1.18$
1	.59	.62	.65	.67	.69
2	.29	.31	.32	.34	.35
3	.19	.21	.22	.22	.23

* May be taken up with Chapter 10 instead of Chapter 4.

8. The rel-variance of the estimated standard deviation for a simple random sample drawn from a binomial distribution (Vol. I, Ch. 4, Sec. 12). *To prove:* For a binomial distribution, the coefficient of variation of the estimated standard deviation, s, in a simple random sample of n elements drawn with replacement is given approximately by

$$V_s \doteq \sqrt{\frac{1}{4n}\left(\frac{1}{PQ} - \frac{4n-6}{n-1}\right)} \tag{8.1}$$

where P is the proportion in the population having a specified characteristic and $Q = 1 - P$.

Proof. We saw in Sec. 6 that the coefficient of variation squared of the estimated standard deviation, s, in a simple random sample of n elements drawn with replacement is given approximately by

$$V_s^2 \doteq \frac{\beta - \dfrac{n-3}{n-1}}{4n}$$

where $\beta = \mu_4/\sigma^4$. For a binomial distribution:

$$\mu_4 = \frac{\sum^{N}(X_i - P)^4}{N} = P(1-P)^4 + Q(0-P)^4$$

$$= PQ - 3P^2Q^2 \tag{8.2}$$

$$\sigma^2 = PQ \tag{8.3}$$

and therefore

$$\beta = \frac{1}{PQ} - 3 \tag{8.4}$$

substituting for β in V_s^2 gives

$$V_s^2 \doteq \frac{1}{4n}\left(\frac{1}{PQ} - \frac{4n-6}{n-1}\right) \tag{8.5}$$

9. Size of sample required to estimate the standard deviation of a proportion with a prescribed precision, say $V_s = .1$ (Vol. I, Ch. 4, Sec. 12, p. 131). *a. To prove:* For n greater than 60, V_s is less than .1 if $.20 \leq P \leq .80$, and it is reasonable to assume that V_s will be less than .1 when the sample proportion, p, lies between .30 and .70.

Proof. For a binomial distribution, from Eq. 8.5,

$$V_s^2 \doteq \frac{1}{4n}\left(\frac{1}{PQ} - \frac{4n-6}{n-1}\right) \tag{9.1}$$

Note that V_s^2 decreases as PQ increases, for a fixed n. If we fix the size of sample at $n = 60$, it follows from Eq. 9.1 that V_s will be less than .1 for $.20 \leq P \leq .80$. Moreover, when the sample proportion, p, lies between .30 and .70, the population proportion, P, is likely to lie between .20 and .80. Thus, when $p = .30$ for a sample of 60, then

$$p - 2\sigma_p = .30 - 2\sqrt{\frac{PQ}{n}} = .30 - 2\sqrt{\frac{.20 \times .80}{60}} = .20$$

and therefore, when $p = .30$, it is likely that P is greater than .20. Similarly, when $p = .70$, then $.70 + 2\sigma_p = .70 + .10 = .80$ and it is likely that P is less than .80. Therefore, we can be reasonably sure that a sample of 60 will be sufficient to estimate the standard deviation of a sample proportion or a total which is between 30 and 70 per cent of the population with a coefficient of variation of less than 10 per cent.

b. To prove: For nP greater than 25, V_s is less than .1, and it is reasonable to assume that V_s is less than .1 if np is 35 or greater.

Proof. From Eq. 9.1 above, solving for n, we obtain

$$n = \frac{\dfrac{1}{PQ} + 4V_s^2 - 4 + \sqrt{\left(4 - 4V_s^2 - \dfrac{1}{PQ}\right)^2 - 16V_s^2\left(\dfrac{1}{PQ} - 6\right)}}{8V_s^2}$$

$$\lim_{P \to 0} nP = \frac{1}{4V_s^2}$$

since $Q \to 1$ and we have $\lim nP = 25$ for $V_s = .1$. Further it is reasonable to expect that nP will be greater than 25 when np is greater than 35, since

$$np - 2\sigma_{(np)} = 35 - 2\sqrt{nPQ} \doteq 35 - 2\sqrt{nP} = 35 - 10 = 25$$

and therefore nP is likely to exceed 25.

10. A simple random sample of a population contains a simple random sample of any subset of the population (Vol. I, Ch. 4, Sec. 7). Suppose that a simple random sample of n elements is drawn from a population of N elements. Let K be the number of elements belonging to some subset of the N elements in the population, and let k be the number of elements in the sample of n belonging to that subset. Then the sample of k is a simple random sample of the K elements, with the same expected sampling fraction as the sample of n. That is, *to prove:*

a. All combinations of k elements are possible samples and are equally likely.

b. $E\dfrac{k}{K} = \dfrac{n}{N}$.

Proof. *a.* Fix the value of k at k_0. Then any particular combination of k_0 elements among the $C_{k_0}^K$ possible samples of k_0 from the subset will occur exactly $C_{n-k_0}^{N-K}$ times among the C_n^N possible samples of n elements from N. Therefore, all samples of k_0 are equally likely.

b. Let $X_i = 1$ if the ith element belongs to the subset;

$= 0$ otherwise.

Then $\sum^n x_i = k$ and $\sum^N X_i = K$. Hence,

$$E\,\frac{k}{K} = E\,\frac{\sum^n x_i}{K} = \frac{n\,\dfrac{\sum^N X_i}{N}}{K} = \frac{n}{N}$$

11. The rel-variance of the ratio of two random variables (Vol. I, Ch. 4, Sec. 18). *To prove:* The rel-variance of a ratio of two random variables u and w is given approximately by

$$V_{(u/w)}^2 \doteq V_u^2 + V_w^2 - 2\rho_{uw} V_u V_w \qquad (11.1 \text{ or I-4.18.12})$$

Proof. An approximation to the variance of a ratio of random variables u/w may be arrived at as follows. Let

$$\Delta u = \frac{u - Eu}{Eu}$$

then

$$u = (Eu)\{1 + \Delta u\}$$

Similarly,

$$w = (Ew)\{1 + \Delta w\}$$

$$V_{(u/w)}^2 = \frac{\sigma_{u/w}^2}{\left(\dfrac{Eu}{Ew}\right)^2} = \frac{E\left\{\dfrac{u}{w} - \dfrac{Eu}{Ew}\right\}^2}{\left(\dfrac{Eu}{Ew}\right)^2} = E\left\{\frac{1 + \Delta u}{1 + \Delta w} - 1\right\}^2$$

$$= E\{(1 + \Delta u)(1 + \Delta w)^{-1} - 1\}^2 \qquad (11.2)$$

Now

$$\frac{1}{1 + \Delta w} = 1 - \Delta w + (\Delta w)^2 - \cdots$$

which will converge to $(1 + \Delta w)^{-1}$ if Δw is less than 1 in absolute value. When Δw converges in probability to zero, it is not necessary that Δw be less than 1 for all possible values of w in order for the initial terms in this expansion, when substituted for $(1 + \Delta w)^{-1}$ in Eq. 11.2, to yield a useful approximation to the rel-variance. Thus, Eq. 11.2 becomes approximately

$$E\{(1+\Delta u)[1-\Delta w+(\Delta w)^2]-1\}^2$$
$$= E\{\Delta u-\Delta w+(\Delta w)^2-\Delta u\Delta w+\Delta u(\Delta w)^2\}^2$$

Retaining terms of the second order or less, and noting that $E\Delta u = E\Delta w = 0$, we have

$$E\{(\Delta u)^2+(\Delta w)^2-2\Delta u\Delta w\} \doteq V_u^2+V_w^2-2\rho_{uw}V_uV_w \qquad (11.3)$$

since

$$E\Delta u\Delta w = \frac{\sigma_{uw}}{(Eu)(Ew)} = \rho_{uw}V_uV_w \qquad (11.4)$$

To see that Eq. 11.3 provides a useful approximation to the rel-variance of the ratio of sample means, for n sufficiently large, we note first that Eq. 11.3 with $u=\bar{x}$, $Eu=\bar{X}$, $w=\bar{y}$, $Ew=\bar{Y}$ may be rewritten

$$E\left\{\frac{(\bar{x}-\bar{X})^2}{\bar{X}^2}+\frac{(\bar{y}-\bar{Y})^2}{\bar{Y}^2}-2\frac{(\bar{x}-\bar{X})(\bar{y}-\bar{Y})}{\bar{X}\bar{Y}}\right\}$$
$$= E\left\{(\bar{x}-\bar{X})-\frac{\bar{X}}{\bar{Y}}(\bar{y}-\bar{Y})\right\}^2\frac{1}{\bar{X}^2} \qquad (11.5)$$

Also

$$V^2_{(\bar{x}/\bar{y})} = E\left(\frac{\bar{x}}{\bar{y}}-\frac{\bar{X}}{\bar{Y}}\right)^2\Big/\frac{\bar{X}^2}{\bar{Y}^2}$$
$$= E\left[(\bar{x}-\bar{X})-\frac{\bar{X}}{\bar{Y}}(\bar{y}-\bar{Y})\right]^2\frac{1}{\bar{X}^2}$$
$$-\frac{1}{\bar{X}^2}E\left[(\bar{x}-\bar{X})-\frac{\bar{X}}{\bar{Y}}(\bar{y}-\bar{Y})\right]^2\left(\frac{\bar{y}^2-\bar{Y}^2}{\bar{y}^2}\right) \qquad (11.6)$$

where the first term is Eq. 11.5, the approximation to the rel-variance. We need merely show that, for some value of n sufficiently large, the second term of Eq. 11.6 becomes very small relative to the first term. To do this we note that (see Remark in Sec. 4, Ch. 3, p. 56)

$$E\left[(\bar{x}-\bar{X})-\frac{\bar{X}}{\bar{Y}}(\bar{y}-\bar{Y})\right]^2\left(\frac{\bar{y}^2-\bar{Y}^2}{\bar{y}^2}\right)$$
$$\leq\left\{E\left[(\bar{x}-\bar{X})-\frac{\bar{X}}{\bar{Y}}(\bar{y}-\bar{Y})\right]^4E\left[\frac{\bar{y}^2-\bar{Y}^2}{\bar{y}^2\bar{Y}^2}\right]^2\right\}^{1/2} \qquad (11.7)$$

Now find a value $\bar{y}$, which represents the smallest of the sample averages from the population, and assume all $y_i > 0$, and let us call that value $\bar{Y}_L$. By substituting $\bar{Y}_L^4$ for $\bar{y}^4$ in the denominator of the right-hand side of the inequality in Eq. 11.7, the inequality is strengthened and the right-hand side becomes equal to

$$\left\{E\left[(\bar{x}-\bar{X})-\frac{\bar{X}}{\bar{Y}}(\bar{y}-\bar{Y})\right]^4 E\frac{(\bar{y}^2-\bar{Y}^2)^2}{\bar{Y}_L^4\bar{Y}^4}\right\}^{1/2} \tag{11.8}$$

By evaluating the expected values of the terms in Eq. 11.8 we find that the order of the first factor* is $1/n^2$, and of the second factor is $1/n$, and therefore the order of Eq. 11.8 is $(1/n^3)^{1/2} = (1/n)^{3/2}$. Since the first term of Eq. 11.6 is of the order $1/n$, it follows that, for n sufficiently large, Eq. 11.8 will be small relative to the approximation to the variance.

A similar proof would hold for estimated totals and other sample estimates.

***12. An indication of a sufficient condition for the approximation to the standard deviation (or coefficient of variation) of an estimated ratio to be reasonably satisfactory** (Vol. I, Ch. 4, Sec. 18).† *To prove:* The approximation to the standard deviation (or coefficient of variation) of an estimated ratio, $r = u/w$, will be reasonably satisfactory provided that $V_w \leq .05$, or provided that $\rho \doteq V_u/V_w$ and $V_w \leq .15$, where V_w is the coefficient of variation of the denominator of the ratio, V_u is the coefficient of variation of the numerator, and ρ is the correlation of u and w.

Proof. Fieller‡ has shown that exact confidence limits, R_1 and R_2, for the ratio $R = U/W$ are given by the solution of

$$(u^2 - t^2s_u^2) - 2R(uw - t^2s_{uw}) + R^2(w^2 - t^2s_w^2) = 0 \tag{12.1}$$

where t determines the appropriate probabilities of the normal distribution, u and w are normally distributed, and s_u^2, s_w^2, and s_{uw} are the sample estimates of the variances and covariance of u and w. Then, at the significance level t, ΔR_1 is the relative length of the upper part of the confidence interval, and ΔR_2 is the relative length of the lower part, where

$$\left.\begin{aligned}\Delta R_1 &= \frac{R_1 - r}{r}\\ \Delta R_2 &= \frac{r - R_2}{r}\end{aligned}\right\} \tag{12.2}$$

We can estimate the rel-variance of a ratio r by substituting sample estimates for each term in Eq. 11.1; that is, an estimate of V_r^2 is

$$v_r^2 = v_u^2 + v_w^2 - 2\rho' v_u v_w \tag{12.3}$$

* See Theorem C.1 in Appendix C to Ch. 3, p. 87.

† By Margaret Gurney, Bureau of the Census.

‡ E. C. Fieller, *Supplement to J. Roy. Stat. Soc.*, **VII** (1940–41), 51.

⋆ May be deferred.

where v_u^2, v_w^2, and ρ' are the sample estimates of the rel-variances and correlation of u and w. With this estimate of the rel-variance, symmetric relative confidence limits are obtained in the usual manner by computing

$$r \pm tv_r \tag{12.4}$$

Thus, tv_r plays a role in defining confidence limits similar to ΔR_1 and ΔR_2, and we can get a measure of the closeness of the approximation of Eq. 12.4 to the limits obtained from Eq. 12.1 by comparing tv_r with ΔR_1 and ΔR_2. Let

$$\varepsilon_1 = \frac{\Delta R_1 - tv_r}{tv_r} = \frac{t(v_w^2 - \rho' v_u v_w) + \sqrt{v_r^2 - t^2 v_u^2 v_w^2(1 - \rho'^2)}}{(1 - t^2 v_w^2)v_r} - 1$$

$$= \frac{tv_w x + \sqrt{1 - t^2 v_w^2(1 - x^2)}}{1 - t^2 v_w^2} - 1 \tag{12.5}$$

where

$$x = \frac{b}{v_r v_w} = \frac{1 - k\rho'}{\sqrt{1 + k^2 - 2k\rho'}} \tag{12.6}$$

$k = v_u/v_w$, and $b = v_w^2 - \rho' v_u v_w$ is the sample estimate of the relative bias of r as an estimate of R (see Eq. 14.1 and Sec. 14–16).

ε_2 is defined analogously for the lower part of the confidence interval and is obtained by replacing t by $-t$.

Note that

$$x^2 = \frac{(1 - k\rho')^2}{(1 - k\rho')^2 + k^2(1 - \rho'^2)} \tag{12.7}$$

and cannot exceed 1.

If we let t and x be positive in Eq. 12.5, we have the maximum of the absolute values of ε_1 and ε_2, which we shall call ε. Then ε is monotonically increasing with increasing x when $tv_w < 1$, and for any specified $tv_w < 1$ the maximum possible value of ε is obtained by setting $x = 1$ in Eq. 12.5. We then have

$$\varepsilon\ (\text{max.}) = \frac{1}{1 - tv_w} - 1 \tag{12.8}$$

Since Eq. 12.8 is monotonically increasing with tv_w, we can set a reasonable maximum value on ε, say ε (max.) $= .15$, and determine for a particular probability level (i.e., a specified t) the maximum value of v_w such that $\varepsilon \leq .15$.

For example, if we specify $t = 2.5$, we have

$$1.15 \geq \frac{1}{1 - 2.5v_w} \quad \text{or} \quad v_w \leq .05$$

Thus, for $v_w \leq .05$, the use of $\pm\, tv_r$ as confidence limits will give approximately the same results as obtained (for normally distributed variates) from Eq. 12.5. This is interpreted as evidence that v_r provides a reasonably good approximation to the rel-variance of r.

Remark 1. If $b = 0$, then $x = 0$ and for $t = 2.5$ and ε (max.) $= .15$ we have, by substituting these values in Eq. 12.5,

$$1.15 \geq \frac{1}{\sqrt{1 - 6.25v_w^2}}$$

or $v_w \leq .19$ is sufficient for the approximation to be good.

Remark 2. By substituting $\pm\, t$ and the observed values of x and v_w in Eq. 12.5, the value of ε can be determined for any particular sample results.

Remark 3. The results above apply also to the population variance $V_r^2 = V_u^2 + V_w^2 - 2\rho V_u V_w$, by substituting V_u^2, V_w^2, and ρ for v_u^2, v_w^2, and ρ' in Eq. 12.5.

13. The variance of the ratio of two random variables estimated from a simple random sample drawn without replacement (Vol. I, Ch. 4, Sec. 18, Eq. 18.1). *To prove:* The variance, σ_r^2, of a ratio r for a simple random sample of n units drawn without replacement from a population of N units is given approximately by

$$\sigma_r^2 \doteq R^2(1-f)\left(\frac{V_X^2 + V_Y^2 - 2\rho_{XY}V_XV_Y}{n}\right) \qquad \text{(13.1 or I-4.18.1)}$$

where

$$r = \frac{\bar{x}}{\bar{y}} = \frac{x}{y} \text{ is an estimate of } R = \frac{\bar{X}}{\bar{Y}} = \frac{X}{Y}$$

$$V_X^2 = \frac{S_X^2}{\bar{X}^2} = \frac{\sum^N (X_i - \bar{X})^2}{(N-1)\bar{X}^2}$$

$$V_Y^2 = \frac{S_Y^2}{\bar{Y}^2} = \frac{\sum^N (Y_i - \bar{Y})^2}{(N-1)\bar{Y}^2}$$

$$\rho_{XY}V_XV_Y = \rho_{XY}\frac{S_XS_Y}{\bar{X}\bar{Y}} = \frac{\sum^N (X_i - \bar{X})(Y_i - \bar{Y})}{(N-1)\bar{X}\bar{Y}} = \frac{S_{XY}}{\bar{X}\bar{Y}}$$

Proof. By Sec. 11 with $u = \bar{x}$ and $v = \bar{y}$ we have

$$\sigma_r^2 = R^2V_r^2 \doteq R^2[V_{\bar{x}}^2 + V_{\bar{y}}^2 - 2\rho_{\bar{x}\bar{y}}V_{\bar{x}}V_{\bar{y}}] \qquad (13.2)$$

By Sec. 2

$$V_{\bar{x}}^2 = \frac{S_{\bar{x}}^2}{\bar{X}^2} = \frac{(1-f)}{\bar{X}^2}\frac{S_X^2}{n} = \frac{(1-f)V_X^2}{n}$$

and similarly for $V_{\bar{y}}^2$. By Sec. 3

$$\rho_{\bar{x}\bar{y}} V_{\bar{x}} V_{\bar{y}} = \rho_{\bar{x}\bar{y}} \frac{S_{\bar{x}}}{\bar{X}} \frac{S_{\bar{y}}}{\bar{Y}} = \rho_{XY} \frac{\sqrt{\frac{1-f}{n}} S_X}{\bar{X}} \frac{\sqrt{\frac{1-f}{n}} S_Y}{\bar{Y}}$$

$$= \frac{(1-f)}{n} \rho_{XY} V_X V_Y$$

Substitute these values into σ_r^2 (Eq. 13.2) to obtain the desired result.

14. An approximation to the bias of a ratio estimate (Vol. I, Ch. 4, Sec. 18). The bias of $r = u/w$ as an estimate of $R = U/W$, where u and w are random variables and $Eu = U$, $Ew = W$, is given approximately by

$$R(V_w^2 - \rho_{uw} V_u V_w) \tag{14.1}$$

To prove:

$$E(r - R) \doteq R(V_w^2 - \rho_{uw} V_u V_w)$$

Proof. The procedure for generating an approximation to the bias is the same as that given in Sec. 11. Here we let

$$\Delta u = \frac{u - Eu}{Eu}$$

$$u = (Eu)(1 + \Delta u) = U(1 + \Delta u)$$

Similarly

$$w = (Ew)(1 + \Delta w) = W(1 + \Delta w)$$

then

$$E\left(\frac{u}{w} - \frac{U}{W}\right) = E\frac{U}{W}[(1 + \Delta u)(1 + \Delta w)^{-1} - 1]$$

$$\doteq E\frac{U}{W}[(1 + \Delta u)\{1 - \Delta w + (\Delta w)^2 - (\Delta w)^3 + \cdots\} - 1]$$

$$\doteq \frac{U}{W} E[1 - \Delta w + (\Delta w)^2 - (\Delta w)^3 + \Delta u$$

$$- \Delta u \Delta w + \Delta u (\Delta w)^2 - 1]$$

and, ignoring the terms of order $1/n^2$ or higher, we have

$$\frac{U}{W}[V_w^2 - \rho_{uw} V_u V_w] \tag{14.2}$$

In a manner similar to that given in Sec. 11, the remainder term in the approximation can be shown to approach 0 faster than Eq. 14.2. Hence, for sufficiently large n, Eq. 14.2 is a satisfactory approximation to the bias.

Remark 1. Whenever the regression line of u on w passes through the origin, the approximation to the bias is zero.

This follows from the considerations below. The regression line of u on w is given by

$$u - U = \rho_{uw} \frac{\sigma_u}{\sigma_w}(w - W) \tag{14.3}$$

If this line goes through the origin, then

$$U = \rho_{uw} \frac{\sigma_u}{\sigma_w} W$$

and

$$\rho_{uw} = \frac{V_w}{V_u}$$

Substituting this value of ρ_{uw} in Eq. 14.1, we see that the approximation to the bias is 0 when the regression of u on w is through the origin.

Remark 2. For simple random sampling with $r = x/y$ and $x = \sum^n x_i$ and $y = \sum^n y_i$, the approximation to the bias of r as an estimate of $R = X/Y$ is given by

$$\frac{1-f}{n} R(V_Y^2 - \rho_{XY} V_X V_Y) \tag{14.4}$$

15. Decreases in the bias of the ratio estimate relative to the standard deviation with increasing sample size (Vol. I, Ch. 4, Sec. 18). *To prove:* With simple random sampling, the bias of a ratio estimate $r = x/y$ decreases faster than the standard error of r, and with a moderately large sample the bias of r will be negligible in relation to its standard error.

Proof. From Eq. 14.4 the approximation to the bias of r is given by

$$\Delta = R \frac{1-f}{n}(V_Y^2 - \rho_{XY} V_X V_Y) \tag{15.1}$$

and from Eq. 13.1, the approximation to the variance of r is

$$\sigma_r^2 \doteq R^2 \frac{1-f}{n}(V_X^2 + V_Y^2 - 2\rho_{XY} V_X V_Y) \tag{15.2}$$

Consequently

$$\frac{\Delta^2}{\sigma_r^2} = \frac{1-f}{n} \frac{(V_Y^2 - \rho_{XY} V_X V_Y)^2}{V_X^2 + V_Y^2 - 2\rho_{XY} V_X V_Y}$$

or

$$\frac{\Delta}{\sigma_r} = \sqrt{\frac{1-f}{n}} \sqrt{\frac{(V_Y^2 - \rho_{XY} V_X V_Y)^2}{V_X^2 + V_Y^2 - 2\rho_{XY} V_X V_Y}}$$

which decreases with increasing n.

16. Two conditions under which the sample ratio is an unbiased estimate of the population ratio (Vol. I, Ch. 4, Sec. 18).

a. The ratio $r = u/w$ of two random variables u and w is an unbiased estimate of $R = Eu/Ew$ when u/w and w are uncorrelated.

b. If $u = \sum^n x_i$ and $w = \sum^n y_i$, the ratio $r = \sum^n x_i / \sum^n y_i$ is an unbiased estimate of $R = X/Y$ when the conditional expected value of x_i is equal to Ry_i for any given y_i. This is a special case of a above.

a. To prove:

$$E\frac{u}{w} = \frac{Eu}{Ew} \text{ when } \sigma_{(u/w),w} = 0$$

Proof. By definition

$$\sigma_{(u/w),w} = Eu - E\frac{u}{w}Ew$$

If $\sigma_{(u/w),w} = 0$, then

$$Eu - E\frac{u}{w}Ew = 0$$

and

$$E\frac{u}{w} = \frac{Eu}{Ew}$$

b. To prove:

$$E\left(\frac{\sum^n x_i}{\sum^n y_i}\right) = \frac{X}{Y} \text{ when } E(x_i|y_i) = Ry_i$$

Proof. By Theorems 13 and 14 of Ch. 3 (p. 61) we may write

$$E\frac{\sum^n x_i}{\sum^n y_i} = E\frac{1}{\sum^n y_i}\sum^n E(x_i|y_i)$$

When

$$E(x_i|y_i) = Ry_i$$

it follows that

$$E\frac{\sum^n x_i}{\sum^n y_i} = E\frac{1}{\sum^n y_i}\sum^n Ry_i = ER = R = \frac{X}{Y}$$

17. The variance of an average and of a total for a subset of the population (Vol. I, Ch. 4, Sec. 10 and 16). For the class of populations for which

$$X_i = 0 \text{ whenever } Y_i = 0$$
$$Y_i = 1 \text{ or } 0$$

a. To prove: The conditional variance of $r = \bar{x}/\bar{y}$ for a particular sample of n units is

$$\frac{N_g - n_g}{N_g n_g} S_g^2$$

and the expected value of this conditional variance over all samples of n units is given approximately by

$$\sigma_r^2 \doteq (1 - f + V_{n_g}^2) \frac{S_g^2}{f N_g} \tag{17.1}$$

where $f = n/N$;

N_g is the number of units for which $Y_i = 1$;

n_g is the number of such units in the sample;

S_g^2 is the variance among such units;

i.e.,

$$S_g^2 = \frac{\sum^{N_g} (X_i - \bar{X}_g)^2}{N_g - 1} \tag{17.2}$$

where

$$\bar{X}_g = \frac{\sum^{N_g} X_i}{N_g}$$

and $V_{n_g}^2$ is given by Eq. 17.9.

An unbiased estimate of $\bar{X}_g$ is

$$r = \frac{\bar{x}}{\bar{y}} = \frac{\sum^{n} x_i}{\sum^{n} y_i} = \frac{\sum^{n_g} x_i}{n_g} \tag{17.3}$$

In $\sum^{n_g} x_i$, the x_i is the value of the X-characteristic for the ith element in the sample having $y_i = 1$.

Proof. By definition

$$\sigma_r^2 = E(r - Er)^2$$

$$= E\left(\frac{\sum^{n_g} x_i}{n_g} - \frac{\sum^{N_g} X_i}{N_g}\right)^2 \tag{17.4}$$

By Theorem 14, Ch. 3 (p. 61), we may write

$$\sigma_r^2 = E E_{n_g}\left(\frac{\sum^{n_g} x_i}{n_g} - \bar{X}_g\right)^2 \tag{17.5}$$

where E_{n_g} means the conditional expected value for a fixed n_g of the expression immediately following it. By Sec. 2 and 10

$$E_{n_g}\left(\frac{\sum^{n_g} x_i}{n_g} - \bar{X}_g\right)^2 = \frac{N_g - n_g}{N_g}\frac{1}{n_g}\frac{\sum^{N_g}(X_i - \bar{X}_g)^2}{N_g - 1} \tag{17.6}$$

From Eq. 17.2, 17.5, and 17.6, we have

$$\sigma_r^2 = E\left(\frac{S_g^2}{n_g} - \frac{S_g^2}{N_g}\right) = S_g^2\left(E\frac{1}{n_g} - \frac{1}{N_g}\right) \tag{17.7}$$

Since

$$n_g = (En_g)(1 + \Delta n_g), \text{ where } \Delta n_g = \frac{n_g - En_g}{En_g}$$

and

$$\frac{1}{1 + \Delta n_g} = 1 - \Delta n_g + (\Delta n_g)^2 - \cdots$$

it follows that

$$E\frac{1}{n_g} = \frac{1}{En_g}E\{1 - \Delta n_g + (\Delta n_g)^2 - \cdots\}$$

$$\doteq \frac{1}{En_g}(1 + V_{n_g}^2) = \frac{1}{fN_g}(1 + V_{n_g}^2) \tag{17.8}$$

since, by Sec. 10, $En_g = (n/N)N_g = fN_g$.

Furthermore,

$$V_{n_g}^2 = \frac{N}{N-1}(1 - f)\frac{Q}{Pn} \doteq \frac{Q}{Pn} \tag{17.9}$$

where

$$P = \frac{N_g}{N}, \quad Q = 1 - P$$

Substitute the approximation of $E(1/n_g)$ in Eq. 17.7 to obtain the result given by Eq. 17.1.

b. To prove: The rel-variance of x', an estimate of a total for a subset of the population, where

$$x' = \frac{N}{n}\sum^{n} x_i = \frac{N}{n}\sum^{n_g} x_i$$

is approximately

$$V_{x'}^2 \doteq (1 - f)\frac{V_g^2 + Q}{Pn} \tag{17.10 or I–4.10.6}$$

where

$$P = \frac{N_g}{N}, \quad Q = 1 - P$$

and

$$V_g^2 = \frac{\sum^{N_g}(X_i - \bar{X}_g)^2}{(N_g - 1)\bar{X}_g^2}$$

Proof. By definition

$$V_{x'}^2 = \frac{\sigma_{x'}^2}{(Ex')^2}$$

Now

$$Ex' = X$$

and

$$\sigma_{x'}^2 = N^2 \frac{(1-f)}{n} \frac{\sum^{N}(X_i - \bar{X})^2}{N-1}$$

where

$$\bar{X} = \frac{\sum^{N} X_i}{N} = \frac{\sum^{N_g} X_i}{N}$$

and if N is large, so that we may assume $N/(N-1) \doteq 1$,

$$V_{x'}^2 \doteq \frac{1-f}{n}\left(\frac{\sum^{N} X_i^2}{N\bar{X}^2} - 1\right) = \frac{1-f}{n}\left(\frac{\sum^{N_g} X_i^2}{NP^2\bar{X}_g^2} - 1\right)$$

$$= \frac{1-f}{n}\left[\frac{1}{P}\left(\frac{\sum^{N_g} X_i^2}{N_g\bar{X}_g^2} - 1\right) + \frac{1}{P} - 1\right]$$

$$= (1-f)\frac{V_g^2 + Q}{Pn}$$

the result given by Eq. 17.10.

18. The rel-variance of the estimated variance of a ratio estimate* (Vol. I, Ch. 4, Sec. 21, Eq. 21.6). *To prove:* The rel-variance, $V_{s^2_r}^2$, of the estimated variance, s_r^2, of the ratio $r = x/y$ for a simple random sample of n units drawn with replacement is given approximately by

$$V_{s^2_r}^2 \doteq \frac{\beta_Z - 1}{n} + \frac{4V_Y^2}{n} - \frac{4\rho_{Z^2Y} V_Y \sqrt{\beta_Z - 1}}{n} \tag{18.1}$$

where ρ_{Z^2Y} is the correlation between Z_i^2 and Y_i, $Z_i = X_i - RY_i$, β_Z

* May be taken up with Chapter 10 instead of Chapter 4.

and V_Y are defined in Sec. 5 and 13, respectively, and either

$$s_r^2 = \frac{\bar{x}^2}{\bar{y}^2}(v_{\bar{x}}^2 + v_{\bar{y}}^2 - 2v_{\bar{x}\bar{y}}), \text{ with } v_{\bar{x}}^2 = \frac{s_X^2}{n\bar{x}^2}, v_{\bar{y}}^2 = \frac{s_Y^2}{n\bar{y}^2}$$

where s_X^2 and s_Y^2 are given by Eq. 4.1, and

$$v_{\bar{x}\bar{y}} = \frac{s_{XY}}{\bar{x}\bar{y}} = \frac{\sum^n (x_i - \bar{x})(y_i - \bar{y})}{(n-1)n\bar{x}\bar{y}}$$

or

$$s_r^2 = \frac{1}{n}\frac{\sum^n y_i^2(r_i - r)^2}{(n-1)\bar{y}^2}, \text{ with } r_i = \frac{x_i}{y_i} \text{ and } \bar{y} = \frac{\sum^n y_i}{n}$$

Consider

$$s_r'^2 = \frac{1}{n}\frac{\sum^n y_i^2(r_i - R)^2}{n\bar{y}^2} = \frac{1}{n}\frac{\sum^n z_i^2}{n\bar{y}^2} \text{ if } z_i = x_i - Ry_i$$

The rel-variance of $s_r'^2$ will be a good approximation to the rel-variance of s_r^2 for large enough n. The derivation here substitutes $r_i - R$ for $r_i - r$. For the effect in an analogous situation where $\bar{X}$ is substituted for $\bar{x}$, see Sec. 5.

Therefore, for n large,

$$V_{s^2_r}^2 \doteq V_{s_r'^2}^2 = V_{(\bar{u}/\bar{y}^2)}^2, \text{ where } \bar{u} = \frac{\sum^n z_i^2}{n}$$

Sampling with replacement has been assumed for simplicity and will be a good approximation to sampling without replacement whenever N is large relative to n.

Proof. By Eq. 11.1, with $u = \bar{u}$, $w = \bar{y}^2$,

$$V_{(\bar{u}/\bar{y}^2)} \doteq V_{\bar{u}}^2 + V_{\bar{y}^2}^2 - 2V_{\bar{u},\bar{y}^2} \tag{18.2}$$

Consider the first term in the right-hand side of Eq. 18.2,

$$V_{\bar{u}}^2 = \left[E\left(\frac{\sum^n z_i^2}{n}\right)^2 - \left(E\frac{\sum^n z_i^2}{n}\right)^2\right] \Big/ \sigma_Z^4$$

By the procedure used in Sec. 5, we find

$$V_{\bar{u}}^2 = \frac{1}{n}\left(\frac{\mu_{4Z}}{\sigma_Z^4} - 1\right) = \frac{1}{n}(\beta_Z - 1) \tag{18.3}$$

Similarly,

$$V_{\bar{y}^2}^2 \doteq \frac{\sigma_{\bar{y}^2}^2}{\bar{Y}^4} \doteq \frac{4V_Y^2}{n} \tag{18.4}$$

where terms of order $1/n^2$ or higher are ignored (see Sec. 6). Finally,

$$V_{\bar{u}\bar{y}^2} = \left[E\left(\frac{\sum^n z_i^2}{n}\right)\left(\frac{\sum^n y_i}{n}\right)^2 - E\,\frac{\sum^n z_i^2}{n}\,E\bar{y}^2 \right] \Big/ \sigma_Z^2 \bar{Y}^2$$

$$\doteq \frac{2\rho_{Z^2Y} V_Y \sqrt{\beta_Z - 1}}{n} \tag{18.5}$$

which is left to the reader to be developed as an exercise. Substituting Eq. 18.3, 18.4, and 18.5 in Eq. 18.2, we obtain Eq. 18.1.

Exercise. Prove that $\rho_{u^2w} \doteq \rho_{uw}$ when

$$u = \frac{1}{n}\sum_i^n x_i \quad \text{and} \quad w = \frac{1}{n}\sum_i^n y_i$$

and n is very large.

19. Condition for the ratio estimate to have a smaller rel-variance than the simple unbiased estimate (Vol. I, Ch. 4, Sec. 19). *To prove:* The approximation to the rel-variance V_r^2 of the ratio of two random variables u/w will be smaller than V_u^2, the rel-variance of an unbiased estimate of a mean or total from the same sample, when

$$\rho_{uw} > \frac{V_w}{2V_u} \tag{19.1}$$

The proof is left to the reader.

Special case of Eq. 19.1. Let

$$u = x' = N\bar{x}$$

$$w = \frac{y'}{Y} = \frac{N\bar{y}}{Y}$$

and

$$\frac{u}{w} = x''$$

From Sec. 2 it follows that

$$V_{x'}^2 = (1-f)\frac{V_X^2}{n} \tag{19.2}$$

Similarly, it follows that

$$V_{(y'/Y)}^2 \doteq (1-f)\frac{V_Y^2}{n} \tag{19.3}$$

From Eq. 3.3 of Sec. 3, it follows that

$$\rho_{x',(y'/Y)} = \rho_{XY} \tag{19.4}$$

And, substituting Eq. 19.2, 19.3, and 19.4 into 19.1, we see that the condition for gain through the use of the ratio estimate x'' over the simple unbiased estimate x' is

$$\rho_{XY} > \frac{V_Y}{2V_X}$$

★20. Consistent estimates of averages and variances (Vol. I, Ch. 4, Sec. 12). *To prove:* The sample estimate $\bar{x}$ of $\bar{X}$ and the sample estimate s^2 of S^2 are unbiased and consistent estimates of $\bar{X}$ and S^2, respectively, where $\bar{x}$ and $\bar{X}$ are defined by Eq. 1.1, s^2 is given by Eq. 4.1, and S^2 by Eq. 2.1.

Proof. The proof that $\bar{x}$ is an unbiased estimate of $\bar{X}$ is given in Sec. 1, and that s^2 is an unbiased estimate of S^2 is given in Sec. 4. Since $\sigma_{\bar{x}}^2$ and σ_{s^2} approach 0 as n increases, it follows from the corollary of Theorem 19, Ch. 3 (p. 75), that $\bar{x}$ and s^2 are consistent estimates of $\bar{X}$ and S^2.

★21. Functions of random variables which are consistent estimates of the same function of population characteristics (Vol. I, Ch. 4, Sec. 18 and 21).

a. $\bar{x}/\bar{y}$ is a consistent estimate of $\bar{X}/\bar{Y}$, where $\bar{x}$ and $\bar{y}$ are defined in Sec. 1, and $\bar{Y} \neq 0$.

b. $s/\bar{x}$ is a consistent estimate of $S/\bar{X} = V$, where s is given by Eq. 4.1 and S^2 by Eq. 2.1, and $\bar{X} \neq 0$.

c. $s^2/\bar{x}^2$ is a consistent estimate of $S^2/\bar{X}^2 = V^2$, and $\bar{X} \neq 0$.

d. $s_r^2 = (\bar{x}^2/\bar{y}^2)(v_x^2 + v_y^2 - 2v_{xy})$ is a consistent estimate of σ_r^2, where s_r^2 is defined in Sec. 18, σ_r^2 is given in Sec. 13, and $\bar{Y} \neq 0$.

The proofs follow immediately from Corollary 2 of Theorem 20, Ch. 3 (p. 75), which states that a rational function of consistent estimates is a consistent estimate of the same rational function of the quantities being estimated, if the denominator does not vanish when the quantities estimated are substituted in the rational function.

★ May be deferred.

REFERENCES

(1) E. C. Fieller, "The Biological Standardization of Insulin," *J. Roy. Stat. Soc. Supplement*, **VII** (1940–1941), 1951.

(2) W. G. Cochran, *Sampling Techniques*, John Wiley & Sons, New York, 1953, Chapters 2, 3, and 4.

(3) W. Edwards Deming, *Some Theory of Sampling*, John Wiley & Sons, New York, 1950, Chapter 4.

CHAPTER 5

Stratified Simple Random Sampling

DERIVATIONS, PROOFS, AND SOME EXTENSIONS OF THEORY FOR CH. 5 OF VOL. I*

NOTE. A stratified simple random sampling plan is one in which the elements (sampling units) of the population are divided into groups, referred to as strata, such that each element is contained in one and only one stratum. The sample is then chosen by selecting a simple random sample of elements from each stratum. The sampling fraction may vary from stratum to stratum or may be uniform in all strata. If the sampling fraction is uniform the sampling plan is referred to as proportionate stratified sampling.

Note that the theory of stratified simple random sampling is equally applicable, once the sampling units have been defined, whether the sampling units are elementary units, as assumed above, or clusters of elementary units. Specific consideration of the theory of cluster sampling is deferred to succeeding chapters.

The notation in this chapter is the same as that followed in Chapter 4 except that a subscript (h) is added to designate the strata.

1. The expected value, variance, covariance, and correlation of unbiased estimates from a stratified sample (Vol. I, Ch. 5, Sec. 3). *To prove:*

a. Estimates of the form

$$\bar{x} = \frac{\sum_{h}^{L} N_h \bar{x}_h}{N} \qquad (1.1 \text{ or I–5.3.2})$$

are unbiased estimates of $\bar{X} = \sum_{h}^{L} N_h \bar{X}_h / N$, where $\bar{x}_h = x_h/n_h = \sum_{i}^{n_h} x_{hi}/n_h$ is a sample mean based on a simple random sample of n_h units from the hth stratum, x_{hi} is the value of a characteristic for the ith unit in the sample from the hth stratum, N_h is the number of units in the hth stratum, L is the number of strata, $N = \sum_{h}^{L} N_h$, $\bar{X}_h = X_h/N_h = \sum_{i}^{N_h} X_{hi}/N_h$ is the population mean for the hth stratum, and X_{hi} is the value of the characteristic for the ith unit in the population in the hth stratum.

---

* Appropriate references to Vol. I are shown in parentheses after section or subsection headings. The number following I– after some equations gives the chapter, section, and number of that particular equation in Vol. I.

b. The variance of $\bar{x}$ is

$$\sigma_{\bar{x}}^2 = \frac{1}{N^2} \sum_h^L N_h^2 \frac{1-f_h}{n_h} S_{hX}^2 \qquad \text{(1.2 or I–5.3.9)}$$

where

$$S_{hX}^2 = \frac{\sum_i^{N_h} (X_{hi} - \bar{X}_h)^2}{N_h - 1} \qquad \text{(1.3 or I–5.1.2)}$$

and

$$f_h = \frac{n_h}{N_h}$$

c. The covariance of $\bar{x}$ and $\bar{y}$ is

$$\sigma_{\bar{x}\bar{y}} = \frac{1}{N^2} \sum_h^L N_h^2 \frac{1-f_h}{n_h} S_{hXY} \qquad (1.4)$$

where

$$S_{hXY} = \sum_i^{N_h} \frac{(X_{hi} - \bar{X}_h)(Y_{hi} - \bar{Y}_h)}{N_h - 1} \qquad (1.5)$$

and $\bar{x}$ and $\bar{y}$ are sample means for two different characteristics of the units included in the sample.

It follows, by definition of the coefficient of correlation, that the correlation between $\bar{x}$ and $\bar{y}$ is

$$\rho_{\bar{x}\bar{y}} = \frac{\sigma_{\bar{x}\bar{y}}}{\sigma_{\bar{x}}\sigma_{\bar{y}}} \qquad (1.6)$$

Note that the variance is a special case of the covariance, i.e.,

$$\sigma_{\bar{x}}^2 = \sigma_{\bar{x}\bar{x}}$$

where $\sigma_{\bar{x}\bar{x}}$ is given by Eq. 1.4 with $\bar{x}$ substituted for $\bar{y}$, and

$$S_{hX}^2 = S_{hXX}$$

where S_{hXX} is given by Eq. 1.5 with X substituted for Y.

Proof. *a*. For the hth stratum: $\bar{x}_h$ depends on the sample and is a random variable, and N_h/N is a constant. Consequently $\bar{x}$ is a linear combination of random variables.* If we take the expected value of $\bar{x}_h$

* A linear combination of random variables is a sum of the form $u = \Sigma a_h u_h$, where the u_h are random variables and the a_h are any constants. Thus, if $u_h = x_h/y_h$, where x_h and y_h are random variables, we say that u is a linear combination of the u_h's but not a linear combination of either the x_h's or the y_h's. Similarly, if $u_h = x_h^b$, where b is different from 1, we again say that u is a linear combination of the u_h's but not of the x_h's. In the case of $\bar{x} = \sum_h^L N_h\bar{x}_h/N$, we let $a_h = N_h/N$ and $u_h = \bar{x}_h$, and we say that $\bar{x}$ is a linear combination of the $\bar{x}_h$'s.

we have $E\bar{x}_h = \bar{X}_h$, which follows from Sec. 1, Ch. 4. Consequently, $E\bar{x} = \bar{X}$, and $\bar{x}$ is an unbiased estimate of $\bar{X}$. This follows since the expected value of a linear combination of random variables is the same linear combination of the expected values of the random variables (Theorem 6, Ch. 3, p. 49), and by substituting $\bar{X}_h$ for $\bar{x}_h$ in Eq. 1.1 we obtain $\bar{X}$.

b. Since the sample selection is carried out independently in each stratum, $\bar{x}_h$ is independent of $\bar{x}_k$, where h and k designate any pair of strata, and $\bar{x}$ is a linear combination of independent random variables. It follows from Theorem 11, Corollary 1, Ch. 3 (p. 56), that if u is a linear combination of independent random variables of the form

$$u = \Sigma a_h u_h \tag{1.7}$$

where the a_h are constants and the u_h are independent random variables, the variance of u is

$$\sigma_u^2 = \Sigma a_h^2 \sigma_{u_h}^2 \tag{1.8}$$

Since $\bar{x}$ is such a linear combination with $a_h = N_h/N$ and $u_h = \bar{x}_h$, the variance of $\bar{x}$ is

$$\sigma_{\bar{x}}^2 = \frac{\sum^L N_h^2 \sigma_{\bar{x}_h}^2}{N^2} \tag{1.9}$$

where $\sigma_{\bar{x}_h}^2$ is the variance of $\bar{x}_h$.

From Sec. 2, Ch. 4, we have (since $\bar{x}_h$ is based on a simple random sample of n_h units from the hth stratum)

$$\sigma_{\bar{x}_h}^2 = (1 - f_h) \frac{S_{hX}^2}{n_h} \tag{1.10}$$

where S_{hX}^2 is given by Eq. 1.3; and by substituting this result in Eq. 1.9 we obtain Eq. 1.2.

c. Since $\bar{x}_h$ is the mean of a first characteristic of the n_h units included in a sample from the hth stratum, and $\bar{y}_h$ is the mean of a second characteristic for the same n_h sampled units, then, from Sec. 3, Ch. 4, the covariance of $\bar{x}_h$ and $\bar{y}_h$ is

$$\sigma_{\bar{x}_h \bar{y}_h} = (1 - f_h) \frac{S_{hXY}}{n_h} \tag{1.11}$$

and, since the samples are selected independently in the respective strata, the covariances $\sigma_{\bar{x}_h \bar{y}_k} = 0$ for h different from k. It follows from the corollary to Theorem 12, Ch. 3 (p. 58), that since $\bar{x}$ is a linear combination of the random variables $\bar{x}_h$, and $\bar{y}$ is the same linear combination of the random variables $\bar{y}_h$,

$$\sigma_{\bar{x}\bar{y}} = \frac{1}{N^2} \sum^L N_h^2 \sigma_{\bar{x}_h \bar{y}_h} \tag{1.12}$$

and, substituting for $\sigma_{\bar{x}_h \bar{y}_h}$ (Eq. 1.11) in Eq. 1.12, we obtain Eq. 1.4.

The proof in Part *b* is seen to be a special case of this result, obtained by substituting $\bar{x}$ for $\bar{y}$.

Remark 1. When a proportionate sample is selected, i.e., $n_h/N_h = n/N = f$, the covariance becomes

$$\sigma_{\bar{x}\bar{y}} = \frac{1-f}{n} \frac{\sum^L N_h S_{hXY}}{N} \tag{1.13}$$

and the variance becomes

$$\sigma_{\bar{x}}^2 = \frac{1-f}{n} S_w^2, \quad \text{with} \quad S_w^2 = \frac{\sum^L N_h S_{hX}^2}{N} \tag{1.14}$$

Remark 2. For estimated totals $x' = \sum_h^L N_h \bar{x}_h$ and $y' = \sum_h^L N_h \bar{y}_h$ the results corresponding to those given in (*a*), (*b*), and (*c*) become

$$Ex' = X = N\bar{X}$$

$$\sigma_{x'y'} = N^2 \sigma_{\bar{x}\bar{y}}$$

$$\sigma_{x'}^2 = N^2 \sigma_{\bar{x}}^2$$

$$\rho_{x'y'} = \frac{\sigma_{x'y'}}{\sigma_{x'}\sigma_{y'}} = \frac{\sigma_{\bar{x}\bar{y}}}{\sigma_{\bar{x}}\sigma_{\bar{y}}} = \rho_{\bar{x}\bar{y}}$$

2. The variance of the ratio estimate (Vol. I, Ch. 5, Sec. 4). *To prove:* The variance of $r = \bar{x}/\bar{y}$ is approximately

$$\sigma_r^2 \doteq \frac{1}{N^2 \bar{Y}^2} \sum^L N_h^2 \frac{1-f_h}{n_h} S_{hZ'}^2 \tag{2.1 or 1-5.4.5}$$

where

$$S_{hZ'}^2 = S_{hX}^2 + R^2 S_{hY}^2 - 2R\rho_{hXY} S_{hX} S_{hY} \tag{2.2 or 1-5.4.6}$$

and where S_{hX}^2 and S_{hY}^2 are given by Eq. 1.3, $\rho_{hXY} = S_{hXY}/S_{hX}S_{hY}$ is the correlation between X_{hi} and Y_{hi} in the *h*th stratum, S_{hXY} is given by Eq. 1.5, and $R = X/Y = \bar{X}/\bar{Y}$. The other terms are defined in Sec. 1.

Proof. From Eq. 11.1, Ch. 4,

$$\sigma_r^2 \doteq R^2 (V_{\bar{x}}^2 + V_{\bar{y}}^2 - 2\rho_{\bar{x}\bar{y}} V_{\bar{x}} V_{\bar{y}}) \tag{2.3}$$

where $V_{\bar{x}} = \sigma_{\bar{x}}/\bar{X}$ and $V_{\bar{y}} = \sigma_{\bar{y}}/\bar{Y}$. Then

$$\sigma_r^2 \doteq R^2 \left(\frac{1}{N^2 \bar{X}^2} \sum^L N_h^2 \frac{1-f_h}{n_h} S_{hX}^2 + \frac{1}{N^2 \bar{Y}^2} \sum^L N_h^2 \frac{1-f_h}{n_h} S_{hY}^2 \right.$$
$$\left. - 2 \frac{1}{N^2 \bar{X}\bar{Y}} \sum^L N_h^2 \frac{1-f_h}{n_h} \rho_{hXY} S_{hX} S_{hY} \right) \tag{2.4}$$

which follows from Sec. 1, above. Assembling terms, we have

$$\sigma_r^2 \doteq \frac{1}{N^2\bar{Y}^2}\sum^L N_h^2 \frac{1-f_h}{n_h}(S_{hX}^2 + R^2S_{hY}^2 - 2R\rho_{hXY}S_{hX}S_{hY}) \quad (2.5)$$

which is equivalent to Eq. 2.1.

Exercise. Show that $S_{hZ'}^2$ is the variance of $Z'_{hi} = X_{hi} - RY_{hi}$, i.e.,

$$S_{hZ'}^2 = \frac{\sum^{N_h}(Z'_{hi} - \bar{Z}'_h)^2}{N_h - 1}$$

3. The variance of the ratio estimate based on the weighted average of ratios of random variables (Vol. I, Ch. 5, Sec. 4). *To prove:* The variance of the estimate

$$r' = \frac{\sum^L Y_h r_h}{Y}$$

is given approximately by

$$\sigma_{r'}^2 \doteq \frac{1}{N^2\bar{Y}^2}\sum^L N_h^2 \frac{1-f_h}{n_h} S_{hZ}^2 \quad (3.1 \text{ or I–5.4.12})$$

where

$$S_{hZ}^2 = S_{hX}^2 + R_h^2S_{hY}^2 - 2R_h\rho_{hXY}S_{hX}S_{hY} \quad (3.2 \text{ or I–5.4.13})$$

the terms in S_{hZ}^2 are defined in Sec. 1 of this chapter, and $r_h = x_h/y_h$, the ratio of the sample aggregates in the hth stratum.

Proof. It is seen from Sec. 13, Ch. 4, that the variance of r_h is given approximately by $\frac{1-f_h}{n_h}\frac{S_{hZ}^2}{\bar{Y}_h^2}$. Consequently, by Eq. 1.7 and 1.8 with

$a_h = Y_h/Y$, $u_h = r_h$, the variance of

$$\frac{\sum^L Y_h r_h}{Y} \quad \text{is} \quad \frac{1}{Y^2}\sum^L Y_h^2 \frac{1-f_h}{n_h}\frac{S_{hZ}^2}{\bar{Y}_h^2}$$

and, since $Y_h = N_h\bar{Y}_h$ and $Y = N\bar{Y}$, we have Eq. 3.1.

Exercise. Show that S_{hZ}^2 is the variance of $Z_{hi} = X_{hi} - R_hY_{hi}$, i.e.,

$$S_{hZ}^2 = \frac{\sum^{N_h}(Z_{hi} - \bar{Z}_h)^2}{N_h - 1}$$

$$= \frac{\sum^{N_h}(X_{hi} - R_hY_{hi})^2}{N_h - 1}$$

4. Comparison of biases of two ratio estimates, using proportionate stratified sampling (Vol. I, Ch. 5, Sec. 4). *To prove:* If Δ_1 and Δ_2 are the approximations to the biases of

$$r = \frac{\sum^L N_h \bar{x}_h}{\sum^L N_h \bar{y}_h} \quad \text{and} \quad r' = \frac{\sum^L Y_h r_h}{Y}$$

respectively, and if σ_r^2 and $\sigma_{r'}^2$ are the approximations to the variances, then for proportionate stratified sampling:

a. Both Δ_1^2/σ_r^2 and $\Delta_2^2/\sigma_{r'}^2$ decrease for increasing size of sample provided the number of strata, L, is held constant.

b. If the size of sample is increased by increasing the number of strata (and if the total size of sample is small relative to the population), then Δ_1^2/σ_r^2 decreases, but Δ_2^2/σ_r^2 may increase with increasing sample size.

c. If the number of strata is L, and if the $\bar{X}_h$ and $\bar{Y}_h$ do not vary widely between strata, then Δ_2 is of the order of L times as large as Δ_1.

d. It follows from Parts *a* and *b* that a sufficient condition for Δ_1 to be small relative to σ_r^2 is that the total sample be large, no matter how small the average size of sample per stratum. It will be zero if $\rho_{wXY} = V_{wX}/V_{wY}$. A sufficient condition for Δ_2 to be small with any size of sample in a stratum is that $\rho_{hXY} \doteq V_{hX}/V_{hY}$ for each stratum (which is the condition for the regression line for X on Y to go through the origin for each stratum). The bias of r' can be small for small samples per stratum under less stringent conditions, but unless the conditions given are approximately met there is a risk of serious bias if r' is used.

Proof. It follows directly from Sec. 14, Ch. 4, that the bias of the ratio estimate r is approximately

$$\Delta_1 = R(V_{y'}^2 - \rho_{x'y'} V_{x'} V_{y'}) \tag{4.1}$$

and the bias of r' is approximately

$$\Delta_2 = \frac{1}{Y} \sum^L Y_h R_h (V_{hy}^2 - \rho_{hxy} V_{hx} V_{hy}) \tag{4.2}$$

where

$$V_{hx}^2 = \frac{1 - f_h}{n_h} V_{hX}^2, \quad V_{x'}^2 = \frac{1}{X^2} \sum^L N_h^2 \frac{1 - f_h}{n_h} S_{hX}^2$$

$$x' = \sum^L N_h \bar{x}_h, \qquad r_h = \frac{x_h}{y_h}$$

$$R = \frac{X}{Y}, \qquad R_h = \frac{X_h}{Y_h}$$

and the other terms are as defined in Sec. 1, 2, and 3. For proportionate stratified sampling Eq. 4.1 and 4.2 become, respectively,

$$\Delta_1 = \frac{1-f}{n} R(V_{wY}^2 - \rho_{wXY} V_{wX} V_{wY}) \tag{4.3}$$

$$= \frac{1-f}{n} R \frac{\sum^L N_h \bar{Y}_h^2 \left(V_{hY}^2 - \frac{R_h}{R} \rho_{hXY} V_{hX} V_{hY}\right)}{N\bar{Y}^2} \tag{4.4}$$

$$= \frac{1-f}{n} A \tag{4.5}$$

$$\Delta_2 = \frac{1}{Y} \sum^L Y_h R_h \frac{1-f_h}{n_h} (V_{hY}^2 - \rho_{hXY} V_{hX} V_{hY})$$

and, since $n_h/N_h = f_h = f$, $Y_h R_h = X_h = N_h \bar{X}_h$, we have

$$\Delta_2 = \frac{1-f}{n\bar{Y}} \sum^L \bar{X}_h (V_{hY}^2 - \rho_{hXY} V_{hX} V_{hY}) \tag{4.6}$$

$$= \frac{L(1-f)}{n} \frac{\sum^L \bar{X}_h}{\bar{Y}L} \frac{\sum^L \bar{X}_h (V_{hY}^2 - \rho_{hXY} V_{hX} V_{hY})}{\sum^L \bar{X}_h} \tag{4.7}$$

$$= \frac{L(1-f)}{n} A' \tag{4.8}$$

The values of A and A' are constant for any fixed set of strata.

From Eq. 4.5 and 4.8, and substituting values for σ_r^2 and $\sigma_{r'}^2$ for proportionate stratified sampling obtained in Sec. 2 and 3, we have

$$\frac{\Delta_1^2}{\sigma_r^2} = \frac{\frac{(1-f)^2}{n^2} A^2}{\frac{1-f}{n} \frac{\sum^L N_h S_{hZ'}^2}{N}} = \frac{\frac{1-f}{n} A^2 N}{\sum^L N_h S_{hZ'}^2} \tag{4.9}$$

$$\frac{\Delta_2^2}{\sigma_{r'}^2} = \frac{\frac{L^2(1-f)^2}{n^2} A'^2}{\frac{1-f}{n} \frac{\sum^L N_h S_{hZ}^2}{N}} = \frac{L^2\left(\frac{1-f}{n}\right) A'^2 N}{\sum^L N_h S_{hZ}^2} \tag{4.10}$$

both of which decrease with increasing size of n if L, A, and A' are fixed. Thus, with a fixed set of strata and for samples large enough, the approximate bias of either r or r' will be trivial relative to the approximate standard deviation.

Note that, although $A^2N/\sum^L N_h S_{hZ'}^2$ and $A'^2N/\sum^L N_h S_{hZ}^2$ depend on the stratification, they are ratios of averages that may not be sensitive to altering the stratification, and sometimes may be affected but slightly as the number of strata is increased.

We shall consider the case where $A^2N/\sum^L N_h S_{hZ'}^2$ and $A'^2N/\sum^L N_h S_{hZ}^2$ are about constant as the number of strata is increased. Suppose, now, that the size of sample is increased by keeping the same average sample take per stratum, but increasing the number of strata, so that $n/L = \bar{n} =$ constant. This does not affect Eq. 4.9, which still decreases with increasing n, but now Eq. 4.10 becomes

$$\frac{\Delta_2^2}{\sigma_{r'}^2} \doteq \frac{n(1-f)A'^2N}{\bar{n}^2\Sigma N_h S_{hZ}^2} \tag{4.11}$$

which increases with increasing n provided f is small for all sizes of sample considered.

Also, if the approximate biases are not zero and if we take the ratio of Δ_2 to Δ_1, we have

$$\frac{\Delta_2}{\Delta_1} = L\frac{A'}{A} \tag{4.12}$$

which will be approximately equal to L when $A' \doteq A$, which latter conditions will hold approximately (from Eq. 4.4 and 4.7) at least when the $\bar{X}_h$, the $\bar{Y}_h$, and the N_h do not differ widely between strata.

Finally, it follows from Eq. 4.9 that Δ_1^2 will be small relative to the variance with a large enough sample (for any number of strata), but from Eq. 4.7 it is seen, with $\bar{n}$ fixed, and with $\rho_{hXY} \neq V_{hX}/V_{hY}$, that Δ_2 need not decrease as n increases unless the sampling fraction becomes large; but a sufficient condition for Δ_2 to be zero or small is that $\rho_{hXY} \doteq V_{hX}/V_{hY}$ for all h.

It appears reasonable to assume that relationships essentially similar to those given above may hold in practice for disproportionate sampling designs, although the exact relationships have not been developed.

5. Difference between the variances of two ratio estimates (Vol. I, Ch. 5, Sec. 4). It is shown below that the difference between the variance of

$$r = \frac{\sum^L N_h \bar{x}_h}{\sum^L N_h \bar{y}_h}$$

and the variance of

$$r' = \frac{\sum^L Y_h r_h}{Y}$$

is given by

$$\sigma_r^2 - \sigma_{r'}^2 \doteq \frac{1}{Y^2} \sum_h^L Y_h^2 \frac{1-f_h}{n_h} [V_{hY}^2(R_h - R)^2$$

$$- 2R_h(V_{hY}^2 - \rho_{hXY} V_{hX} V_{hY})(R_h - R)] \quad (5.1)$$

Note that the first term in brackets in Eq. 5.1 is positive and will increase as the variation in the R_h increases, and the second term involves the approximations to the bias of the r_h and will be small when the approximate biases are small, i.e., when the $\rho_{hXY} \doteq V_{hX}/V_{hY}$. This is an indication that the larger the difference among the stratum ratios, the greater the gain in using r' rather than r, so long as the second term is small. However, if the second term is large relative to the first, then the use of r' should be avoided.

Derivation of the difference $\sigma_r^2 - \sigma_{r'}^2$ is as follows:

The variance of r is given by Eq. 2.1. The variance of r' is given by Eq. 3.1. The difference is

$$\sigma_r^2 - \sigma_{r'}^2 = \frac{1}{N^2 \bar{Y}^2} \sum_h^L N_h^2 \frac{1-f_h}{n_h} (S_{hZ'}^2 - S_{hZ}^2) \quad (5.2)$$

Now

$$S_{hZ'}^2 - S_{hZ}^2 = S_{hX}^2 + R^2 S_{hY}^2 - 2R\rho_{hXY} S_{hX} S_{hY}$$

$$- (S_{hX}^2 + R_h^2 S_{hY}^2 - 2R_h \rho_{hXY} S_{hX} S_{hY})$$

$$= S_{hY}^2(R^2 - R_h^2) - 2\rho_{hXY} S_{hX} S_{hY}(R - R_h)$$

$$= \bar{Y}_h^2 V_{hY}^2 (R_h - R)^2$$

$$- 2\bar{Y}_h^2 R_h (V_{hY}^2 - \rho_{hXY} V_{hX} V_{hY})(R_h - R) \quad (5.3)$$

Substituting Eq. 5.3 into Eq. 5.2 and substituting $Y = N\bar{Y}$ and $Y_h = N_h \bar{Y}_h$, we obtain Eq. 5.1.

6. Total population variance expressed as a sum of components (Vol. I, Ch. 5, Eq. 5.6). *To prove:* For a stratified sampling design the variance between elementary units in the population, σ^2, can be written as the sum of the variance between the stratum means, σ_b^2, and the variance between elementary units within strata, σ_w^2, i.e.,

$$\sigma^2 = \sigma_b^2 + \sigma_w^2 \quad \text{(6.1 or I–5.5.6)}$$

where

$$\sigma^2 = \frac{\sum^L \sum^{N_h} (X_{hi} - \bar{X})^2}{N} \quad (6.2)$$

$$\sigma_b^2 = \frac{\sum^L N_h(\bar{X}_h - \bar{X})^2}{N} \qquad (6.3 \text{ or I–5.5.4})$$

$$\sigma_w^2 = \frac{\sum^L \sum^{N_h}(X_{hi} - \bar{X}_h)^2}{N} \qquad (6.4 \text{ or I–5.5.7})$$

Proof.

$$N\sigma^2 = \sum^L \sum^{N_h}(X_{hi} - \bar{X})^2 = \sum^L \sum^{N_h}\{(X_{hi} - \bar{X}_h) + (\bar{X}_h - \bar{X})\}^2$$

$$= \sum^L \sum^{N_h}(X_{hi} - \bar{X}_h)^2 + 2\sum^L \sum^{N_h}(X_{hi} - \bar{X}_h)(\bar{X}_h - \bar{X}) + \sum^L \sum^{N_h}(\bar{X}_h - \bar{X})^2$$

and, since $\sum^{N_h}(X_{hi} - \bar{X}_h) = 0$ for each stratum, and since

$$\sum^L \sum^{N_h}(\bar{X}_h - \bar{X})^2 = \sum^L N_h(\bar{X}_h - \bar{X})^2$$

we have

$$\sigma^2 = \frac{\sum^L \sum^{N_h}(X_{hi} - \bar{X}_h)^2}{N} + \frac{\sum^L N_h(\bar{X}_h - \bar{X})^2}{N} = \sigma_w^2 + \sigma_b^2$$

7. Gain due to stratification using proportionate sampling (Vol. I, Ch. 5, Eq. 5.3). *To prove:* The absolute gain due to stratification with proportionate sampling is

$$\sigma_{\bar{x}}^2 - \sigma_{\bar{x}_s}^2 \doteq \frac{1-f}{n}\left(\sigma_b^2 - \frac{S_w^2}{\bar{N}}\right) \qquad (7.1 \text{ or I–5.5.3})$$

where σ_b^2 is given by Eq. 6.3, and

$$S_w^2 = \frac{1}{N}\sum^L \frac{N_h}{N_h - 1}\sum^{N_h}(X_{hi} - \bar{X}_h)^2 \qquad (7.2)$$

where $\sigma_{\bar{x}}^2$ represents the variance of $\bar{x}$ for a simple random sample, and $\sigma_{\bar{x}_s}^2$ represents the variance of $\bar{x}$ based on a proportionate stratified sample.

Proof. The variance of a sample mean based on a simple random sample of n elements is (from Ch. 4, Sec. 2)

$$\sigma_{\bar{x}}^2 = (1-f)\frac{S^2}{n} = \frac{1-f}{n}\frac{N}{N-1}\sigma^2 = \frac{1-f}{n}\frac{N}{N-1}(\sigma_b^2 + \sigma_w^2) \qquad (7.3)$$

where σ_b^2 is given by Eq. 6.3 and σ_w^2 is given by Eq. 6.4.

The variance of a sample mean based on a proportionate stratified sample of n elements is (from Sec. 1, Remark 1)

$$\sigma_{\bar{x}_s}^2 = (1-f)\frac{S_w^2}{n} \qquad (7.4 \text{ or I–5.3.10})$$

Therefore,

$$\sigma_{\bar{x}}^2 - \sigma_{\bar{x}_s}^2 = \frac{1-f}{n}(S^2 - S_w^2) \tag{7.5}$$

Assume that N is large enough that the approximation $N/(N-1) \doteq 1$ is good, and the N_h are either so large, or so nearly constant, that the approximation $N_h/(N_h - 1) = \bar{N}/(\bar{N} - 1)$ is satisfactory. Then

$$\sigma^2 \doteq S^2, \; S_w^2 \doteq \frac{\bar{N}}{\bar{N}-1}\sigma_w^2$$

and

$$\sigma_{\bar{x}}^2 - \sigma_{\bar{x}_s}^2 \doteq \frac{1-f}{n}\left(\sigma^2 - \frac{\bar{N}}{\bar{N}-1}\sigma_w^2\right) = \frac{1-f}{n}\left(\sigma_b^2 + \sigma_w^2 - \frac{\bar{N}}{\bar{N}-1}\sigma_w^2\right)$$

$$= \frac{1-f}{n}\left(\sigma_b^2 - \frac{\sigma_w^2}{\bar{N}-1}\right)$$

$$= \frac{1-f}{n}\left(\sigma_b^2 - \frac{S_w^2}{\bar{N}}\right) \tag{7.1}$$

8. Variance between the stratum means with random grouping of elements into strata (Vol. I, Ch. 5, Sec. 5). *To prove:* If strata are formed by a random grouping of elements, the variance, σ_b^2, between the stratum means is approximately equal to $S_w^2/\bar{N}$.

Proof. If the N_h elements in the hth stratum were obtained by distributing all N elements of the population at random among the L strata, so that N_h elements were put into the hth stratum, $h = 1, \cdots, L$, then $E\bar{X}_h = \bar{X}$, and

$$\sigma_b^2 = E\frac{\sum^L N_h(\bar{X}_h - \bar{X})^2}{N} = E\frac{\sum^L N_h \bar{X}_h^2}{N} - \bar{X}^2 = \frac{\sum^L N_h E\bar{X}_h^2}{N} - \bar{X}^2$$

$$= \frac{\sum^L N_h\left(S^2\dfrac{N-N_h}{NN_h} + \bar{X}^2\right)}{N} - \bar{X}^2$$

$$= \frac{\sum^L S^2 \dfrac{N-N_h}{N}}{N} + \frac{\sum^L N_h \bar{X}^2}{N} - \bar{X}^2$$

$$= \frac{S^2(L-1)}{N} = \frac{S^2}{\bar{N}}\frac{L-1}{L}$$

$$\doteq \frac{S^2}{\bar{N}}, \text{ if } L \text{ is large.}$$

Also, from Sec. 4, Ch. 4, $ES_h^2 = S^2$, where S_h^2 is given by Eq. 1.3, provided the strata are made up by random assignment of the units in the population, and therefore $ES_w^2 = S^2$, where S_w^2 is given by Eq. 7.2, or $S_w^2 \doteq S^2$ for a reasonably large population.

It follows that $\sigma_b^2 \doteq S_w^2/\bar{N}$ provided there are enough strata that $L/(L-1) \doteq 1$.

9. Optimum allocation to strata (Vol. I, Ch. 5, Eq. 8.1). *To prove:* The values of n_h which minimize the variance

$$\tilde{\sigma}^2 = \frac{1}{N^2} \sum^L N_h^2 \frac{1-f_h}{n_h} \tilde{S}_h^2 \tag{9.1}$$

subject to the condition

$$\sum^L n_h = n \quad \text{or} \quad \sum^L n_h - n = 0 \tag{9.2}$$

are

$$n_h = \frac{N_h \tilde{S}_h}{\sum^L N_h \tilde{S}_h} n \tag{9.3 or I–5.8.1}$$

where $\tilde{S}_h$ is given by S_h, $S_{hZ'}/\bar{Y}$, or $S_{hZ}/\bar{Y}$, depending on the form of the estimate.

Equation 9.1 is the variance of $\bar{x}$ (Eq. 1.1) when $\tilde{S}_h^2 = S_h^2$ as defined by Eq. 1.2; it is the variance of

$$r = \frac{\sum^L N_h \bar{x}_h}{\sum^L N_h \bar{y}_h}$$

when $\tilde{S}_h^2 = S_{hZ'}^2/\bar{Y}^2$ and $S_{hZ'}^2$ is defined by Eq. 2.2; and it is the variance of $r' = \sum^L Y_h r_h / Y$ when $\tilde{S}_h^2 = S_{hZ}^2/\bar{Y}^2$ and S_{hZ}^2 is defined by Eq. 3.2. The form of the variance, and thus of the optimum values of n_h, is the same for totals estimated by multiplying the above estimates by a constant.

Proof. To obtain the minimum $\tilde{\sigma}^2$ subject to a fixed size of sample we set up the Lagrangian F,* i.e., we define the function

$$F = F_0 + \lambda_1 F_1 + \lambda_2 F_2 + \cdots + \lambda_k F_k \tag{9.4}$$

where F_0 is the function to be minimized, F_1 is the relationship determined by a first condition to be imposed, F_2 is the relationship determined by a

* For a discussion of the Lagrange method of obtaining a relative minimum or maximum value of a function subject to conditions see E. Goursat and E. R. Hedrick, *Mathematical Analysis*, Vol. I, Ginn and Co., 1940, p. 128, Sec. 61, or other texts.

second condition to be imposed, etc.; and λ_1 and λ_2, etc., are Lagrange multipliers whose values are obtained as a part of the solution.

In the present problem F_0 is the variance to be minimized and is given by Eq. 9.1; $F_1 = \sum^L n_h - n$ is the condition to be imposed (from Eq. 9.2). There are no other conditions. Consequently, in this particular problem, we have

$$F = \frac{1}{N^2} \sum^L N_h^2 \left(1 - \frac{n_h}{N_h}\right) \frac{\tilde{S}_h^2}{n_h} + \lambda\left(\sum^L n_h - n\right) \tag{9.5}$$

We ascertain the optimum values of the n_h by taking derivatives of F with respect to the n_h and setting each of the derivatives equal to zero. This gives L equations in $L + 1$ unknowns (the n_h's and λ). The condition given by Eq. 9.2 makes $L + 1$ equations. With as many equations as unknowns we can solve them simultaneously to obtain the values for the n_h that minimize (or maximize) F_0 subject to the condition F_1. One can examine the results in cases where there is any doubt to ascertain whether the solution gives a minimum or a maximum value for F_0. Thus,

$$\frac{\partial F}{\partial n_h} = -\frac{N_h^2 \tilde{S}_h^2}{N^2 n_h^2} + \lambda = 0 \quad (h = 1, \cdots, L)$$

or

$$n_h = \frac{N_h \tilde{S}_h}{N\sqrt{\lambda}} \tag{9.6}$$

Summing over the L equations, we have

$$\sum^L n_h = \frac{1}{\sqrt{\lambda}} \frac{\sum^L N_h \tilde{S}_h}{N}$$

and from Eq. 9.2 we substitute n for $\sum^L n_h$ and solve for $\sqrt{\lambda}$. By substituting this value for $\sqrt{\lambda}$ in Eq. 9.6, we obtain

$$n_h = \frac{N_h \tilde{S}_h}{\sum^L N_h \tilde{S}_h} n \tag{9.3}$$

To see that the substitution of n_h given by Eq. 9.3 yields a minimum we note that the variance

$$N^2 \tilde{\sigma}^2 = \sum^L N_h^2 \left(1 - \frac{n_h}{N_h}\right) \frac{\tilde{S}_h^2}{n_h}$$

can be written as

$$N^2\tilde{\sigma}^2 = -\sum^L N_h\tilde{S}_h^2 + \sum^L n_h\left(\frac{N_h\tilde{S}_h}{n_h}\right)^2$$

$$= -\sum^L N_h\tilde{S}_h^2 + \sum^L n_h\left(\frac{N_h\tilde{S}_h}{n_h} - K\right)^2 + 2K\sum^L N_h\tilde{S}_h - nK^2 \qquad (9.7)$$

where K is a constant. The second term of the right-hand member of Eq. 9.7 is the only term involving the n_h's. Therefore, $\tilde{\sigma}^2$ will be at a minimum when this term is zero, i.e., when $K = N_h\tilde{S}_h/n_h$ or

$$n_h = \frac{N_h\tilde{S}_h}{K} \qquad (9.8)$$

and, summing both sides for $h = 1, 2, \cdots, L$,

$$\sum^L n_h = n = \frac{\sum^L N_h\tilde{S}_h}{K}$$

or

$$K = \frac{\sum^L N_h\tilde{S}_h}{n}$$

When this value of K is substituted into Eq. 9.8, we obtain Eq. 9.3, the values of n_h which make $\tilde{\sigma}^2$ a minimum.

The variance at the optimum is given by

$$\sigma^2\,(\text{opt.}) = \frac{1}{N^2}\left\{\frac{(\sum^L N_h\tilde{S}_h)^2}{n} - \sum^L N_h\tilde{S}_h^2\right\} \qquad (9.9 \text{ or I–5.8.5})$$

which is obtained by substituting Eq. 9.3 in Eq. 9.1.

10. Gain of optimum allocation over proportionate stratified sampling (Vol. I, Ch. 5, Eq. 8.6 and 8.7). *To prove:* The relative gain over proportionate stratified sampling of optimum allocation of a fixed size of sample to the strata is given by

$$\frac{\sigma^2\,(\text{prop.}) - \sigma^2\,(\text{opt.})}{\sigma^2\,(\text{prop.})} = \frac{V_{\tilde{S}_h}^2}{(1-f)(1 + V_{\tilde{S}_h}^2)} \qquad (10.1 \text{ or I–5.8.6})$$

$$\doteq \frac{V_{\tilde{S}_h}^2}{1 + V_{\tilde{S}_h}^2} \qquad (10.2)$$

where

$$\sigma^2_{\bar{S}_h} = \frac{1}{N}\sum^L N_h(\bar{S}_h - \bar{S})^2 = \frac{\sum^L N_h\bar{S}_h^2}{N} - \bar{S}^2$$

$$V^2_{\bar{S}_h} = \frac{\sigma^2_{\bar{S}_h}}{\bar{S}^2}$$

$$\bar{S} = \frac{\sum^L N_h\bar{S}_h}{N}$$

Proof. We have already seen that

$$\sigma^2\,(\text{prop.}) = \frac{1-f}{n}\frac{\sum^L N_h\bar{S}_h^2}{N} = \frac{\sum^L N_h\bar{S}_h^2}{Nn} - \frac{\sum^L N_h\bar{S}_h^2}{N^2}$$

$$\sigma^2\,(\text{opt.}) = \frac{(\sum^L N_h\bar{S}_h)^2}{N^2n} - \frac{\sum^L N_h\bar{S}_h^2}{N^2}$$

Then, it follows that

$$\frac{\sigma^2\,(\text{prop.}) - \sigma^2\,(\text{opt.})}{\sigma^2\,(\text{prop.})} = \frac{\frac{\sum^L N_h\bar{S}_h^2}{Nn} - \frac{\bar{S}^2}{n}}{\frac{N-n}{Nn}\frac{\sum^L N_h\bar{S}_h^2}{N}}$$

$$= \frac{\sigma^2_{\bar{S}_h}}{(1-f)\sum^L N_h\bar{S}_h^2/N}$$

Substituting $\sigma^2_{\bar{S}_h} + \bar{S}^2$ for $\sum^L N_h\bar{S}_h^2/N$ and dividing numerator and denominator by $\bar{S}^2$, we get Eq. 10.1.

11. Optimum allocation with variable costs between strata (Vol. I, Ch. 5, Eq. 12.1, 12.2, and 12.3). *To prove:* The optimum allocation to strata of a sample of elements when the cost of including a unit in the sample varies from stratum to stratum is

$$n_h = \frac{N_h\bar{S}_h/\sqrt{C_h}}{\sum^L (N_h\bar{S}_h/\sqrt{C_h})}\,n \qquad (11.1 \text{ or I–5.12.1})$$

where n, the total size of sample, is determined to yield (1) a minimum variance when the total expenditure is fixed, or (2) a minimum expenditure when the precision is specified at ε^2.

For case 1:

$$n = \frac{C}{\sum^{L} N_h \tilde{S}_h \sqrt{C_h}} \sum^{L} \frac{N_h \tilde{S}_h}{\sqrt{C_h}} \qquad \text{(11.2 or I–5.12.2)}$$

For case 2:

$$n = \frac{\sum^{L}(N_h \tilde{S}_h \sqrt{C_h}) \sum^{L}(N_h \tilde{S}_h / \sqrt{C_h})}{N^2 \varepsilon^2 + \sum^{L} N_h \tilde{S}_h^2} \qquad \text{(11.3 or I–5.12.3)}$$

Proof. The variance of an estimated average or ratio from a stratified random sample is (see Eq. 9.1)

$$\tilde{\sigma}^2 = \frac{1}{N^2} \sum^{L} N_h^2 (1 - f_h) \frac{\tilde{S}_h^2}{n_h} = \frac{1}{N^2} \sum^{L} N_h^2 \tilde{S}_h^2 \left(\frac{1}{n_h} - \frac{1}{N_h} \right)$$

The cost of the survey is assumed to be of the form

$$C = \sum^{L} C_h n_h$$

(1) To determine values of n_h which minimize the variance subject to a fixed total cost, C, we set up the Lagrangian F:

$$F = \frac{1}{N^2} \sum^{L} N_h^2 \tilde{S}_h^2 \left(\frac{1}{n_h} - \frac{1}{N_h} \right) + \lambda \left(\sum^{L} C_h n_h - C \right)$$

$$\frac{\partial F}{\partial n_h} = - \frac{1}{N^2} \frac{N_h^2 \tilde{S}_h^2}{n_h^2} + \lambda C_h = 0$$

Solving for n_h, we have

$$n_h = \frac{N_h \tilde{S}_h}{N \sqrt{C_h}} \cdot \frac{1}{\sqrt{\lambda}}$$

and

$$n = \sum^{L} n_h = \frac{\sum^{L}(N_h \tilde{S}_h / \sqrt{C_h})}{N \sqrt{\lambda}}$$

Substituting for $\sqrt{\lambda}$, we have

$$n_h = \frac{N_h \tilde{S}_h / \sqrt{C_h}}{\sum^{L}(N_h \tilde{S}_h / \sqrt{C_h})} n$$

When the total cost is fixed at $C = \sum^{L} C_h n_h$, we have

$$C = \frac{\sum^{L} N_h \tilde{S}_h \sqrt{C_h}}{\sum^{L}(N_h \tilde{S}_h / \sqrt{C_h})} n$$

and the optimum n is given by Eq. 11.2.

(2) To determine the values of n_h that minimize the cost subject to a prescribed precision, ε, we set up the Lagrangian F:

$$F = \sum^{L} C_h n_h + \lambda \left[\frac{1}{N^2} \sum^{L} N_h^2 \tilde{S}_h^2 \left(\frac{1}{n_h} - \frac{1}{N_h} \right) - \varepsilon^2 \right]$$

Proceeding as in (1) above, we find that the optimum values for the n_h are, again, as given by Eq. 11.1.

When the precision is fixed at

$$\varepsilon^2 = \frac{1}{N^2} \left(\sum^{L} \frac{N_h^2 \tilde{S}_h^2}{n_h} - \sum^{L} N_h \tilde{S}_h^2 \right)$$

we have, by substituting for n_h its optimum value,

$$N^2 \varepsilon^2 = \sum^{L} \left(\frac{N_h^2 \tilde{S}_h^2 \sum^{L} N_h \tilde{S}_h / \sqrt{C_h}}{n N_h \tilde{S}_h / \sqrt{C_h}} \right) - \sum^{L} N_h \tilde{S}_h^2$$

and the optimum n is given by Eq. 11.3.

12. Sample estimates of population variances (Vol. I, Ch. 4, Sec. 14). *To prove:*

a. A consistent and unbiased estimate of S_{hX}^2 as given by Eq. 1.3 is

$$s_{hX}^2 = \frac{\sum^{n_h} (x_{hi} - \bar{x}_h)^2}{n_h - 1} \qquad (12.1 \text{ or I–5.14.1})$$

b. A consistent estimate of $S_{hZ'}^2$ as given by Eq. 2.2 is

$$s_{hZ'}^2 = s_{hX}^2 + r^2 s_{hY}^2 - 2r s_{hXY} \qquad (12.2 \text{ or I–5.14.2})$$

where s_{hX}^2 and s_{hY}^2 are given by Eq. 12.1, and

$$s_{hXY} = \frac{\sum^{n_h} (x_{hi} - \bar{x}_h)(y_{hi} - \bar{y}_h)}{n_h - 1} \qquad (12.3)$$

c. A consistent estimate of S_{hZ}^2 as given by Eq. 3.2 is

$$s_{hZ}^2 = s_{hX}^2 + r_h^2 s_{hY}^2 - 2r_h s_{hXY} \qquad (12.4 \text{ or I–5.14.3})$$

d. An unbiased estimate of σ_{bX}^2 as given by Eq. 6.3 based on a proportionate stratified sample is

$$\frac{1}{n} \sum^{L} n_h (\bar{x}_h - \bar{x})^2 - \frac{1-f}{\bar{n}} \left[\frac{1}{L} \sum^{L} \left(1 - \frac{N_h}{N} \right) s_{hX}^2 \right] \qquad (12.5)$$

where s_{hX}^2 is given by Eq. 12.1 and $\bar{n} = n/L$. Note that for large $\bar{n}$ the second term of Eq. 12.5 can be neglected.

e. An unbiased estimate of σ_X^2 as given by Eq. 6.2 based on a proportionate stratified sample is

$$\frac{1}{n}\sum_h^L\sum_i^{n_h}(x_{hi} - \bar{x})^2 + \frac{1-f}{n}s_{wX}^2 \tag{12.6}$$

where

$$s_{wX}^2 = \frac{1}{n}\sum_h^L n_h s_{hX}^2$$

Note that for large n the second term of Eq. 12.6 can be neglected.

Proof. The proofs are left to the reader.

Hint: The proofs for Parts *a*, *b*, and *c* follow the same reasoning as given in Sec. 20 and 21 of Ch. 4. The proof for Part *d* follows from noting that

$$E\sum_h^L n_h(\bar{x}_h - \bar{x})^2 = E\sum_h^L n_h\bar{x}_h^2 - nE\bar{x}^2$$

$$= \sum_h^L n_h(\sigma_{\bar{x}_h}^2 + \bar{X}_h^2) - n(\sigma_{\bar{x}}^2 + \bar{X}^2)$$

and the proof for Part *e* follows from noting that

$$E\sum_h^L\sum_i^{n_h}(x_{hi} - \bar{x})^2 = \sum_h^L\frac{n_h}{N_h}\sum_i^{N_h}X_{hi}^2 - n(\sigma_{\bar{x}}^2 + \bar{X}^2)$$

13. Variance for stratification after sampling (Vol. I, Ch. 5, Sec. 16, Remark 3). Suppose that we have a simple random sample of n units. Classify the units of the sample into L classes. Let x_{hi} be the value of the ith unit in the hth class. Let N_h be the known number of units in the hth class in the population, and let n_h be the number of units in the hth class in the sample.

Construct the estimate

$$\bar{x} = \frac{1}{N}\sum_h^L\frac{N_h}{n_h}\sum_i^{n_h}x_{hi} \tag{13.1}$$

It is easy to show that $\bar{x}$ is an unbiased estimate of the population mean $\bar{X}$, if we exclude all samples in which one or more of the n_h is zero.

The variance of $\bar{x}$ may be obtained in the following way. First, we note that

$$\sigma_{\bar{x}}^2 = E(\bar{x} - \bar{X})^2 = E\{E_{n_h}(\bar{x} - \bar{X})^2\}$$

where E_{n_h} denotes the expected value for a fixed set of values $n_1, n_2, \cdots,$

n_L. But $E_{n_h}(\bar{x} - \bar{X})^2$ is then the variance of a sample mean based on a stratified sample with n_h elements from stratum h. Therefore,

$$E_{n_h}(\bar{x} - \bar{X})^2 = \frac{1}{N^2}\sum_h^L N_h^2 \frac{N_h - n_h}{N_h}\frac{S_h^2}{n_h}$$

$$= \frac{1}{N^2}\sum_h^L N_h^2 S_h^2 \frac{1}{n_h} - \frac{1}{N^2}\sum_h^L N_h S_h^2$$

To find the expected value of this expression, we need to evaluate $E(1/n_h)$. This can be evaluated exactly,* but the exact value has a complicated expression. It can be shown, however, that

$$E\frac{1}{n_h} \doteq \frac{1}{fN_h}\left\{1 + \frac{1-f}{n}\frac{N - N_h}{N_h}\right\}$$

This follows from Eq. 17.8 and 17.9 of Ch. 4 with $P = P_h = N_h/N$ and $Q = Q_h = 1 - P$. Hence,

$$\sigma_{\bar{x}}^2 = E\{E_{n_h}(\bar{x} - \bar{X})^2\}$$

$$\doteq \frac{1}{N^2}\sum^L \frac{N_h^2 S_h^2}{fN_h} + \frac{1}{N^2}\frac{1-f}{n}\sum^L \frac{N_h^2 S_h^2}{fN_h}\frac{N - N_h}{N_h} - \frac{1}{N^2}\sum^L N_h S_h^2$$

and, by substituting $S_w^2 = \Sigma N_h S_h^2/N$, $f = n/N$, and $Q_h = (N - N_h)/N$,

$$\sigma_{\bar{x}}^2 = \frac{1-f}{n}\left(S_w^2 + \frac{1}{\bar{n}}\frac{\sum^L S_h^2 Q_h}{L}\right) \qquad (13.2 \text{ or I–5.16.1})$$

We note that the first term is precisely the variance for a proportionate stratified sample selected from the strata defined earlier, and that the second term is positive. Hence, the variance for stratification after sampling is larger than the variance for proportionate stratified sampling.

We note further that for a sufficiently large average sample size per stratum, $\bar{n}$, the second term will be small compared with the first term, since the second term is of the order of $1/\bar{n}$ as large as the first. Unless $\bar{n}$ is sufficiently large, however, the net effect of such an approach by "stratification after sampling" may substantially increase the variance over the usual estimate with simple random sampling.

14. Increase in variance arising from duplication of a subset of elements (Vol. I, Ch. 5, Sec. 16, Remark 4). *To prove:* Suppose that from a random

---

* Frederick F. Stephan, "The Expected Value and the Variance of the Reciprocal and Other Negative Powers of a Positive Bernoullian Variate," *Annals Math. Stat.*, **XVI** (1945).

sample of n elements we select a random subsample of n_1 elements, duplicate these n_1 elements, and add them to the original sample. Then the mean based on the $n + n_1$ elements is an unbiased estimate of the population mean, and its variance is greater than the variance of the mean based on the original n elements by the approximate factor

$$\frac{1 + 3\dfrac{n_1}{n}}{\left(1 + \dfrac{n_1}{n}\right)^2}$$

Proof. Let the elements in the random subsample have the values $x_1, x_2, \cdots, x_{n_1}$, and the remaining elements in the sample have the values $x_{n_1+1}, \cdots, x_n$. The estimate of the mean based on the sample of $n + n_1$ elements is then

$$\bar{x}' = \frac{2\sum_i^{n_1} x_i + \sum_{j=n_1+1}^{n} x_j}{n + n_1} = \frac{2u + v}{n + n_1}, \text{ say.}$$

Now,

$$E\bar{x}' = \frac{1}{n + n_1}(2Eu + Ev)$$

$$= \frac{1}{n + n_1}[2n_1\bar{X} + (n - n_1)\bar{X}]$$

$$= \bar{X}$$

so that $\bar{x}'$ is unbiased. Also

$$\sigma_{\bar{x}'}^2 = \frac{1}{(n + n_1)^2}(4\sigma_u^2 + \sigma_v^2 + 4\sigma_{uv})$$

Clearly,

$$\sigma_u^2 = \frac{N - n_1}{N} n_1 S_X^2$$

and

$$\sigma_v^2 = \frac{N - (n - n_1)}{N}(n - n_1)S_X^2$$

It is readily shown that

$$\sigma_{uv} = -\frac{n_1(n - n_1)}{N} S_X^2$$

Hence, we have

$$\sigma_{\bar{x}'}^2 = \frac{S_X^2}{(n + n_1)^2} \left\{ \frac{Nn\left(1 + 3\dfrac{n_1}{n}\right) - (n + n_1)^2}{N} \right\}$$

For the original sample of n, the variance of the mean is

$$\sigma_{\bar{x}}^2 = \frac{N - n}{N} \frac{S_X^2}{n}$$

so that

$$\frac{\sigma_{\bar{x}'}^2}{\sigma_{\bar{x}}^2} = \frac{N}{N - n} \left(\frac{n}{n + n_1}\right)^2 \left(1 + 3\frac{n_1}{n}\right) - \frac{n}{N - n}$$

If N is large compared to n, we have

$$\frac{\sigma_{\bar{x}'}^2}{\sigma_{\bar{x}}^2} \doteq \frac{1 + 3\dfrac{n_1}{n}}{\left(1 + \dfrac{n_1}{n}\right)^2}$$

The relative loss in efficiency is thus

$$\frac{\sigma_{\bar{x}'}^2 - \sigma_{\bar{x}}^2}{\sigma_{\bar{x}}^2} \doteq \frac{\dfrac{n_1}{n}\left(1 - \dfrac{n_1}{n}\right)}{\left(1 + \dfrac{n_1}{n}\right)^2}$$

This has its maximum value (for $0 < n_1/n < 1$) when $n_1/n = \frac{1}{3}$, and for this value of n_1/n, the loss in efficiency is .125.

REFERENCES

(1) Tore Dalenius and Margaret Gurney, "The Problem of Optimum Stratification, II," *Skandinavisk Aktuarietidskrift*, 1951, pp. 133–148.

(2) L. R. Frankel and J. S. Stock, "The Allocation of Samplings among Several Strata," *Annals Math. Stat.*, **10** (1939), 288–293.

(3) J. Neyman, "On the Two Different Aspects of the Representative Method: The Method of Stratified Sampling and the Method of Purposive Selection," *J. Roy. Stat. Soc.*, **109** (1934), 558–606.

(4) P. V. Sukhatme, "Contributions to the Theory of the Representative Method," *J. Roy. Stat. Soc. Supplement*, 1935, pp. 253–268.

(5) A. A. Tschuprow, "On the Mathematical Expectation of the Moments of Frequency Distributions in the Case of Correlated Observations," *Metron*, **2** (1923), 646–680.

(6) W. G. Cochran, *Sampling Techniques*, John Wiley & Sons, New York, 1953, Chapter 5.

(7) W. Edwards Deming, *Some Theory of Sampling*, John Wiley & Sons, New York, 1950, Chapter 6.

CHAPTER 6

Simple One- or More Stage Cluster Sampling

DERIVATIONS, PROOFS, AND SOME EXTENSIONS OF THEORY FOR CH. 6 OF VOL. I*

NOTE. A *simple cluster sampling* plan is a sampling plan in which (*a*) the elementary units of the population to be sampled are grouped into clusters, such that each elementary unit is associated with one and only one cluster; and (*b*) a sample is drawn by using the clusters as sampling units and selecting a simple random sample of the clusters. The clusters are referred to as primary sampling units (psu's) or as first-stage sampling units.

If all elementary units in the selected clusters are included in the sample, the sampling plan is a one-stage sampling plan. If a subsample is selected from each of the selected psu's, with a uniform fraction of the second-stage sampling units selected from each primary unit included in the sample, the sampling plan is referred to as a *simple two-stage cluster sampling plan.* Additional stages of sampling can be introduced.

In this chapter we develop the theory for these simple cluster sampling designs and for certain extensions of them. Attention is given to optimum sample design with different cost functions. Measures of homogeneity are defined, and their effect on optimum two-stage sample design is considered.

For simplification in some proofs sampling with replacement is assumed. The results obtained are to be regarded as approximations to those for sampling without replacement in cases where the sampling fractions involved are not too large.

Some notation used in this chapter. The notation in this chapter is an extension of that in Chapter 4, with the listing units (second-stage units) in this chapter corresponding to the sampling units in Chapter 4. Thus, N in this chapter is the total number of second-stage units, and N_i is the number in the ith primary unit. The N units are here regarded as grouped

* Appropriate references to Vol. I are shown in parentheses after section or subsection headings. The number following I– after some equations gives the chapter, section, and number of that particular equation in Vol. I.

into M clusters that serve as primary sampling units. The basic notation of this chapter is as follows:

M = Number of first-stage units (or primary sampling units) in the population.

m = Number of first-stage units in the sample.

N_i = Number of second-stage units (or listing units) in ith first-stage unit in the population.

n_i = Number of second-stage units in the sample from ith first-stage unit in the sample.

X_{ij} = Value of X-characteristic for jth second-stage unit in ith first-stage unit, $i = 1, \cdots, M$, and $j = 1, \cdots, N_i$.

x_{ij} = Value of X-characteristic for jth second-stage unit in the sample from ith first-stage unit in the sample, $i = 1, \cdots, m$, and $j = 1, \cdots, n_i$.

Y_{ij} and y_{ij} are defined similarly for a second characteristic.

Sums are indicated by dropping subscripts.

$$X_i = \sum_j^{N_i} X_{ij}, \quad X = \sum_i^{M} X_i = \sum_i^{M} \sum_j^{N_i} X_{ij}$$

$$x_i = \sum_j^{n_i} x_{ij}, \quad x = \sum_i^{m} x_i = \sum_i^{m} \sum_j^{n_i} x_{ij}$$

$$N = \sum_i^{M} N_i, \quad n = \sum_i^{m} n_i$$

Average values per second-stage unit are

$$\bar{\bar{X}} = X/N, \quad \bar{\bar{X}}_i = X_i/N_i$$

$$\bar{\bar{x}} = x/n, \quad \bar{\bar{x}}_i = x_i/n_i$$

Average values per first-stage unit are

$$\bar{X} = X/M, \quad \bar{x} = x/m$$

$$\bar{N} = N/M, \quad \bar{n} = n/m$$

Ratios are

$$R = X/Y, \quad R_i = X_i/Y_i$$

$$r = x/y, \quad r_i = x_i/y_i$$

Simple unbiased estimates of totals are x' and x'_i, where

x' is the estimate of X ⎫ See Sec. 1 for further
x'_i is the estimate of X_i ⎭ definitions of these estimates.

Furthermore

$$f_1 = m/M, \quad f_{2i} = n_i/N_i$$

and when f_{2i} is constant for all i

$$f_2 = f_{2i} = E\bar{n}/\bar{N}$$

and

$$f = f_1 f_2 = En/N$$

1. The variance and covariance for a two-stage and for a multi-stage sampling design (Vol. I, Ch. 6, Sec. 6). *a. Variance of simple unbiased estimate for two-stage sample. To prove:* Let us first consider the simple case of a population consisting of M first-stage units with N_i second-stage units within the ith first-stage unit. Assume that a simple random sample of m first-stage units is selected, and a simple random sample of n_i second-stage units is selected from the ith selected first-stage unit. Suppose that the estimate made from the sample is

$$x' = \frac{M}{m}\sum_i^m x'_i = \frac{M}{m}\sum_i^m \frac{N_i}{n_i}\sum_j^{n_i} x_{ij} \tag{1.1}$$

where x_{ij} is the value for the jth selected second-stage unit from the ith selected first-stage unit. We shall show below that $Ex' = X$, where

$$X = \sum_i^M \sum_j^{N_i} X_{ij}$$

and X_{ij} is the value of the jth second-stage unit of the ith first-stage unit in the population. We shall also show that the variance of x' is

$$\sigma_{x'}^2 = \frac{M^2}{m}\frac{M-m}{M}S_{1X}^2 + \frac{M}{m}\sum_i^M \frac{N_i^2}{n_i}\frac{N_i - n_i}{N_i}S_{2iX}^2 \tag{1.2}$$

where

$$S_{1X}^2 = \frac{\sum_i^M (X_i - \bar{X})^2}{M-1} \tag{1.3}$$

$$S_{2iX}^2 = \frac{\sum_j^{N_i} (X_{ij} - \bar{\bar{X}}_i)^2}{N_i - 1} \tag{1.4}$$

and

$$X_i = \sum_j^{N_i} X_{ij}; \quad \bar{X} = \frac{\sum_i^M X_i}{M} = \frac{X}{M}; \quad \bar{\bar{X}}_i = \frac{\sum_j^{N_i} X_{ij}}{N_i} = \frac{X_i}{N_i} \tag{1.5}$$

Proof. We can write

$$x' = \frac{M}{m} \sum_i^m x_i'$$

where

$$x_i' = N_i \bar{\bar{x}}_i$$

Then, by Theorem 6 of Ch. 3 (p. 49),

$$Ex' = \frac{M}{m} \sum_i^m Ex_i'$$

Now Ex_i', by Theorem 14 of Ch. 3 (p. 61), is equal to $E(Ex_i' | b^\star = B_j)$, where $E(x_i' | b^\star = B_j)$ means the conditional expected value of x_i', knowing that the ith sampled first-stage sampling unit is the jth first-stage unit in the population. For short we shall call $E(x_i' | b^\star = B_j) = E_i x_i'$. Hence

$$Ex' = \frac{M}{m} \sum_i^m Ex_i' = \frac{M}{m} \sum_i^m E(E_i x_i') = \frac{M}{m} \sum_i^m E\left(E_i \frac{N_i}{n_i} \sum_j^{n_i} x_{ij}\right)$$

$$= \frac{M}{m} \sum_i^m E \frac{N_i}{n_i} \frac{n_i}{N_i} \sum_j^{N_i} X_{ij}' = \frac{M}{m} \sum_i^m EX_i'$$

where X_{ij}' and X_i' are to be distinguished from X_{ij} and X_i defined above in that they are random variables and take on different values dependent on which first-stage unit is selected in the sample. The values of X_i and X_{ij} are uniquely associated with the ith first-stage sampling unit in the population. Finally, from Sec. 1 of Ch. 4, we have

$$Ex' = \frac{M}{m} \sum_i^m EX_i' = \frac{M}{m} \sum_i^m \frac{1}{M} \sum_j^M X_j = M\bar{X} = X$$

We now wish to express the variance of x' in terms of the two components, one being the contribution arising from the first-stage sampling and the other from the second-stage sampling.

By Theorem 15, Ch. 3 (p. 65), with $u = x'$ and $b^\star = [1]$, where the expression $|[1]$ is used to indicate a fixed sample of first-stage units, we may write

$$\sigma_{x'}^2 = E\sigma_{x'|[1]}^2 + \sigma_{E(x'|[1])}^2 \tag{1.6}$$

In this case, $\sigma_{x'|[1]}^2$ is the conditional variance of x', holding the first-stage units constant; $E(x'|[1])$ is the conditional expected value of x', holding the first-stage units constant; and $\sigma^2_{E(x'|[1])}$ is the variance of these conditional expected values over all possible samples of first-stage units.*

Consider the first term in the right-hand member of Eq. 1.6. The variance of x' for a fixed set of first-stage units in the sample is the variance of $(M/m)x''$, where x'' is the estimated total for the fixed set of m psu's, with each of the m psu's now serving as a stratum. Consequently, by Sec. 1, Ch. 5, the variance of x' for the fixed set of primary units is

$$\sigma^2_{x'|[1]} = \frac{M^2}{m^2} \sum^{m} N_i^2 \frac{N_i - n_i}{n_i N_i} S^2_{2iX}$$

Hence, by Theorem 6, Ch. 3 (p. 49),

$$E\sigma^2_{x'|[1]} = \frac{M^2}{m^2} E \sum^{m} N_i^2 \frac{N_i - n_i}{n_i N_i} S^2_{2iX}$$

$$= \frac{M^2}{m^2} mEN_i^2 \frac{N_i - n_i}{n_i N_i} S^2_{2iX}$$

$$= \frac{M^2}{m^2} m \frac{1}{M} \sum^{M} N_i^2 \frac{N_i - n_i}{n_i N_i} S^2_{2iX} \tag{1.7}$$

since

$$u_i = N_i^2 \frac{N_i - n_i}{n_i N_i} S^2_{2iX}$$

is a random variable having M possible values, each with probability $1/M$. Equation 1.7 is equal to the second term in the right-hand member of Eq. 1.2 and represents the contribution to the variance due to sampling second-stage units within first-stage units.

Consider now $\sigma^2_{E(x'|[1])}$ in Eq. 1.6, which represents the first-stage contribution to the variance. Now,

$$E(x'|[1]) = \frac{M}{m} E\left\{\left(\sum_i^m N_i \bar{\bar{x}}_i\right)|[1]\right\}$$

where $\bar{\bar{x}}_i = \sum_j^{n_i} x_{ij}/n_i$ is the sample average per second-stage unit from the ith first-stage unit in the sample. Since the first-stage units are held constant, they can be regarded as strata, and from Sec. 1 of Ch. 5 we have

$$E(x'|[1]) = \frac{M}{m} \sum_i^m N_i \bar{\bar{X}}_i' = \frac{M}{m} \sum_i^m X_i'$$

* Note that Theorem 16 (p. 68) could have been applied in this case, also, and with exactly the same steps as with Theorem 15. Theorem 16 has an advantage when more than two stages of sampling are involved (as in Sec. 4 of Ch. 7).

where the primes are used to indicate that $\bar{\bar{X}}_i'$ and X_i' are random variables (values for the ith first-stage unit in the sample). Hence, by Sec. 2 of Ch. 4

$$\sigma^2_{E(x'|[1])} = \sigma^2_{\left(\frac{M}{m}\sum_i^m X'_i\right)} = M^2 \frac{M-m}{Mm} \frac{\sum_i^M (X_i - \bar{X})^2}{M-1} \tag{1.8}$$

which is equal to the first term in the right-hand member of Eq. 1.2 and represents the contribution to the variance due to sampling first-stage units. Therefore, substituting Eq. 1.7 and 1.8 into Eq. 1.6, we obtain Eq. 1.2.

If the second-stage sampling fractions are uniform, i.e., $n_i/N_i = f_2 = E\bar{n}/\bar{N}$, for all first-stage units, the variance of x' becomes

$$\sigma^2_{x'} = M^2 \frac{M-m}{Mm} S^2_{1X} + N^2 \frac{\bar{N}-\bar{n}}{\bar{N}m\bar{n}} S^2_{2X} \tag{1.9}$$

where S^2_{1X} is given by Eq. 1.3, and

$$S^2_{2X} = \frac{1}{N} \sum_i^M \frac{N_i}{N_i - 1} \sum_j^{N_i} (X_{ij} - \bar{\bar{X}}_i)^2 = \frac{1}{N} \sum_i^M N_i S^2_{2iX} \tag{1.10}$$

and the rel-variance of x' is given by

$$V^2_{x'} = \frac{\sigma^2_{x'}}{X^2} = \frac{M-m}{Mm} B^2_X + \frac{\bar{N}-\bar{n}}{\bar{N}m\bar{n}} W^2_X \tag{1.11 or I-6.6.4}$$

where

$$B^2_X = \frac{S^2_{1X}}{\bar{X}^2}, \quad W^2_X = \frac{S^2_{2X}}{\bar{\bar{X}}^2}, \quad \bar{X} = \frac{X}{M}, \quad \bar{\bar{X}} = \frac{X}{N}$$

and where, in Eq. 1.9 and 1.11, $\bar{n}$ is the expected number of listing units in the sample per psu in the sample.

b. Covariance of simple unbiased estimate for two-stage sample. To prove: The covariance of x' and y' for a two-stage sampling design is

$$\begin{aligned} \sigma_{x'y'} &= E(x' - X)(y' - Y) \\ &= \frac{M^2}{m} \frac{M-m}{M} S_{1XY} + \frac{M}{m} \sum_i^M \frac{N_i^2}{n_i} \frac{N_i - n_i}{N_i} S_{2iXY} \end{aligned} \tag{1.12}$$

where x' and y' are defined by Eq. 1.1, and where

$$S_{1XY} = \frac{\sum_i^M (X_i - \bar{X})(Y_i - \bar{Y})}{M-1} \tag{1.13}$$

$$S_{2iXY} = \frac{\sum_{j}^{N_i}(X_{ij} - \bar{\bar{X}}_i)(Y_{ij} - \bar{\bar{Y}}_i)}{N_i - 1} \tag{1.14}$$

with X_{ij}, $\bar{\bar{X}}_i$, $\bar{X}$, and X defined as in Eq. 1.5, and with Y_{ij}, $\bar{\bar{Y}}_i$, $\bar{Y}$, and Y similarly defined for the Y-characteristic.

The component of the covariance of x' and y' due to first-stage sampling is

$$\frac{M^2}{m}\frac{M - m}{M} S_{1XY} \tag{1.15}$$

and the component of the covariance of x' and y' due to second-stage sampling is

$$\frac{M}{m}\sum_{i}^{M}\frac{N_i^2}{n_i}\frac{N_i - n_i}{N_i} S_{2iXY} \tag{1.16}$$

Proof. The proof that the covariance, $\sigma_{x'y'}$, for a two-stage sampling design is given by Eq. 1.12 follows the same steps as used in Part *a* to prove that $\sigma_{x'}^2$ is given by Eq. 1.2 and is left to the reader. Note that Eq. 1.12 becomes Eq. 1.2 when we substitute values of X for the corresponding values of Y. Thus, by this substitution S_{1XX} in Eq. 1.13 equals S_{1X}^2 in Eq. 1.3, and S_{2iXX} in Eq. 1.14 equals S_{2iX}^2 in Eq. 1.4. It follows that

$$\sigma_{y'y'} = \sigma_{y'}^2 \tag{1.17}$$

As in Eq. 1.9, when uniform sampling fractions are used, the covariance becomes

$$\sigma_{x'y'} = M^2 \frac{M - m}{Mm} S_{1XY} + N^2 \frac{\bar{N} - \bar{n}}{\bar{N}m\bar{n}} S_{2XY} \tag{1.18}$$

where S_{1XY} is given by Eq. 1.13,

$$S_{2XY} = \frac{1}{N}\sum_{i}^{M} N_i S_{2iXY} \tag{1.19}$$

and $\bar{n}$ in Eq. 1.18 is the expected number of listing units in the sample per psu in the sample.

c. Rel-variance of a ratio for two-stage sample. To prove: Let the estimate of the ratio be

$$r = \frac{x'}{y'} \tag{1.20}$$

where x' is defined by Eq. 1.1 and y' is similarly defined for the Y-characteristic. Then the rel-variance of r can be written as

$$V_r^2 \doteq \frac{1}{X^2}\frac{M^2}{m}\frac{M-m}{M}S_1^2 + \frac{1}{X^2}\frac{M}{m}\sum_i^M \frac{N_i^2}{n_i}\frac{N_i-n_i}{N_i}S_{2i}^2 \qquad (1.21)$$

where

$$S_1^2 = S_{1X}^2 + R^2S_{1Y}^2 - 2RS_{1XY} \qquad (1.22)$$

with S_{1X}^2 defined by Eq. 1.3, S_{1Y}^2 similarly defined, S_{1XY} defined by Eq. 1.13, and $R = X/Y$, and where

$$S_{2i}^2 = S_{2iX}^2 + R^2S_{2iY}^2 - 2RS_{2iXY} \qquad (1.23)$$

with S_{2iX}^2 defined by Eq. 1.4, S_{2iY}^2 similarly defined, and S_{2iXY} defined by Eq. 1.14.

Proof. By Sec. 11 of Ch. 4, we have

$$V_r^2 \doteq \frac{\sigma_{x'}^2}{X^2} + \frac{\sigma_{y'}^2}{Y^2} - 2\,\frac{\sigma_{x'y'}}{XY} \qquad (1.24)$$

where $\sigma_{x'}^2$ and $\sigma_{y'}^2$ are given by Eq. 1.2, and $\sigma_{x'y'}$ is given by Eq. 1.12. By making the substitutions indicated above for $\sigma_{x'}^2$, $\sigma_{y'}^2$, and $\sigma_{x'y'}$ in Eq. 1.24 and combining the first-stage contributions to the variance and covariance for X and Y, we obtain the first term of Eq. 1.21. Similarly, combining the second-stage contributions, we obtain the second term of Eq. 1.21.

When uniform sampling fractions are used, i.e., $n_i/N_i = f_2 = E\bar{n}/\bar{N}$, the rel-variance of r becomes

$$V_r^2 \doteq \frac{M-m}{M}\frac{B^2}{m} + \frac{\bar{N}-\bar{n}}{\bar{N}}\frac{W^2}{m\bar{n}} \qquad (1.25 \text{ or I–6.6.10})$$

where

$$B^2 = B_X^2 + B_Y^2 - 2B_{XY} \qquad (1.26 \text{ or I–6.6.11})$$

with

$$B_{XY} = \frac{S_{1XY}}{\bar{X}\bar{Y}} \qquad (1.27 \text{ or I–6.6.8})$$

S_{1XY} given by Eq. 1.13,

$$B_X^2 = B_{XX}, \quad B_Y^2 = B_{YY}$$

and

$$W^2 = W_X^2 + W_Y^2 - 2W_{XY} \qquad (1.28)$$

with

$$W_{XY} = \frac{S_{2XY}}{\bar{\bar{X}}\bar{\bar{Y}}} = \frac{\sum^{M} N_i S_{2iXY}}{N\bar{\bar{X}}\bar{\bar{Y}}} \qquad (1.29 \text{ or I-6.6.9})$$

S_{2iXY} given by Eq. 1.14,

$$W_X^2 = W_{XX}, \quad W_Y^2 = W_{YY}$$

and where in Eq. 1.25, $\bar{n}$ is the expected number of listing units in the sample per psu in the sample. Equation 1.25 follows directly from Eq. 1.21 with $n_i/N_i = E\bar{n}/\bar{N}$.

d. Variance of simple unbiased estimate for multi-stage sample. The variance for a simple random sample of m first-stage units selected without replacement from M first-stage units can be expressed in terms of the contribution to the variance from the first-stage units and the combined contribution from all subsequent stages of sampling.

Let $\sum^{m} u_i/m$ be an unbiased estimate of $\bar{U}$, where u_i is an estimate obtained from the ith selected first-stage unit of a sample of m first-stage units, $i = 1, 2, \cdots, m$; and also let

$$E_i u_i = U_j \qquad j = 1, 2, \cdots, M$$

where E_i designates a conditional expected value for the ith selected primary unit (which is, say, the jth unit in the population of psu's).

To prove:

$$\sigma_{\bar{u}}^2 = \frac{M-m}{(M-1)m}\sigma_U^2 + \frac{1}{Mm}\sum_i^M \sigma_{iu_i}^2 \tag{1.30}$$

where

$$\sigma_U^2 = \frac{1}{M}\sum_i^M (U_i - \bar{U})^2 \tag{1.31}$$

$$\sigma_{iu_i}^2 = E_i(u_i - E_i u_i)^2 \tag{1.32}$$

and $\sum_i^M \sigma_{iu_i}^2/Mm$ represents the contribution to the variance from all subsequent stages of sampling.

Proof. By Theorem 15, Ch. 3 (p. 65),

$$\sigma_{\bar{u}}^2 = E\sigma_{\bar{u}|[1]}^2 + \sigma_{E(\bar{u}|[1])}^2 \tag{1.33}$$

where the second term represents the contribution to the variance from first-stage sampling, and the first term represents the contribution from all subsequent stages of sampling. From Sec. 1*a*,

$$\sigma_{E(\bar{u}|[1])}^2 = \frac{M-m}{(M-1)m}\sigma_U^2 \tag{1.34}$$

Also, since the subsampling is carried out independently in each psu, and since the first-stage units can be regarded as m strata when the m first-stage units are held constant, it follows that

$$\sigma^2_{\bar{u}|[1]} = \frac{1}{m^2}\sum_i^m E_i(u_i - E_i u_i)^2 = \frac{1}{m^2}\sum_i^m \sigma^2_{iu_i}$$

From Theorem 6, Ch. 3 (p. 49), we have

$$E\sigma^2_{\bar{u}|[1]} = \frac{1}{m^2} E\sum_i^m \sigma^2_{iu_i} = \frac{1}{m^2}\sum_i^m \frac{1}{M}\sum_i^M \sigma^2_{iu_i}$$

$$= \frac{1}{Mm}\sum_i^M \sigma^2_{iu_i} \tag{1.35}$$

Substituting Eq. 1.34 and 1.35 into Eq. 1.33, we have $\sigma^2_{\bar{u}}$ as given by Eq. 1.30.

2. Estimates of the total variance and of the total covariance for a multi-stage sampling design where the first-stage units are selected with simple random sampling (Vol. I, Ch. 6, Sec. 7, Remark). Assume that it is desired to estimate the total variance and covariance of unbiased estimates from a multi-stage design without estimating the components of the variance and covariance.

a. The estimates $s^2_{\bar{u}}$ and $s_{\bar{u}\bar{w}}$ given below are unbiased estimates of the total variance and covariance as long as there are at least two first-stage units in the sample, selected with replacement. The subsequent stages of sampling are not restricted either as to the number of stages of sampling or the method of sampling, so long as the u_i are unbiased estimates of $\bar{U}$, where u_i is an estimate made from the ith psu in the sample, and the subsampling in any psu is independent of that in any other psu.

Let $u_1, u_2, \cdots, u_m$ be m unbiased estimates of $\bar{U}$ made from each of m independently selected psu's.

To prove:

$$s^2_{\bar{u}} = \frac{\sum_i^m (u_i - \bar{u})^2}{m(m-1)} \tag{2.1}$$

is an unbiased estimate of

$$\sigma^2_{\bar{u}} = E(\bar{u} - \bar{U})^2 \tag{2.2}$$

where

$$\bar{u} = \frac{\sum^m u_i}{m}$$

Proof. From Corollary 1 of Theorem 11, Ch. 3 (p. 56),

$$\sigma_{\bar{u}}^2 = \frac{1}{m}\sigma_u^2 \tag{2.3}$$

where

$$\sigma_u^2 = E(u_i - \bar{U})^2 \tag{2.4}$$

From Sec. 4 of Ch. 4, we have

$$Es_{\bar{u}}^2 = \sigma_{\bar{u}}^2 \tag{2.5}$$

b. Assume that u_i and $\bar{u}$ are unbiased estimates of $\bar{U}$, w_i and $\bar{w}$ are unbiased estimates of $\bar{W}$, $i = 1, 2, \cdots, m$; and m of the M primary units are independently selected by simple random sampling with replacement.

To prove:

$$s_{\bar{u}\bar{w}} = \frac{\sum^{m}(u_i - \bar{u})(w_i - \bar{w})}{m(m-1)} \tag{2.6}$$

is an unbiased estimate of

$$\sigma_{\bar{u}\bar{w}} = E(\bar{u} - \bar{U})(\bar{w} - \bar{W})$$

The proof is left to the reader.

c. Assume that a simple random sample of m first-stage units is selected without replacement.

To prove:

$$Es_{\bar{u}}^2 = \frac{M}{M-1}\frac{\sigma_U^2}{m} + \frac{\sum_{i}^{M}\sigma_{iu_i}^2}{Mm} \tag{2.7}$$

$$Es_{\bar{u}\bar{w}} = \frac{M}{M-1}\frac{\sigma_{UW}}{m} + \frac{\sum_{i}^{M}\sigma_{iu_iw_i}}{Mm} \tag{2.8}$$

The proof is left to the reader. As the following hint indicates, the only complication is the need for using finite sampling corrections.

Hint: Use the following equations:

$$Es_{\bar{u}}^2 = E\frac{\sum_{i}^{m}(u_i - \bar{u})^2}{m(m-1)} = \frac{1}{m(m-1)}\left(\sum^{m}Eu_i^2 - mE\bar{u}^2\right)$$

$$E\bar{u}^2 = \sigma_{\bar{u}}^2 + \bar{U}^2$$

$$\sigma_{\bar{u}}^2 = \frac{M-m}{(M-1)m}\sigma_U^2 + \frac{\sum_{i}^{M}\sigma_{iu_i}^2}{Mm}$$

$$E\sum_{i}^{m}u_i^2 = m(\sigma_u^2 + \bar{U}^2)$$

Note that $Es_{\bar{u}}^2$ is not an unbiased estimate of $\sigma_{\bar{u}}^2$ given by Eq. 1.30, but the bias will be small if m/M is small.

d. Assume that a simple random sample of m first-stage units is selected, and let $\bar{u} = \sum^{m} u_i/m$. Assume, further, that a simple random sample of m' units is selected from the m first-stage units originally selected. Let

$$\bar{u}' = \frac{\sum^{m'} u_i}{m'} \quad \text{and} \quad s_{\bar{u}}'^2 = \frac{\sum^{m'}(u_i - \bar{u}')^2}{m(m'-1)}$$

To prove: If the m and m' units are selected without replacement,

$$Es_{\bar{u}}'^2 = Es_{\bar{u}}^2 = \hat{\sigma}_{\bar{u}}^2 \tag{2.9}$$

where $s_{\bar{u}}^2$ is given by Eq. 2.1 and $\hat{\sigma}_{\bar{u}}^2$ is given by Eq. 2.7.

Proof. The proof follows from the fact that

$$E\frac{\sum^{m'}(u_i - \bar{u}')^2}{m'-1} = E\frac{\sum^{m}(u_i - \bar{u})^2}{m-1}$$

3. Estimates of the total variance and rel-variance for a two-stage design (Vol. I, Ch. 6, Sec. 7).

a. Let x' be given by Eq. 1.1; then, if M is large relative to m, a consistent estimate of $\sigma_{x'}^2$ (Eq. 1.2) is

$$s_{x'}'^2 = \frac{\sum_i^{m'}(\hat{x}_i - \hat{x})^2}{m(m'-1)} \tag{3.1}$$

where

$$\left.\begin{aligned} \hat{x}_i &= M\frac{N_i}{n_i}\sum_j^{n_i} x_{ij} = Mx_i' \\ \hat{x} &= \frac{\sum^{m'}\hat{x}_i}{m'} \end{aligned}\right\} \tag{3.2}$$

and m' is the number of first-stage units selected by simple random sampling to estimate the variance. The remaining terms in Eq. 3.2 are defined in Sec. 1.

To prove: $Es_{x'}'^2 \doteq \sigma_{x'}^2$, when M is large relative to m.

Proof. The proof follows from Eq. 2.7 and 2.9, where it has been shown that

$$Es_{\bar{u}}'^2 = \frac{M}{M-1}\frac{\sigma_U^2}{m} + \frac{\sum_i^{M}\sigma_{iu_i}^2}{Mm} \tag{3.3}$$

where

$$\sigma_U^2 = \frac{\sum^{M}(U_i - \bar{U})^2}{M}$$

and

$$\sigma_{iu_i}^2 = E_i(u_i - E_iu_i)^2$$

Now let

$$u_i = \hat{x}_i \quad \text{and} \quad \bar{u} = \hat{x}$$

Then

$$U_i = MX_i$$

since

$$\hat{x}_i = Mx_i' \quad \text{and} \quad E_ix_i' = X_i$$

and $\bar{U} = M\bar{X} = \Sigma X_i$. Also,

$$s_{\bar{u}}'^2 = s_{x'}'^2 = \frac{\sum^{m'}(\hat{x}_i - \hat{x})^2}{m(m' - 1)} \tag{3.4}$$

Remark 1. It will be convenient for use in subsequent sections to let

$$s_X^2 = \frac{m}{M^2} s_{x'}'^2 = \frac{\sum^{m'}(x_i' - \bar{x}')^2}{(m' - 1)} \tag{3.5}$$

and

$$\bar{x}' = \frac{\sum^{m'} x_i'}{m'}$$

From $U_i = MX_i$ it follows that

$$\sigma_U^2 = \frac{M^2\sum^{M}(X_i - \bar{X})^2}{M} \tag{3.6}$$

Hence, $M\sigma_U^2/(M - 1) = M^2S_{1X}^2$, where S_{1X}^2 is given by Eq. 1.3.

From $u_i = \hat{x}_i$ we have

$$\sigma_{iu_i}^2 = E_i(u_i - E_iu_i)^2 = M^2E_i(x_i' - X_i)^2$$

$$= M^2N_i^2 \frac{N_i - n_i}{N_in_i} S_{2iX}^2 \tag{3.7}$$

where S_{2iX}^2 is given by Eq. 1.4. Substituting Eq. 3.6 and 3.7 in Eq. 3.3, we have

$$Es_{\bar{u}}'^2 = Es_{x'}'^2 = M^2 \frac{S_{1X}^2}{m} + \frac{M}{m} \sum^{M} N_i^2 \frac{N_i - n_i}{N_in_i} S_{2iX}^2 \tag{3.8}$$

Note that $Es_{x'}'^2$ approaches $\sigma_{x'}^2$ as given by Eq. 1.2 when M is large relative to m. The second term of Eq. 3.8 is the same as the second term of

Eq. 1.2. The ratio of the first term of Eq. 3.8 to that of Eq. 1.2 is $M/(M - m)$, which is near 1 for M large relative to m. Then, for M large relative to m,

$$v_{x'}^2 = \frac{s_{x'}^{\prime 2}}{x'^2} \tag{3.9}$$

is a consistent estimate of $V_{x'}^2$ (Eq. 1.11), where $s_{x'}^{\prime 2}$ is given by Eq. 3.1 and x' by Eq. 1.1.

Remark 2. The proof that $s_{x'}^{\prime 2}$ is a consistent estimate of $\sigma_{x'}^2$, with M large relative to m follows immediately from Sec. 2*a*, when it is recognized that having M large relative to m is equivalent to sampling the m units with replacement. Thus, it was shown in Sec. 2*a* that

$$Es_{\bar{u}}^2 = E\frac{\sum_i^{m'}(u_i - \bar{u})^2}{m(m' - 1)} = \sigma_{\bar{u}}^2$$

If we let

$$u_i = M\frac{N_i}{n_i}\sum_j^{n_i} x_{ij}$$

and

$$\bar{u} = \frac{M}{m'}\sum_i^{m'}\frac{N_i}{n_i}\sum_j^{n_i} x_{ij} = x'$$

then it follows that

$$Es_{x'}^{\prime 2} = \sigma_{\bar{u}}^2$$

b. When the sampling fractions within first-stage units are uniform for all first-stage units and $m' = m$, the estimate $v_{x'}^2$ (Eq. 3.9) becomes

$$\frac{\sum^m (x_i - \bar{x})^2}{m(m - 1)\bar{x}^2} = \frac{s_{cX}^2}{m\bar{x}^2} \tag{3.10}$$

where

$$s_{cX}^2 = \frac{\sum^m (x_i - \bar{x})^2}{m - 1}$$

This can readily be verified by making the following substitutions which hold in this special case:

$$\frac{n_i}{N_i} = f_2, \quad \frac{m}{M} = f_1, \quad f_1 f_2 = f$$

$$\hat{x}_i = \frac{m x_i}{f_1 f_2} \tag{3.11}$$

$$x' = \frac{x}{f_1 f_2}, \quad \bar{x} = \frac{x}{m}$$

For this case a more accurate estimate of the rel-variance is

$$v_{x'}^2 = (1 - f)\frac{s_{cX}^2}{m\bar{x}^2} \tag{3.12}$$

The reader can verify that this estimate is a closer approximation of $V_{x'}^2$ than is $s_{cX}^2/m\bar{x}^2$.

Note that s_X^2 as given by Eq. 3.5 with $m' = m$ and a uniform subsampling fraction is related to s_{cX}^2 as follows:

$$s_{cX}^2 = s_X^2 f_2^2$$

Note also that s_{cX}^2 as defined by Eq. 3.10 is referred to as the variance between ultimate clusters, and $v_{x'}^2$ as given by Eq. 3.12 or 3.10 is referred to as an ultimate cluster estimate of the rel-variance of x'. The term *ultimate cluster* is used to denote the units in the sample from a primary sampling unit.

c. We shall next consider an estimate of the rel-variance of $r = x'/y'$ for a two-stage design. The rel-variance of r is given by

$$V_r^2 \doteq \frac{\sigma_{x'}^2}{X^2} + \frac{\sigma_{y'}^2}{Y^2} - \frac{2\sigma_{x'y'}}{XY}$$

From Sec. 2*b*, and with terms defined in Part *a* above, an unbiased estimate of $\sigma_{x'y'}$ is

$$s'_{x'y'} = \frac{\sum^{m'}(\hat{x}_i - \hat{x})(\hat{y}_i - \hat{y})}{m(m' - 1)} = \frac{M^2}{m}\frac{\sum^{m'}(x'_i - \bar{x}')(y'_i - \bar{y}')}{m' - 1} = \frac{M^2}{m}s_{XY} \tag{3.13}$$

Similarly

$$s'^2_{y'} = \frac{\sum^{m'}(\hat{y}_i - \hat{y})^2}{m(m' - 1)} = \frac{M^2}{m}\frac{\sum^{m'}(y'_i - \bar{y}')^2}{m' - 1} = \frac{M^2}{m}s_Y^2 \tag{3.14}$$

Therefore, a consistent estimate of V_r^2 is

$$v_r^2 = \frac{s'^2_{x'}}{x'^2} + \frac{s'^2_{y'}}{y'^2} - 2\frac{s'_{x'y'}}{x'y'} = \frac{1}{x'^2}\frac{M^2}{m}s^2 \tag{3.15}$$

where

$$s^2 = s_X^2 + r^2 s_Y^2 - 2rs_{XY} \tag{3.16}$$

with the $s'^2_{x'}$, $s'^2_{y'}$, and $s'_{x'y'}$ defined by Eq. 3.1, 3.14, and 3.13, respectively.

When the sampling fractions within first-stage units are uniform for all first-stage units and $m' = m$, the estimate above (Eq. 3.15) becomes

$$v_r^2 = \frac{s_{cX}^2}{m\bar{x}^2} + \frac{s_{cY}^2}{m\bar{y}^2} - 2\frac{s_{cXY}}{m\bar{x}\bar{y}} \tag{3.17}$$

where $s_{cX}^2/m\bar{x}^2$ is defined in Eq. 3.10 above, $s_{cY}^2/m\bar{y}^2$ is similarly defined, and

$$\frac{s_{cXY}}{m\bar{x}\bar{y}} = \frac{\sum^{m}(x_i - \bar{x})(y_i - \bar{y})}{m(m-1)\bar{x}\bar{y}} \tag{3.18}$$

An improved estimate is

$$v_r^2 = \frac{1-f}{m}\left(\frac{s_{cX}^2}{\bar{x}^2} + \frac{s_{cY}^2}{\bar{y}^2} - \frac{2s_{cXY}}{\bar{x}\bar{y}}\right) \tag{3.19}$$

which should be used when the over-all sampling fraction, f, is large.

Remark 3. An estimate of S^2, the variance between listing units in the population, may be made from a cluster subsample as follows:

$$s^2 = \frac{\sum^{m}\sum^{n_i}(x_{ij} - \bar{\bar{x}})^2}{n-1} \tag{3.20}$$

Although s^2 is not an unbiased estimate of S^2, its bias will be trivial whenever the number of primary units in the sample is large. Thus, when the primary units are equal in size, i.e., $N_i = \bar{N}$,

$$Es^2 = S^2 - \frac{S^2}{N}\left\{1 - \frac{N-1}{n-1}\left[\frac{\bar{n}}{\bar{N}}\frac{m-1}{M-1}(1-\delta) - \delta\bar{n}\frac{M-m}{M-1}\right]\right\} \tag{3.21}$$

and the remainder term will be small for m (and thus N and n) large. (The terms in Eq. 3.21 are defined below.) Equation 3.21 follows readily from the fact that

$$E\sum^{m}\sum^{n_i}x_{ij}^2 = n\sigma^2 + n\bar{\bar{X}}^2$$

and

$$E\bar{\bar{x}}^2 = \sigma_{\bar{\bar{x}}}^2 + \bar{\bar{X}}^2$$

where

$$\sigma_{\bar{\bar{x}}}^2 = \frac{M-m}{M-1}\frac{\sigma_b^2}{m} + \frac{\bar{N}-\bar{n}}{\bar{N}-1}\frac{\sigma_w^2}{m\bar{n}}$$

$$\sigma^2 = \sigma_b^2 + \sigma_w^2$$

$$\sigma_b^2 = \frac{\sum^{M}(\bar{X}_i - \bar{\bar{X}})^2}{M}$$

$$\sigma_w^2 = \frac{\sum^{M}\sum^{N_i}(X_{ij} - \bar{X}_i)^2}{M\bar{N}}$$

$$\delta = \frac{\sigma_b^2 - \sigma^2/\bar{N}}{(\bar{N}-1)\sigma^2/\bar{N}}$$

$$\sigma^2 = \frac{N}{N-1}S^2 = \frac{\sum^{M}\sum^{N_i}(X_{ij} - \bar{\bar{X}})^2}{N}$$

4. Estimates of the components of the rel-variance of a ratio estimate for a two-stage sampling design (Vol. I, Ch. 6, Sec. 7). Consistent estimates of the components of the rel-variance of a ratio, r, for a two-stage sampling design may be made as follows: The estimate, r, is given by Eq. 1.20. The rel-variance of r is given by Eq. 1.21.

a. To prove: A consistent estimate of the within-psu component of the rel-variance, i.e., of the second term in Eq. 1.21, is

$$\frac{1}{x'^2}\frac{M^2}{m}\frac{\sum^{m'}\frac{N_i^2}{n_i}\frac{N_i - n_i}{N_i}s_{2i}^2}{m'} \tag{4.1}$$

where m' is the number of units selected from the m first-stage units for estimating the variance,

$$s_{2i}^2 = s_{2iX}^2 + r^2 s_{2iY}^2 - 2rs_{2iXY} \tag{4.2}$$

with

$$s_{2iXY} = \frac{\sum^{n_i}(x_{ij} - \bar{\bar{x}}_i)(y_{ij} - \bar{\bar{y}}_i)}{n_i - 1} \tag{4.3}$$

$$s_{2iX}^2 = s_{2iXX} \quad \text{and} \quad s_{2iY}^2 = s_{2iYY}$$

Proof. The within component of the rel-variance may be rewritten

$$\frac{1}{X^2}\frac{M^2}{m}\left[\frac{\sum^{M}\frac{N_i^2}{n_i}\frac{N_i - n_i}{N_i}S_{2i}^2}{M}\right] \tag{4.4}$$

Now,

$$\frac{\sum^{m'}\frac{N_i^2}{n_i}\frac{N_i - n_i}{N_i}S_{2i}^2}{m'} \tag{4.5}$$

is an unbiased estimate of the term in brackets in Eq. 4.4. We need to find a consistent estimate of S_{2i}^2, for Eq. 4.5 to be a consistent estimate of the term in brackets in Eq. 4.4. Since the n_i elements are a simple random sample from the N_i elements, it follows from Sec. 4, Ch. 4, that

$$E_i s_{2iXY} = E_i\frac{\sum^{n_i}(x_{ij} - \bar{\bar{x}}_i)(y_{ij} - \bar{\bar{y}}_i)}{n_i - 1} = S_{2iXY} \tag{4.6}$$

where S_{2iXY} is defined by Eq. 1.14. Similarly, $E_i s_{2iX}^2 = S_{2iX}^2$ and $E_i s_{2iY}^2 = S_{2iY}^2$. Since, also by Sec. 21, Ch. 4, r is a consistent estimate of

R, s_{2i}^2 is a consistent estimate of S_{2i}^2, and

$$\hat{s}^2 = \frac{\sum^{m'} \frac{N_i^2}{n_i} \frac{N_i - n_i}{N_i} s_{2i}^2}{m'} \tag{4.7}$$

is a consistent estimate of the term in brackets in Eq. 4.4. Also, since x' is an unbiased estimate of X, Eq. 4.1 is a consistent estimate of the within component, namely, Eq. 4.4.

For a simple two-stage design with uniform sampling fractions, the rel-variance whose components we wish to estimate is given by Eq. 1.25. In this case $m/M = f_1$, $n_i/N_i = f_2$, $x' = x/f_1f_2 = n\bar{\bar{x}}/f_1f_2$, $y' = y/f_1f_2 = n\bar{\bar{y}}/f_1f_2$, and $n = m\bar{n}$. When $m' = m$, the estimate of the within component (Eq. 4.1), with the above substitutions, becomes

$$\frac{1 - f_2}{m\bar{n}} w^2 \tag{4.8}$$

where

$$w^2 = w_X^2 + w_Y^2 - 2w_{XY} \tag{4.9 or I–6.7.6}$$

$$w_{XY} = \frac{s_{2XY}}{\bar{\bar{x}}\bar{\bar{y}}}$$

$$s_{2XY} = \frac{1}{n} \sum_i^m \frac{n_i}{n_i - 1} \sum_j^{n_i} (x_{ij} - \bar{\bar{x}}_i)(y_{ij} - \bar{\bar{y}}_i) \tag{4.10}$$

$$w_X^2 = w_{XX} \quad \text{and} \quad w_Y^2 = w_{YY}$$

$$s_{2X}^2 = s_{2XX} \quad \text{and} \quad s_{2Y}^2 = s_{2YY}$$

b. To prove: A consistent estimate of the between-psu component of the rel-variance, i.e., of the first term in Eq. 1.21, is

$$\frac{1}{x'^2} \frac{M^2}{m} \frac{M - m}{M} (s^2 - \hat{s}^2) \tag{4.11}$$

where $\hat{s}^2$ is given by Eq. 4.7,

$$s^2 = s_X^2 + r^2 s_Y^2 - 2rs_{XY} \tag{4.12}$$

s_X^2, s_Y^2, and s_{XY} are defined in Eq. 3.5 and 3.13.

Proof. From Eq. 3.8 it follows that

$$Es_X^2 = S_{1X}^2 + \frac{\sum^M \frac{N_i^2}{n_i} \frac{N_i - n_i}{N_i} S_{2iX}^2}{M} \tag{4.13}$$

Since also

$$Es_Y^2 = S_{1Y}^2 + \frac{\sum^M \frac{N_i^2}{n_i} \frac{N_i - n_i}{N_i} S_{2iY}^2}{M} \tag{4.14}$$

$$Es_{XY} = S_{1XY} + \frac{\sum^M \frac{N_i^2}{n_i} \frac{N_i - n_i}{N_i} S_{2iXY}}{M} \tag{4.15}$$

and r is a consistent estimate of R, it follows that s^2 is a consistent estimate of

$$S_1^2 + \frac{\sum^M \frac{N_i^2}{n_i} \frac{N_i - n_i}{N_i} S_{2i}^2}{M} \tag{4.16}$$

where S_1^2 is given by Eq. 1.22 and S_{2i}^2 is given by Eq. 1.23. Hence, the term in parentheses in Eq. 4.11 is a consistent estimate of S_1^2, and, since x' is an unbiased estimate of X, Eq. 4.11 is a consistent estimate of the between-psu component.

For a simple two-stage design, with uniform sampling fractions, the estimate of the between-psu component (Eq. 4.11) when $m' = m$ becomes

$$\frac{1 - f_1}{m} b^2$$

where

$$b^2 = b_X^2 + b_Y^2 - 2b_{XY} \qquad \text{(4.17 or I–6.7.11)}$$

with

$$b_{XY} = \frac{1}{\bar{x}\bar{y}} [s_{cXY} - \bar{n}(1 - f_2)s_{2XY}] \qquad \text{(4.18 or I–6.7.13)}$$

s_{cXY} defined by Eq. 3.18,

$$\bar{x} = \frac{x}{m}, \quad \bar{y} = \frac{y}{m}$$

$$b_X^2 = b_{XX}, \quad b_Y^2 = b_{YY}$$

Remark. When the number of second-stage units used to estimate the variance is n_i', not necessarily equal to n_i, the estimate of the within-psu component is given by Eq. 4.1 with

$$s_{2iXY} = \frac{\sum^{n'_i} (x_{ij} - \bar{\bar{x}}_i)(y_{ij} - \bar{\bar{y}}_i)}{n_i' - 1} \tag{4.19}$$

$$\bar{\bar{x}}_i = \frac{\sum^{n'_i} x_{ij}}{n} \quad \text{and} \quad \bar{\bar{y}}_i = \frac{\sum^{n'_i} y_{ij}}{n_i'}$$

The estimate of the between-psu component is given by

$$\frac{1}{x'^2}\frac{M^2}{m}\frac{M-m}{M}(s'^2-\hat{s}'^2) \tag{4.20}$$

where s'^2 is given by s^2 with $x_i' = (N_i/n_i')\sum^{n'_i} x_{ij}$ and y_i' is similarly defined.

$$\hat{s}'^2 = \frac{\sum^{m'} \frac{N_i^2}{n_i'}\frac{N_i-n_i'}{N_i} s_{2i}^2}{m'} \tag{4.21}$$

The estimate of the total variance is obtained by adding the estimates of the components.

5. Rel-variance of a ratio estimate expressed in terms of δ, the measure of homogeneity; an estimate of δ from the sample (Vol. I, Ch. 4, Sec. 8). *a. To prove:* For a simple two-stage sampling design (see Note on p. 142) the rel-variance of r given by Eq. 1.25 is approximately equal to

$$\frac{1-f}{m\bar{n}}\hat{V}^2[1+\delta(\bar{n}-1)] \tag{5.1 or I-6.8.6}$$

where $\hat{V}^2$ and δ are defined by Eq. 5.4 and 5.5 below, and in Eq. 5.1 through Eq. 5.10, $\bar{n}$ is the expected number of listing units in the sample per psu in the sample.

Proof. From Eq. 1.25, the rel-variance of r is

$$V_r^2 \doteq \frac{M-m}{M}\frac{B^2}{m}+\frac{\bar{N}-\bar{n}}{\bar{N}}\frac{W^2}{m\bar{n}} \tag{5.2 or I-6.6.10}$$

and if m is small relative to M

$$V_r^2 \doteq \frac{M-1}{M}\frac{B^2}{m}+\frac{\bar{N}-\bar{n}}{\bar{N}}\frac{W^2}{m\bar{n}} \tag{5.3}$$

where B^2 and W^2 are defined by Eq. 1.26 and 1.28. By definition

$$\hat{V}^2 = \frac{M-1}{M}B^2+\frac{\bar{N}-1}{\bar{N}}W^2 \tag{5.4 or I-6.8.10}$$

and

$$\delta = \frac{\frac{M-1}{M}B^2-\frac{\hat{V}^2}{\bar{N}}}{(\bar{N}-1)\hat{V}^2/\bar{N}} \tag{5.5 or I-6.8.11}$$

From Eq. 5.4 and 5.5 we obtain

$$\frac{M-1}{M}B^2 = \frac{\hat{V}^2}{\bar{N}}[1+\delta(\bar{N}-1)] \tag{5.6}$$

and

$$W^2 = \hat{V}^2(1 - \delta) \tag{5.7}$$

If we substitute in Eq. 5.3 for B^2 from Eq. 5.6 and for W^2 from Eq. 5.7 and simplify, we obtain

$$V_r^2 \doteq \frac{\hat{V}^2}{m\bar{n}}[1 + \delta(\bar{n} - 1)] \tag{5.8}$$

It is easy to show, also, that

$$V_r^2 \doteq \frac{1-f}{m\bar{n}}\hat{V}^2[1 + \delta(\bar{n} - 1)] \tag{5.1}$$

and ordinarily Eq. 5.1 is a closer approximation to Eq. 5.2 than is Eq. 5.8. The reader can readily verify this by showing that the differences are

$$\text{Eq. 5.8} - \text{Eq. 5.2} = \frac{B^2}{M}\frac{m-1}{m} \tag{5.9}$$

$$\text{Eq. 5.1} - \text{Eq. 5.2} = \frac{\bar{N} - \bar{n}}{\bar{N}M}\left(B^2 - \frac{W^2}{\bar{N}}\right) - \frac{1}{m-1}(1-f)\left(\frac{B^2}{M}\frac{m-1}{m}\right) \tag{5.10}$$

and that Eq. 5.10 is smaller in absolute value than Eq. 5.9 provided $m > 1$ and $B^2 - (W^2/\bar{N}) \geq 0$, as will commonly be the case.

Exercises

5.1. When $N_i = \bar{N}$, show that

$$\frac{N-1}{N}V^2 = \hat{V}^2$$

where

$$V^2 = V_X^2 + V_Y^2 - 2V_{XY} \tag{5.11}$$

and

$$V_{XY} = \frac{\sum^M \sum^{N_i} (X_{ij} - \bar{\bar{X}})(Y_{ij} - \bar{\bar{Y}})}{(N-1)\bar{\bar{X}}\bar{\bar{Y}}} \tag{5.12}$$

$$V_X^2 = V_{XX}, \quad V_Y^2 = V_{YY} \tag{5.13}$$

5.2. Assume that $N_i = \bar{N}$ for all i and define

$$\bar{\bar{x}} = \frac{\sum^m \sum^{n_i} x_{ij}}{m\bar{n}}$$

Show that

$$V_{\bar{\bar{x}}}^2 \doteq \frac{1-f}{m\bar{n}}V_X^2[1 + \delta(\bar{n} - 1)] \tag{5.14 or I–6.8.1}$$

where

$$V_X^2 = \frac{\sum^M \sum^{N_i}(X_{ij} - \bar{\bar{X}})^2}{(N-1)\bar{\bar{X}}^2} = \frac{N}{N-1}\frac{\sigma^2}{\bar{\bar{X}}^2} \tag{5.15}$$

$$\delta = \frac{\sigma_b^2 - \sigma^2/\bar{N}}{(\bar{N}-1)\sigma^2/\bar{N}} \tag{5.16}$$

and

$$\sigma_b^2 = \frac{\sum^M(\bar{X}_i - \bar{\bar{X}})^2}{M} \tag{5.17}$$

b. A consistent estimate of δ (Eq. 5.5) can be made by substituting consistent sample estimates, term for term, for the population values involved in the definition of δ. A simpler but equivalent estimate (provided M is large) is given by Eq. 5.18.

To prove: A consistent estimate of δ is

$$\delta' = \frac{s_c^2 - \bar{n}s_2^2}{s_c^2 + \bar{n}(\bar{n}-1)s_2^2} \tag{5.18 or I–6.8.14}$$

where

$$s_c^2 = s_{cX}^2 + r^2 s_{cY}^2 - 2rs_{cXY}$$

with s_{cX}^2, s_{cY}^2, and s_{cXY} defined by Eq. 3.10 and 3.18, and where

$$s_2^2 = s_{2X}^2 + r^2 s_{2Y}^2 - 2rs_{2XY} \tag{5.19}$$

with s_{2X}^2, s_{2Y}^2, and s_{2XY} defined by Eq. 4.10.

Proof. We first note that if M is large, Eq. 5.5 can be written as

$$\delta = \frac{S_1^2 - \bar{N}S_2^2}{S_1^2 + \bar{N}(\bar{N}-1)S_2^2} \tag{5.20 or I–6.8.13}$$

since, in Eq. 5.5,

$$B^2 = \frac{S_1^2}{\bar{X}^2}$$

$$\hat{V}^2 \doteq B^2 + \frac{\bar{N}-1}{\bar{N}} W^2$$

with

$$W^2 = \frac{S_2^2}{\bar{\bar{X}}^2}$$

where we assume $(M-1)/M \doteq 1$ and where S_1^2 is given by Eq. 1.22, $S_2^2 = S_{2X}^2 + R^2 S_{2Y}^2 - 2RS_{2XY}$ with S_{2XY} given by Eq. 1.19, $S_{2X}^2 = S_{2XX}$ and $S_{2Y}^2 = S_{2YY}$. We shall now show that the expected value of the numerator of δ' is a consistent estimate of f_2^2 times the numerator of δ and that the expected value of the denominator of δ' is a consistent estimate of f_2^2 times the denominator of δ, and the proof that δ' is a consistent estimate of δ follows immediately from Theorem 20, Ch. 3 (p. 75). Let

$$\frac{n_i}{N_i} = f_2$$

$$s_c^2 = f_2^2 s^2$$

where s^2 is given in Eq. 4.12. From Sec. 4*b* with the substitution $n_i/N_i = f_2$, it follows that s_c^2 is a consistent estimate of

$$f_2^2 \left[S_1^2 + \frac{\bar{N}(1 - f_2)}{f_2} S_2^2\right]$$

It follows from Sec. 4*a* that s_2^2 is a consistent estimate of S_2^2.

It follows from Theorem 20, Ch. 3 (p. 75), that the numerator of δ', $(s_c^2 - \bar{n}s_2^2)$, is a consistent estimate of

$$f_2^2(S_1^2 - \bar{N}S_2^2)$$

and the denominator of δ', $[s_c^2 + \bar{n}(\bar{n} - 1)s_2^2]$, is a consistent estimate of

$$f_2^2[S_1^2 + \bar{N}(\bar{N} - 1)S_2^2]$$

6. The measure of homogeneity when the primary units are equal in size (Vol. I, Ch. 6, Sec. 8, Eq. 8.3). The measure of homogeneity, δ, between listing units (second-stage units) within primary units when each primary unit contains $\bar{N}$ listing units is the intraclass correlation and is given by

$$\delta = \frac{\sigma_b^2 - \sigma^2/\bar{N}}{(\bar{N} - 1)\sigma^2/\bar{N}} \qquad \text{(6.1 or I–6.8.3)}$$

where σ_b^2 and σ^2 are defined below.

When a population is composed of equal-sized primary units (in terms of listing units) and when a single primary unit is drawn at random and two listing units are drawn at random from this primary unit, the intraclass correlation is defined by

$$\delta = \frac{E(x_{ij} - \bar{\bar{X}})(x_{ik} - \bar{\bar{X}})}{\sqrt{E(x_{ij} - \bar{\bar{X}})^2}\sqrt{E(x_{ik} - \bar{\bar{X}})^2}} \qquad (6.2)$$

From Sec. 2, Ch. 4, with $n = 1$, it follows that

$$E(x_{ij} - \bar{\bar{X}})^2 = \frac{\sum^{M}\sum_{j}^{\bar{N}}(X_{ij} - \bar{\bar{X}})^2}{M\bar{N}} = \sigma^2 \qquad (6.3)$$

Substituting Eq. 6.3 into 6.2 gives

$$\delta\sigma^2 = E(x_{ij} - \bar{\bar{X}})(x_{ik} - \bar{\bar{X}})$$

By Theorem 14 of Ch. 3 (p. 61),

$$\delta\sigma^2 = EE_i(x_{ij} - \bar{\bar{X}})(x_{ik} - \bar{\bar{X}}) = E\frac{\sum\limits_{j\neq k}^{\bar{N}} (X_{ij} - \bar{\bar{X}})(X_{ik} - \bar{\bar{X}})}{\bar{N}(\bar{N}-1)}$$

$$= \frac{\sum\limits_{i}^{M} \sum\limits_{j\neq k}^{\bar{N}} (X_{ij} - \bar{\bar{X}})(X_{ik} - \bar{\bar{X}})}{M\bar{N}(\bar{N}-1)}$$

$$= \frac{\sum\limits_{i}^{M} \left[\left\{\sum\limits_{j}^{\bar{N}}(X_{ij} - \bar{\bar{X}})\right\}^2 - \sum\limits_{j}^{\bar{N}} (X_{ij} - \bar{\bar{X}})^2\right]}{M\bar{N}(\bar{N}-1)} \tag{6.4}$$

Equation 6.4 can be rewritten as follows:

$$\delta\sigma^2 = \frac{\sum\limits^{M}(\bar{X}_i - \bar{\bar{X}})^2}{M} - \frac{\sum\limits^{M}\sum\limits^{\bar{N}}(X_{ij} - \bar{X}_i)^2}{M\bar{N}(\bar{N}-1)}$$

$$= \sigma_b^2 - \frac{\sigma_w^2}{\bar{N}-1}$$

where

$$\sigma_b^2 = \frac{1}{M}\sum^{M} (\bar{X}_i - \bar{\bar{X}})^2, \quad \text{and} \quad \sigma_w^2 = \frac{1}{M\bar{N}}\sum^{M}\sum^{\bar{N}} (X_{ij} - \bar{X}_i)^2$$

and hence

$$\delta = \frac{\sigma_b^2 - \sigma_w^2/(\bar{N}-1)}{\sigma^2} \tag{6.5}$$

and since $\sigma^2 = \sigma_b^2 + \sigma_w^2$ it follows that

$$\delta = \frac{\sigma_b^2 - \sigma^2/\bar{N}}{(\bar{N}-1)\sigma^2/\bar{N}} \tag{6.1}$$

***7. Relationship between δ for primary units and δ for ultimate clusters** (Vol. I, Ch. 6, Sec. 8, pp. 262 and 266). *To prove:* The measure of homogeneity, δ_1, for listing units within primary units is approximately equal to the expected value of the measure of homogeneity, δ_2, for listing units within ultimate clusters when the ultimate clusters are formed by proportionate sampling of listing units within primary units. An

* May be deferred. This development is given to illustrate more correctly the properties of ultimate clusters. Section 5 suffices to show that a consistent estimate of δ is readily available from ultimate clusters.

ultimate cluster is defined as all the listing units included in the sample from a selected primary unit.

Proof. Let

$$\delta_1 = \frac{\dfrac{M-1}{M} B^2 - \dfrac{\hat{V}^2}{\bar{N}}}{(\bar{N}-1)\hat{V}^2/\bar{N}} \qquad \text{(7.1 or I–6.8.11)}$$

where $\hat{V}^2$ is defined by Eq. 5.4, B^2 and W^2 are defined as in Sec. 1; and let

$$\delta_2 = E\frac{\dfrac{MK-1}{MK} b_2^2 - \dfrac{\hat{v}_2^2}{E\bar{n}}}{(\bar{n}-1)\hat{v}_2^2/E\bar{n}} \doteq \frac{E\left[\dfrac{MK-1}{MK} b_2^2 - \dfrac{\hat{v}_2^2}{E\bar{n}}\right]}{E[(\bar{n}-1)\hat{v}_2^2/E\bar{n}]} \qquad (7.2)$$

where $K = \bar{N}/E\bar{n}$ is the number of ultimate clusters into which the listing units in each primary unit could be grouped without replacement, where the expected value is over all possible sets of ultimate clusters, and where

$$b_2^2 = b_{2X}^2 + b_{2Y}^2 - 2b_{2XY} \qquad (7.3)$$

with

$$b_{2XY} = \frac{\sum_i^M \sum_\alpha^K \left(\sum_j^{n_i} X_{i\alpha j} - \tilde{X}\right)\left(\sum_j^{n_i} Y_{i\alpha j} - \tilde{Y}\right)}{(MK-1)\tilde{X}\tilde{Y}} \qquad (7.4)$$

$$b_{2X}^2 = b_{2XX}, \quad b_{2Y}^2 = b_{2YY}$$

$X_{i\alpha j}$ is the value of the X-characteristic for the jth listing unit in the αth ultimate cluster of the ith psu, $Y_{i\alpha j}$ is similarly defined for the Y-characteristic,

$$\tilde{X} = \bar{X}/K, \quad \tilde{Y} = \bar{Y}/K$$

$n_i = N_i/K$ is the number of listing units per ultimate cluster in the ith primary unit,

$$\hat{v}_2^2 = \frac{MK-1}{MK} b_2^2 + \frac{\bar{n}-1}{\bar{n}} w_2^2 \qquad (7.5)$$

and

$$w_2^2 = w_{2X}^2 + w_{2Y}^2 - 2w_{2XY} \qquad (7.6)$$

with

$$w_{2XY} = \frac{1}{N\bar{\bar{X}}\bar{\bar{Y}}} \sum_i^M \sum_\alpha^K n_i \frac{\sum_j^{n_i} (X_{i\alpha j} - \bar{\bar{X}}_{i\alpha})(Y_{i\alpha j} - \bar{\bar{Y}}_{i\alpha})}{n_i - 1} \qquad (7.7)$$

$$w_{2X}^2 = w_{2XX}, \quad w_{2Y}^2 = w_{2YY}$$

$$\bar{\bar{X}}_{i\alpha} = \frac{\sum^{n_i} X_{i\alpha j}}{n_i}, \quad \bar{\bar{Y}}_{i\alpha} = \frac{\sum^{n_i} Y_{i\alpha j}}{n_i}$$

We will show

$$Eb_2^2 = \frac{MK}{MK-1}\left[\frac{M-1}{M}B^2 + (1-f_2)\frac{W^2}{\bar{n}}\right] \tag{7.8}$$

and

$$Ew_2^2 = W^2 \tag{7.9}$$

where $f_2 = 1/K$ and, in Eq. 7.8, $\bar{n} = f_2\bar{N}$. Substituting these expected values into Eq. 7.2, making use of Eq. 7.5, and simplifying, we obtain

$$\delta_2 \doteq \frac{\dfrac{M-1}{M}B^2 - \dfrac{\hat{V}^2}{\bar{N}}}{(\bar{N}-1)\hat{V}^2/\bar{N}} = \delta_1$$

Consider first

$$Eb_2^2 = Eb_{2X}^2 + Eb_{2Y}^2 - 2Eb_{2XY} \tag{7.10}$$

Now

$$Eb_{2XY} = \frac{MK}{MK-1}\,\frac{\sum^{M}\sum^{K} EX_{i\alpha}Y_{i\alpha} - MK\tilde{X}\tilde{Y}}{MK\tilde{X}\tilde{Y}} \tag{7.11}$$

Now, $X_{i\alpha}$ and $Y_{i\alpha}$ are ultimate cluster totals for the αth ultimate cluster in the ith primary unit, i.e., $X_{i\alpha} = \sum_j^{n_i} X_{i\alpha j}$ and $Y_{i\alpha} = \sum_j^{n_i} Y_{i\alpha j}$, where $X_{i\alpha j}$ and $Y_{i\alpha j}$ are values for the jth listing unit in the $i\alpha$th ultimate cluster of the particular subdivision into ultimate clusters. Since the ultimate clusters are formed only within primary units and all possible subdivisions are considered, $X_{i\alpha}$ and $Y_{i\alpha}$ are the sample totals for a simple random sample of n_i listing units from the N_i listing units in the ith primary unit. Hence,

$$EX_{i\alpha}Y_{i\alpha} = \frac{N_i - n_i}{N_i} n_i S_{2iXY} + n_i^2 \bar{X}_i \bar{Y}_i \tag{7.12}$$

Substituting Eq. 7.12 in Eq. 7.11, using $n_i = N_i/K$, $\tilde{X} = \bar{X}/K$, $\tilde{Y} = \bar{Y}/K$, and simplifying, we obtain

$$Eb_{2XY} = \frac{MK}{MK-1}\left[\frac{\sum^M (X_i - \bar{X})(Y_i - \bar{Y})}{M\bar{X}\bar{Y}} + \frac{(1-f_2)}{\bar{n}}\,\frac{\sum^M N_i S_{2iXY}}{N\bar{\bar{X}}\bar{\bar{Y}}}\right]$$

$$= \frac{MK}{MK-1}\left[\frac{M-1}{M}B_{XY} + (1-f_2)\frac{W_{XY}}{\bar{n}}\right] \tag{7.13}$$

where in Eq. 7.13 and also in Eq. 7.14 and 7.15, $\bar{n} = f_2\bar{N}$. Substituting X for Y in Eq. 7.13,

$$Eb_{2X}^2 = \frac{MK}{MK-1}\left[\frac{M-1}{M}B_X^2 + (1-f_2)\frac{W_X^2}{\bar{n}}\right] \tag{7.14}$$

Similarly,

$$Eb_{2Y}^2 = \frac{MK}{MK-1}\left[\frac{M-1}{M} B_Y^2 + (1-f_2)\frac{W_Y^2}{\bar{n}}\right] \tag{7.15}$$

Substituting Eq. 7.13, 7.14, and 7.15 into Eq. 7.10, we obtain Eq. 7.8.

Consider now

$$Ew_2^2 = Ew_{2X}^2 + Ew_{2Y}^2 - 2Ew_{2XY} \tag{7.16}$$

Since the n_i elements are a simple random sample from N_i elements, we have, by Sec. 4, Ch. 4,

$$Ew_{2XY} = \frac{\sum_i^M \sum_\alpha^K n_i S_{2iXY}}{N\bar{\bar{X}}\bar{\bar{Y}}} = \frac{\sum_i^M N_i S_{2iXY}}{N\bar{\bar{X}}\bar{\bar{Y}}} = W_{XY} \tag{7.17}$$

When X is substituted for Y, we obtain

$$Ew_{2X}^2 = W_X^2 \tag{7.18}$$

Similarly,

$$Ew_{2Y}^2 = W_Y^2 \tag{7.19}$$

Substituting Eq. 7.17, 7.18, and 7.19 into Eq. 7.16, we obtain Eq. 7.9.

8. Some physical properties of frequently occurring populations, and values of δ under specified conditions (Vol. I, Ch. 9, Sec. 8, and Ch. 6, Sec. 8).

a. Many actual populations are characterized by the following physical properties:

i. The elements within a cluster are positively correlated with regard to a specified characteristic.

ii. Clusters containing large numbers of elements have greater internal heterogeneity than clusters containing small numbers of elements.

iii. Increasing the size of the cluster brings in correlated elements (e.g., in population or agriculture surveys, larger clusters are formed by including households or farms in adjacent areas).

The first of these properties is widely recognized, and the losses of efficiency through the use of large whole clusters as sampling units are frequently cited. The second and third properties hold just as commonly in actual populations, and ordinarily for the same populations for which the first property holds.

The presence of these physical properties leads to the following mathematical relationships which have been found useful in making choices among alternative sample designs.

(1) The sizes of the primary sampling units, N_i, are negatively correlated with the δ_i, the measures of homogeneity among the elements within the primary units, where δ_i is defined as either

$$\delta_i = \frac{E_i(x_{ij} - \bar{X})(x_{ik} - \bar{X})}{E(x_{ij} - \bar{X})^2} = \frac{1}{\sigma^2}\left[(\bar{X}_i - \bar{X})^2 - \frac{\sum^{N_i}(X_{ij} - \bar{X}_i)^2}{N_i(N_i - 1)}\right]$$

or

$$\delta_i = \frac{(\bar{X}_i - \bar{X})^2}{\sigma^2}$$

(2) The N_i and $N_i\delta_i$ are positively correlated.

(3) The N_i and σ_i^2 are positively correlated, where σ_i^2 is the variance among elements in the ith primary unit.

(4) The N_i and σ_i^2/N_i are negatively correlated.

The use of these relationships can often determine the choice among alternative sampling procedures in situations where more specific characteristics of populations are unknown. The relationships, of course, do not necessarily hold, and exceptions to them will be found.

b. The following values of the measure of homogeneity (Eq. 5.5) hold for a population of clusters for the conditions specified:

(1) The maximum possible value for δ is

$$\delta\,(\text{max.}) = 1 \tag{8.1}$$

if all listing units in any cluster are alike in that the values of the characteristics X_{ij} and Y_{ij} are uniform for all listing units in a cluster, but have some different values in different clusters. The clusters need not be uniform in size.

(2) The minimum possible value for δ is

$$\delta\,(\text{min.}) = -\frac{1}{\bar{N} - 1} \tag{8.2}$$

The minimum value of δ is obtained when B^2 in Eq. 5.5 is equal to 0. This will occur

(i) if the clusters are equal in size, the estimate is a simple unbiased estimate of a mean or total, and $X_i = \bar{X}$ for all i; and

(ii) for a ratio estimate when $X_i/Y_i = R$ for all i.

(3) If the clusters are equal in size, i.e., $N_i = \bar{N}$, and if the primary units are formed by randomly grouping the population of listing units into M clusters of $\bar{N}$ listing units each, then

$$\delta = -\frac{1}{N-1} \doteq 0 \tag{8.3}$$

(4) For any specified distribution of cluster sizes, $N_1, N_2, \cdots, N_M$, if the N_i listing units associated with the ith cluster are a simple random sample from all listing units in the population, then, for the ratio estimate x/y,

$$\delta = -\frac{1 + V_N^2}{N - (1 + V_N^2)} \doteq 0 \tag{8.4}$$

For a simple unbiased estimate

$$\delta = \frac{V_N^2\left(1 - \frac{V_X^2}{N}\right) - \frac{V_X^2}{N}}{V_N^2\left(1 - \frac{V_X^2}{N}\right) - \frac{V_X^2}{N} + V_X^2} \tag{8.5}$$

Note that Eq. 8.4 and 8.5 reduce to 8.3 when $N_i = \bar{N}$.

The proof of the above theorems is left to the reader.

***9. Relationship of the measure of homogeneity for second-stage units within primary units to the measures of homogeneity for elementary units within primary units and within second-stage units** (Vol. I, Ch. 6, Sec. 8, Eq. 8.16). *To prove:* The measure of homogeneity, δ_L, for second-stage units within primary units bears the following relationship to the measures of homogeneity, δ_1 and δ_2, for elementary units within primary units and for elementary units within second-stage units, respectively:

$$\delta_L = \frac{\lambda[1 + \delta_1(\bar{K} - 1)] - [1 + \delta_2(\bar{\bar{K}} - 1)]}{(\bar{N} - 1)[1 + \delta_2(\bar{\bar{K}} - 1)]} \tag{9.1}$$

$$\doteq \frac{[1 + \delta_1(\bar{K} - 1)] - [1 + \delta_2(\bar{\bar{K}} - 1)]}{(\bar{N} - 1)[1 + \delta_2(\bar{\bar{K}} - 1)]} \qquad \text{(9.2 or I-6.8.16)}$$

In the population, M is the number of primary units;

$N = \sum^{M} N_i$ is the number of second-stage units;

$K = \sum^{M} K_i = \sum^{M} \sum^{N_i} K_{ij}$ is the number of elementary units.

* May be deferred.

A single bar denotes an average per primary unit, a double bar denotes an average per second-stage unit, and a triple bar denotes an average per elementary unit. By definition,

$$\delta_L = \frac{\dfrac{M-1}{M} B_L^2 - \dfrac{\hat{V}_L^2}{\bar{N}}}{(\bar{N}-1)\hat{V}_L^2/\bar{N}}; \quad \hat{V}_L^2 = \frac{M-1}{M} B_L^2 + \frac{\bar{N}-1}{\bar{N}} W_L^2 \tag{9.3}$$

$$B_L^2 = B_{LX}^2 + B_{LY}^2 - 2B_{LXY} \tag{9.4}$$

$$B_{LXY} = \frac{\sum^{M}(X_i - \bar{X})(Y_i - \bar{Y})}{(M-1)\bar{X}\bar{Y}} \tag{9.5}$$

$$B_{LX}^2 = B_{LXX}, \quad B_{LY}^2 = B_{LYY} \tag{9.6}$$

$$W_L^2 = W_{LX}^2 + W_{LY}^2 - 2W_{LXY} \tag{9.7}$$

$$W_{LXY} = \frac{\sum^{M} \dfrac{N_i}{N_i - 1} \sum^{N_i} (X_{ij} - \bar{\bar{X}}_i)(Y_{ij} - \bar{\bar{Y}}_i)}{N\bar{\bar{X}}\bar{\bar{Y}}} \tag{9.8}$$

$$W_{LX}^2 = W_{LXX}; \quad W_{LY}^2 = W_{LYY} \tag{9.9}$$

Also,

$$\delta_1 = \frac{\dfrac{M-1}{M} B_1^2 - \dfrac{\hat{V}_1^2}{\bar{K}}}{(\bar{K}-1)\hat{V}_1^2/\bar{K}}; \quad \hat{V}_1^2 = \frac{M-1}{M} B_1^2 + \frac{\bar{K}-1}{\bar{K}} W_1^2 \tag{9.10}$$

$$B_1^2 = B_L^2 \tag{9.11}$$

$$W_1^2 = W_{1X}^2 + W_{1Y}^2 - 2W_{1XY} \tag{9.12}$$

$$W_{1XY} = \frac{\sum^{M} \dfrac{K_i}{K_i - 1} \sum^{N_i} \sum^{K_{ij}} (X_{ijl} - \bar{\bar{\bar{X}}}_i)(Y_{ijl} - \bar{\bar{\bar{Y}}}_i)}{K\bar{\bar{\bar{X}}}\bar{\bar{\bar{Y}}}} \tag{9.13}$$

$$W_{1X}^2 = W_{1XX}, \quad W_{1Y}^2 = W_{1YY} \tag{9.14}$$

and

$$\delta_2 = \frac{\dfrac{N-1}{N} B_2^2 - \dfrac{\hat{V}_2^2}{\bar{\bar{K}}}}{(\bar{\bar{K}}-1)\hat{V}_2^2/\bar{\bar{K}}}; \quad \hat{V}_2^2 = \frac{N-1}{N} B_2^2 + \frac{\bar{\bar{K}}-1}{\bar{\bar{K}}} W_2^2 \tag{9.15}$$

$$B_2^2 = B_{2X}^2 + B_{2Y}^2 - 2B_{2XY} \tag{9.16}$$

$$B_{2XY} = \frac{\sum^{M}\sum^{N_i}(X_{ij} - \bar{\bar{X}})(Y_{ij} - \bar{\bar{Y}})}{(N-1)\bar{\bar{X}}\bar{\bar{Y}}} \tag{9.17}$$

$$B_{2X}^2 = B_{2XX}, \quad B_{2Y}^2 = B_{2YY} \tag{9.18}$$

$$W_2^2 = W_{2X}^2 + W_{2Y}^2 - 2W_{2XY} \tag{9.19}$$

$$W_{2XY} = \frac{\sum^{M}\sum^{N_i}\frac{K_{ij}}{K_{ij}-1}\sum^{K_{ij}}(X_{ijl} - \bar{\bar{X}}_{ij})(Y_{ijl} - \bar{\bar{Y}}_{ij})}{K\bar{\bar{X}}\bar{\bar{Y}}} \tag{9.20}$$

$$W_{2X}^2 = W_{2XX}, \quad W_{2Y}^2 = W_{2YY} \tag{9.21}$$

$$\lambda = \frac{\hat{V}_1^2 V_L^2}{\hat{V}_2^2 \hat{V}_L^2}\frac{N-1}{N}, \quad V_L^2 = B_2^2 \tag{9.22}$$

Proof. From the pairs of equations 9.3, 9.10, and 9.15 we readily obtain Eq. 9.23, 9.24, and 9.25 as follows:

$$\frac{M-1}{M}B_L^2 = \frac{\hat{V}_L^2}{\bar{N}}[1 + \delta_L(\bar{N} - 1)]$$

$$= \frac{\hat{V}_L^2}{(N-1)B_2^2/N}\frac{(N-1)B_2^2/N}{\bar{N}}[1 + \delta_L(\bar{N} - 1)] \tag{9.23}$$

$$\frac{M-1}{M}B_1^2 = \frac{\hat{V}_1^2}{\bar{K}}[1 + \delta_1(\bar{K} - 1)] \tag{9.24}$$

$$\frac{N-1}{N}B_2^2 = \frac{\hat{V}_2^2}{\bar{\bar{K}}}[1 + \delta_2(\bar{\bar{K}} - 1)] \tag{9.25}$$

Since $B_1^2 = B_L^2$, the right-hand member of Eq. 9.24 = the right-hand member of Eq. 9.23, and by substituting Eq. 9.25 for $(N-1)B_2^2/N$ in the numerator of the right-hand member of Eq. 9.23, we obtain

$$\frac{\hat{V}_1^2}{\bar{K}}[1 + \delta_1(\bar{K} - 1)] = \frac{\hat{V}_L^2}{(N-1)B_2^2/N}\frac{\hat{V}_2^2}{\bar{N}\bar{\bar{K}}}[1 + \delta_2(\bar{\bar{K}} - 1)][1 + \delta_L(\bar{N} - 1)] \tag{9.26}$$

If we solve Eq. 9.26 for δ_L, we obtain Eq. 9.1. In most cases λ will be close to 1. In fact, when the primary units are equal in size and also the second-stage units are equal in size, then $\lambda = 1$.

10. Optimum values for a simple two-stage sampling design with a simple cost function (Vol. I, Ch. 6, Sec. 16). For a simple two-stage sampling design, the optimum number of primary units, m, and the optimum expected size of ultimate cluster, $\bar{n}$, subject to a fixed expected total expenditure, $C = C_1 m + C_2 m\bar{n}$, are

$$\text{opt. } m = \frac{C}{C_1 + C_2\bar{n}} \tag{10.1}$$

and

$$\text{opt. } \bar{n} = \sqrt{\frac{C_1}{C_2}\frac{W^2}{B^2 - W^2/\bar{N}}} \doteq \sqrt{\frac{C_1}{C_2}\frac{1-\delta}{\delta}} \qquad \text{(10.2 or I-6.16.2)}$$

where δ is given by Eq. 5.5 and B^2 and W^2 are defined in Sec. 1c, and $\bar{n}$ in Eq. 10.1 is the optimum $\bar{n}$.

Derivation. The rel-variance of a simple two-stage design (see Sec. 1, Eq. 1.25) is

$$V_r^2 \doteq \left(1 - \frac{m}{M}\right)\frac{B^2}{m} + \left(1 - \frac{\bar{n}}{\bar{N}}\right)\frac{W^2}{m\bar{n}} \qquad \text{(10.3 or I-6.6.10)}$$

To obtain the values of m and $\bar{n}$ which make V_r^2 a minimum for a fixed total expenditure, set up the Lagrangian $F = V_r^2 + \lambda(C_1 m + C_2 m\bar{n} - C)$. Then the solution of the equations, $\partial F/\partial m = 0$, $\partial F/\partial \bar{n} = 0$, and the cost equation will give the optimum values. Thus

$$\frac{\partial F}{\partial m} = -\frac{B^2}{m^2} - \frac{W^2}{m^2\bar{n}} + \frac{W^2}{m^2\bar{N}} + \lambda(C_1 + C_2\bar{n}) = 0 \tag{10.4}$$

$$\frac{\partial F}{\partial \bar{n}} = -\frac{W^2}{m\bar{n}^2} + \lambda C_2 m = 0 \tag{10.5}$$

Multiplying Eq. 10.4 by m and subtracting Eq. 10.5 multiplied by $\bar{n}$, we obtain

$$\lambda m^2 = \frac{B^2 - W^2/\bar{N}}{C_1} \tag{10.6}$$

From Eq. 10.5

$$\lambda m^2 = \frac{W^2}{C_2\bar{n}^2} \tag{10.7}$$

Equating the right-hand members of Eq. 10.6 and 10.7 and solving for $\bar{n}$, we obtain the optimum $\bar{n}$ as given in Eq. 10.2. Then opt. m is obtained by substitution of this result in the cost equation.

***11. Optimum values for a simple two-stage sampling design with a more general cost function** (Vol. I, Ch. 6, Sec. 18, 19).* For a simple two-stage

* This development and the proof of convergence are due to B. J. Tepping and B. Skalak.

* May be deferred.

sampling design, the optimum expected size of ultimate cluster is

$$\bar{n} = \sqrt{\frac{\frac{C_0}{a} + C_1}{C_2} \frac{W^2}{B^2 - W^2/\bar{N}}} \doteq \sqrt{\frac{\frac{C_0}{a} + C_1}{C_2} \frac{1 - \delta}{\delta}} \quad \text{(11.1 or I-6.18.1)}$$

where the expected total expenditure is $C = C_0\sqrt{m} + C_1 m + C_2 m\bar{n}$ and where m, the optimum number of primary units, is determined to yield either (a) a minimum error when the total expenditure, C, is fixed, in which case

$$m = \frac{a^2}{4}, \quad \text{or} \quad 2\sqrt{m} = a \qquad \text{(11.2 or I-6.18.3)}$$

and

$$a = \frac{\sqrt{1 + 4\frac{C}{C_0}\frac{C_1 + C_2\bar{n}}{C_0}} - 1}{\frac{C_1 + C_2\bar{n}}{C_0}} \qquad \text{(11.3 or I-6.18.2)}$$

or (b) a minimum expenditure when the precision is specified as ε, in which case

$$a = \sqrt{4\frac{B^2 - \frac{W^2}{\bar{N}} + \frac{W^2}{\bar{n}}}{\varepsilon + \frac{B^2}{M}}} \doteq \sqrt{4\frac{\hat{V}^2\delta}{\varepsilon}\left(1 + \frac{1-\delta}{\delta}\frac{1}{\bar{n}}\right)} \quad \text{(11.4 or I-6.19.2)}$$

where δ is given by Eq. 5.5 and B^2 and W^2 are defined in Sec. 1c.

The optimum solution for $\bar{n}$ and m subject to the condition of fixed cost is obtained by substituting any guessed value for a in Eq. 11.1, then substituting the resulting value for $\bar{n}$ in Eq. 11.3, then substituting the new value for a back into Eq. 11.3, and continuing this process until successive solutions for $\bar{n}$ and a yield the same values to the desired accuracy. Then the final a is substituted in Eq. 11.2 to solve for the optimum value of m. The optimum values subject to a fixed variance are obtained by a similar process, using Eq. 11.1, 11.4, and 11.2.

It can be shown that this iterative process will converge, but the proof is beyond the scope of this book.

Derivation of the terms used in the iterative process. The rel-variance of a simple two-stage design (see Sec. 1, Eq. 1.25) is

$$V_r^2 \doteq \left(1 - \frac{m}{M}\right)\frac{B^2}{m} + \left(1 - \frac{\bar{n}}{\bar{N}}\right)\frac{W^2}{m\bar{n}} \qquad \text{(11.5 or I-6.6.10)}$$

To determine the values of m and $\bar{n}$ which minimize the precision subject to a fixed cost we set up the Lagrangian F_C:

$$F_C = V_r^2 + \lambda(C_0\sqrt{m} + C_1 m + C_2 m\bar{n} - C)$$

To determine the values of m and $\bar{n}$ which minimize the cost subject to a prescribed precision, ε, we set up the Lagrangian F_ε:

$$F_\varepsilon = \mu(V_r^2 - \varepsilon) + C$$

Then

$$\frac{\partial F_C}{\partial \bar{n}} = -\frac{W^2}{m\bar{n}^2} + \lambda C_2 m = 0 = \frac{\partial F_\varepsilon}{\partial \bar{n}} \text{ with } \lambda = \frac{1}{\mu} \tag{11.6}$$

$$\frac{\partial F_C}{\partial m} = -\frac{B^2}{m^2} - \frac{W^2}{m^2\bar{n}} + \frac{W^2}{m^2\bar{N}} + \lambda\left(\frac{C_0}{2\sqrt{m}} + C_1 + C_2\bar{n}\right) = 0 = \frac{\partial F_\varepsilon}{\partial m}$$

$$\text{with } \lambda = \frac{1}{\mu} \tag{11.7}$$

From Eq. 11.6 we obtain

$$\lambda m^2 = \frac{W^2}{C_2\bar{n}^2} \tag{11.8}$$

Multiplying Eq. 11.6 by $\bar{n}$ and subtracting Eq. 11.7 multiplied by m, we obtain

$$\lambda m^2 = \frac{B^2 - W^2/\bar{N}}{(C_0/2\sqrt{m}) + C_1} \tag{11.9}$$

Equating Eq. 11.8 and 11.9, solving for $\bar{n}$, and substituting $2\sqrt{m} = a$, we obtain the optimum $\bar{n}$ given in Eq. 11.1. The alternative forms involving δ are obtained by using the rel-variance in terms of δ as given in Eq. 5.1 of Sec. 5. For the case in which the cost is fixed we obtain a as given in Eq. 11.3 above by recognizing the cost equation as a quadratic in $\sqrt{m}$. For the case in which the precision is fixed we obtain a by setting V_r^2 as given in Eq. 11.5 (or for the alternative form V_r^2 as given by Eq. 5.1) equal to ε and solving for m. Then $a = \sqrt{4m}$.

REFERENCES

(1) M. N. Ghosh, "Expected Travel among Random Points in a Region," *Calcutta Stat. Assn. Bull.* **6** (1949), 83–87.
(2) R. J. Jessen, "Statistical Investigation of a Sample Survey for Obtaining Farm Facts," *Iowa Agr. Exp. Stat. Res. Bull.* 304 (1942).

(3) P. C. Mahalanobis, "A Sample Survey of the Acreage under Jute in Bengal," *Sankhya*, **4** (1940), 511–530.
(4) P. C. Mahalanobis, "On Large-Scale Sample Surveys," *Phil. Trans. Roy. Soc.*, Series B, **231** (1946), 329–451.
(5) Eli S. Marks, "A Lower Bound for the Expected Travel among *m* Random Points," *Annals Math. Stat.*, **19** (1948), 419–422.
(6) Garnet E. McCreary, "Cost Functions for Sample Surveys," unpublished thesis, Iowa State College, Ames, Iowa, 1950.
(7) F. Yates and I. Zacopanay, "The Estimation of the Efficiency of Sampling with Special Reference to Sampling for Yield in Cereal Experiments," *J. Agr. Sci.*, **25** (1935), 543–577.
(8) U. S. Bureau of the Census, Sampling Staff, *A Chapter in Population Sampling*, U. S. Government Printing Office, Washington, D. C., 1947.
(9) W. G. Cochran, *Sampling Techniques*, John Wiley & Sons, New York, 1953, Chapters 9 and 10.
(10) W. E. Deming, *Some Theory of Sampling*, John Wiley & Sons, New York, 1950, Chapter 5.
(11) F. X. Schumacher and R. A. Chapman, *Sampling Methods in Forestry and Range Management*, Duke University, School of Forestry, 1942, Chapter 6.

CHAPTER 7

Stratified Single- or Multi-stage Cluster Sampling

DERIVATIONS, PROOFS, AND SOME EXTENSIONS OF THEORY FOR CH. 7 OF VOL. I*

NOTE. The theory for stratified cluster sampling with one or more stages of sampling is presented in this chapter. Topics covered include the estimate of the variance and its components, gains due to stratification with cluster sampling, and the optimum design under selected cost conditions.

It is sometimes necessary to isolate the contribution to the total variance attributable to a particular stage of sampling or to express the variance in terms of the contribution from each stage of sampling. This chapter indicates a procedure which makes it possible to write down the variance of sample estimates for any number of stages of sampling in terms of the components of the variance.

The notation in this chapter is the same as that introduced in Chapter 6 except that here a subscript (h) is added to designate the strata.

1. The rel-variance of a ratio estimate for a two-stage stratified sampling design (Vol. I, Ch. 7, Sec. 5). Assume that we have a population consisting of L primary strata, M_h primary units in the hth stratum, and N_{hi} second-stage units within the hith primary unit. Assume, further, that a simple random sample of m_h first-stage units is selected from M_h, and a simple random sample of n_{hi} second-stage units is selected from the N_{hi} in the hith primary unit. Now let

$$r = \frac{x'}{y'} = \frac{\sum_h^L x'_h}{\sum_h^L y'_h} \tag{1.1}$$

* Appropriate references to Vol. I are shown in parentheses after section or subsection headings. The number following I– after some equations gives the chapter, section, and number of that particular equation in Vol. I.

where

$$x'_h = \frac{M_h}{m_h} \sum_i^{m_h} x'_{hi}, \quad y'_h = \frac{M_h}{m_h} \sum_i^{m_h} y'_{hi}$$

and

$$x'_{hi} = \frac{N_{hi}}{n_{hi}} \sum_j^{n_{hi}} x_{hij}, \quad y'_{hi} = \frac{N_{hi}}{n_{hi}} \sum_j^{n_{hi}} y_{hij}$$

where x_{hij} and y_{hij} are the values of the X- and Y-characteristics for the *hij*th second-stage unit in the sample.

To prove: The rel-variance of r is

$$V_r^2 \doteq \frac{1}{X^2} \sum^L \frac{M_h^2}{m_h} \frac{M_h - m_h}{M_h} S_{1h}^2 + \frac{1}{X^2} \sum^L \frac{M_h}{m_h} \sum^{M_h} \frac{N_{hi}^2}{n_{hi}} \frac{N_{hi} - n_{hi}}{N_{hi}} S_{2hi}^2$$

(1.2 or I–7.5.3)

where

$$S_{1h}^2 = S_{1hX}^2 + R^2 S_{1hY}^2 - 2RS_{1hXY} \qquad \text{(1.3 or I–7.5.4)}$$

with

$$S_{1hXY} = \frac{\sum^{M_h} (X_{hi} - \bar{X}_h)(Y_{hi} - \bar{Y}_h)}{M_h - 1} \qquad \text{(1.4 or I–7.5.6)}$$

$$S_{1hX}^2 = S_{1hXX}, \quad S_{1hY}^2 = S_{1hYY}, \quad R = \frac{X}{Y}$$

$$\bar{X}_h = \frac{X_h}{M_h}, \quad \text{and} \quad \bar{Y}_h = \frac{Y_h}{M_h}$$

Also,

$$S_{2hi}^2 = S_{2hiX}^2 + R^2 S_{2hiY}^2 - 2RS_{2hiXY} \qquad \text{(1.5 or I–7.5.9)}$$

with

$$S_{2hiXY} = \frac{\sum^{N_{hi}} (X_{hij} - \bar{\bar{X}}_{hi})(Y_{hij} - \bar{\bar{Y}}_{hi})}{N_{hi} - 1} \qquad \text{(1.6 or I–7.5.11)}$$

$$S_{2hiX}^2 = S_{2hiXX}, \quad S_{2hiY}^2 = S_{2hiYY}$$

$$\bar{\bar{X}}_{hi} = \frac{X_{hi}}{N_{hi}}, \quad \text{and} \quad \bar{\bar{Y}}_{hi} = \frac{Y_{hi}}{N_{hi}}$$

Proof. From Ch. 4, Sec. 11, the rel-variance of r is

$$V_r^2 \doteq V_{x'}^2 + V_{y'}^2 - 2V_{x'y'} \qquad (1.7)$$

where

$$V_{x'}^2 = \frac{\sigma_{x'}^2}{(Ex')^2}, \quad V_{y'}^2 = \frac{\sigma_{y'}^2}{(Ey')^2}, \quad \text{and} \quad V_{x'y'} = \frac{\sigma_{x'y'}}{(Ex')(Ey')}$$

First, by Theorem 5, Ch. 3 (p. 48), $Ex' = \sum^L Ex'_h$ and by Sec. 1, Ch. 6, it follows that $Ex'_h = X_h$ when we recognize that x'_h is given by x' in Eq. 1.1, Ch. 6, with the subscript h added. Hence, $Ex' = \sum^L X_h = X$. Similarly $Ey' = Y$.

By Eq. 1.8, Ch. 5,

$$\sigma^2_{x'} = \sum^L \sigma^2_{x'_h}, \quad \sigma^2_{y'} = \sum^L \sigma^2_{y'_h}, \quad \text{and} \quad \sigma_{x'y'} = \sum^L \sigma_{x'_h y'_h}$$

Now $\sigma^2_{x'_h}$ is given by Eq. 1.2, Ch. 6, with the subscript h added provided we assume that x' in Eq. 1.1, Ch. 6, is an estimate of a stratum total; $\sigma^2_{y'_h}$ is similarly defined; and $\sigma_{x'_h y'_h}$ is given by Eq. 1.12, Ch. 6, with the subscript h added. Assembling the terms representing the first-stage contribution to the variance gives the first term in Eq. 1.2. Similarly, assembling the terms representing the second-stage contribution to the variance gives the second term in Eq. 1.2.

When the second-stage sampling fractions are the same for all first-stage units in a stratum, i.e., $n_{hi}/N_{hi} = f_{2h}$, and $m_h/M_h = f_{1h}$, then the over-all sampling fraction in the hth stratum is $f_h = f_{1h}f_{2h}$, the estimate (Eq. 1.1) becomes

$$r = \frac{x'}{y'} = \frac{\sum^L \frac{1}{f_h} x_h}{\sum^L \frac{1}{f_h} y_h} \qquad (1.8 \text{ or I-7.5.13})$$

and the rel-variance of r becomes

$$V_r^2 \doteq \frac{1}{X^2} \sum^L \frac{M_h^2}{m_h} \frac{M_h - m_h}{M_h} S_{1h}^2 + \frac{1}{X^2} \sum^L \frac{N_h^2}{m_h \bar{n}_h} \frac{\bar{N}_h - \bar{n}_h}{\bar{N}_h} S_{2h}^2 \qquad (1.9)$$

with S_{1h}^2 given in Eq. 1.3,

$$S_{2h}^2 = \frac{\sum^{M_h} N_{hi} S_{2hi}^2}{M_h \bar{N}_h} \qquad (1.10 \text{ or I-7.5.15})$$

and S_{2hi}^2 given in Eq. 1.5.

Equation 1.9 may also be written

$$V_r^2 \doteq \frac{1}{X^2} \sum^L X_h^2 \frac{M_h - m_h}{M_h} \frac{B_h^2}{m_h} + \frac{1}{X^2} \sum^L X_h^2 \frac{\bar{N}_h - \bar{n}_h}{\bar{N}_h} \frac{W_h^2}{m_h \bar{n}_h} \qquad (1.11 \text{ or I-7.5.16})$$

where

$$B_h^2 = \frac{S_{1h}^2}{\bar{X}_h^2}, \quad W_h^2 = \frac{S_{2h}^2}{\bar{\bar{X}}_h^2}$$

$$\bar{X}_h = \frac{X_h}{M_h}, \quad \bar{\bar{X}}_h = \frac{X_h}{N_h}$$

and where in Eq. 1.9 and 1.11 $\bar{n}_h = f_{2h}\bar{N}_h$ is the expected number of listing units in the sample per psu in the sample for the hth stratum.

2. The estimate of the rel-variance for two-stage stratified sampling (Vol. I, Ch. 7, Sec. 6). An estimate, v_r^2, of the rel-variance, V_r^2 (Eq. 1.2), for a two-stage stratified design in which a simple random sample of at least two first-stage units is selected within each stratum and a simple random subsample of second-stage units is selected is given by

$$v_r^2 = \frac{1}{x'^2} \sum_h^L \frac{M_h^2}{m_h} s_{c'h}^2 \qquad \text{(2.1 or I-7.6.1)}$$

where

$$s_{c'h}^2 = s_{c'hX}^2 + r^2 s_{c'hY}^2 - 2r s_{c'hXY} \qquad \text{(2.2 or I-7.6.2)}$$

$$s_{c'hXY} = \frac{\sum^{m'_h} (x'_{hi} - \bar{x}'_h)(y'_{hi} - \bar{y}'_h)}{m'_h - 1} \qquad \text{(2.3 or I-7.6.5)}$$

$$x'_{hi} = \frac{N_{hi}}{n_{hi}} x_{hi}, \quad \bar{x}'_h = \sum^{m'_h} x'_{hi}/m'_h$$

y'_{hi} and $\bar{y}'_h$ are similarly defined,

$$s_{c'hX}^2 = s_{c'hXX}, \quad s_{c'hY}^2 = s_{c'hYY}$$

m'_h is the number of primary units from the hth stratum used in estimating the variance and may be smaller than m_h. This result follows immediately from Sec. 3, Ch. 6, with the subscript h added to each estimated variance, since

$$V_r^2 \doteq \frac{1}{X^2} \sum^L (\sigma_{x'_h}^2 + R^2 \sigma_{y'_h}^2 - 2R\sigma_{x'_h y'_h})$$

When the sample is self-weighting within strata (i.e., $n_{hi}/N_{hi} = f_{2h}$, $m_h/M_h = f_{1h}$, and $f_h = f_{1h} f_{2h}$), $s_{c'h}^2$, Eq. 2.2, becomes

$$\frac{1}{f_{2h}^2} s_{ch}^2 \qquad \text{(2.4)}$$

where

$$s_{ch}^2 = s_{chX}^2 + r^2 s_{chY}^2 - 2r s_{chXY} \qquad \text{(2.5)}$$

$$s_{chXY} = \frac{\sum^{m'_h} (x_{hi} - \bar{x}_h)(y_{hi} - \bar{y}_h)}{m'_h - 1} \qquad \text{(2.6)}$$

$$\bar{x}_h = \frac{\sum_i^{m'_h} x_{hi}}{m'_h}, \quad \bar{y}_h = \frac{\sum_i^{m'_h} y_{hi}}{m'_h}$$

$$s_{chX}^2 = s_{chXX} \quad \text{and} \quad s_{chY}^2 = s_{chYY}$$

An estimate of V_r^2 which is more accurate than Eq. 2.1 is

$$v_r^2 = \frac{1}{x'^2} \sum^{L} (1 - f_h) \frac{m_h}{f_h^2} s_{ch}^2 \qquad \text{(2.7 or I-7.6.6)}$$

Note that, when $L = 1$, and when s_{ch}^2 is estimated from the m_h units in the sample, Eq. 2.7 reduces to Eq. 3.19, Ch. 6.

3. Estimates of the components of the rel-variance of a ratio for a two-stage stratified sampling design (Vol. I, Ch. 7, Sec. 7). Consistent estimates of the components of the rel-variance of a ratio, r, for a two-stage stratified sampling design may be made as follows: The estimate, r, is given by Eq. 1.1. The rel-variance of r is given by Eq. 1.2.

A consistent estimate of the within-psu component of the rel-variance, i.e., of the second term in Eq. 1.2, is

$$\frac{1}{x'^2} \sum^{L} \frac{M_h^2}{m_h} \frac{\sum^{m'_h} \frac{N_{hi}^2}{n_{hi}} \frac{N_{hi} - n_{hi}}{N_{hi}} s_{2hi}^2}{m'_h} = \frac{1}{x'^2} \sum^{L} \frac{M_h^2}{m_h} \hat{s}_h^2 \qquad (3.1)$$

where m'_h is the number of first-stage units from the hth stratum used in estimating the variance and may be less than m_h,

$$s_{2hi}^2 = s_{2hiX}^2 + r^2 s_{2hiY}^2 - 2r s_{2hiXY} \qquad (3.2)$$

with

$$s_{2hiXY} = \frac{\sum^{n_{hi}} (x_{hij} - \bar{\bar{x}}_{hi})(y_{hij} - \bar{\bar{y}}_{hi})}{n_{hi} - 1} \qquad (3.3)$$

$$s_{2hiX}^2 = s_{2hiXX}, \quad \text{and} \quad s_{2hiY}^2 = s_{2hiYY}$$

Equation 3.1 follows from Sec. 4, Ch. 6, when the estimates in that chapter are assumed to be estimates of the variances for the hth stratum, and from the fact that the variance of a stratified sample sum is the sum of the within-strata variances.

A consistent estimate of the between-psu component of the rel-variance, i.e., of the first term in Eq. 1.2, is

$$\frac{1}{x'^2} \sum^{L} \frac{M_h^2}{m_h} \frac{M_h - m_h}{M_h} (s_{c'h}^2 - \hat{s}_h^2) \qquad (3.4)$$

where $\hat{s}_h^2$ is defined in Eq. 3.1 and $s_{c'h}^2$ is defined by Eq. 2.2. This result follows immediately from Sec. 4, Ch. 6, and the fact that the variance for a stratified sample sum is the sum of the within-strata variances.

4. The rel-variance of a ratio estimate for three- or more stage stratified sampling (Vol. I, Ch. 7, Sec. 12). *To prove:* The rel-variance of a ratio, $r = x'/y'$, for a stratified sampling design with K stages of sampling can be expressed as the sum of K terms each representing the contribution from one of the stages of sampling. In the special case of three-stage sampling, it will be shown that the rel-variance of x'/y' is equal to

$$V_r^2 = \text{Eq. } 1.2 + \frac{1}{X^2}\sum_h^L \frac{M_h}{m_h}\sum_i^{M_h}\frac{N_{hi}}{n_{hi}}\sum_j^{N_{hi}}\frac{Q_{hij}^2}{q_{hij}}\frac{Q_{hij}-q_{hij}}{Q_{hij}}S_{3hij}^2 \tag{4.1}$$

where Eq. 1.2 represents the contribution from each of the first two stages of sampling and where the last term of Eq. 4.1 represents the contribution to the rel-variance from the third stage of sampling. The notation is defined below.

Proof. From Sec. 11, Ch. 4, we have for the rel-variance of a ratio of random variables

$$V_r^2 \doteq V_{x'}^2 + V_{y'}^2 - 2V_{x'y'} = \frac{\sigma_{x'}^2}{(Ex')^2} + \frac{\sigma_{y'}^2}{(Ey')^2} - \frac{2\sigma_{x'y'}}{Ex'Ey'}$$

$$= \frac{1}{(Ex')^2}(\sigma_{x'}^2 + R^2\sigma_{y'}^2 - 2R\sigma_{x'y'}) \tag{4.2}$$

where

$$R = \frac{Ex'}{Ey'} = \frac{X}{Y}$$

For a stratified population and a simple random sample of units at each stage of sampling

$$x' = \sum^L x_h'$$

where x_h' is the simple unbiased estimate of X_h, the hth stratum total. For three stages of sampling the simple unbiased estimate of X_h is given by Eq. 4.8, and this form of estimate is extended to obtain the simple unbiased estimate for any number of stages of sampling. Similarly,

$$y' = \sum^L y_h'$$

From Sec. 1, Ch. 5,

$$\sigma_{x'}^2 = \sum^L \sigma_{x'_h}^2, \quad \sigma_{y'}^2 = \sum^L \sigma_{y'_h}^2, \quad \text{and} \quad \sigma_{x'y'} = \sum^L \sigma_{x'_h y'_h} \tag{4.3}$$

We shall now indicate how to apply Theorem 16 to express $\sigma_{x'_h}^2$ in terms of the contribution to the variance for the hth stratum from each stage of sampling. The developments for $\sigma_{y'_h}^2$ and $\sigma_{x'_h y'_h}$ follow exactly the same steps. By Theorem 16, Ch. 3 (p. 68), with $u = x_h'$, $z_j = E(x_h' \mid j!)$, the contribution to the variance in the hth stratum from the jth stage of sampling is

$$E\sigma^2_{z_j|[1, 2, 3, \cdots, j-1]} \tag{4.4}$$

In Eq. 4.4 the symbol

$$z_j = E(x'_h|[1, 2, 3, \cdots, j]) \tag{4.5}$$

is the expected value of x'_h, considering the units selected at the jth stage of sampling as strata. Since the jth-stage units are selected from the $(j-1)$th-stage units, etc., regarding the jth-stage units as strata implies that the units selected at all previous stages are also fixed. Now

$$\sigma^2_{z_j|[1, 2, 3, \cdots, j-1]} \tag{4.6}$$

means the variance of z_j (Eq. 4.5), where the variance is evaluated within the units selected at the $(j-1)$th stage of sampling. If a simple random sample of jth-stage units is selected from each $(j-1)$th-stage unit in the sample, the units selected at the $(j-1)$th stage of sampling can be regarded as strata; and, from Corollary 1 to Theorem 11, Ch. 3 (p. 56), we can write down the conditional variance of any linear combination of random variables. The final step of taking the expected value of Eq. 4.6 makes use of the theorem (Theorem 5, Ch. 3, p. 48) that the expected value of a sum is equal to the sum of the expected values, or, more generally, Theorem 6, Ch. 3 (p. 49), for a linear combination of random variables. The above steps make it possible to write down the contribution to the variance from any stage of sampling for any linear combination of random variables, as will be illustrated below.

Consider the evaluation of Eq. 4.4, for the contribution to the variance from the third stage of sampling. We must then evaluate

$$E\sigma^2_{z_3|[1, 2]} \tag{4.7}$$

where

$$z_3 = E(x'_h|[1, 2, 3])$$

Suppose that M_h and m_h are the number of first-stage units in the hth stratum in the population and in the sample, respectively; N_{hi} and n_{hi} are the number of second-stage units in the hith psu in the population and in the sample, respectively; and Q_{hij} and q_{hij} are the number of third-stage units in the hijth second-stage unit in the population and in the sample, respectively. Then

$$\left.\begin{aligned} z_3 = x'_h = E(x'_h|[1, 2, 3]) &= \frac{M_h}{m_h}\sum_i^{m_h}\frac{N_{hi}}{n_{hi}}\sum_j^{n_{hi}}\frac{Q_{hij}}{q_{hij}}\sum_k^{q_{hij}} x_{hijk} \\ &= \frac{M_h}{m_h}\sum_i^{m_h}\frac{N_{hi}}{n_{hi}}\sum_j^{n_{hi}} x'_{hij} \end{aligned}\right\} \tag{4.8}$$

where x_{hijk} is the value of X for the *hijk*th third-stage unit in the sample, and

$$x'_{hij} = \frac{Q_{hij}}{q_{hij}} \sum_{k}^{q_{hij}} x_{hijk} \tag{4.9}$$

is the sample estimate of X_{hij}, the value of X for the *hij*th second-stage unit in the sample. By Corollary 1 to Theorem 11 of Ch. 3 (p. 56), since we can regard the second-stage units in the sample as strata,

$$\sigma^2_{z_3|[1,2]} = \frac{M_h^2}{m_h^2} \sum_{i}^{m_h} \frac{N_{hi}^2}{n_{hi}^2} \sum_{j}^{n_{hi}}{}' \sigma^2_{x'_{hij}} \tag{4.10}$$

and since a simple random sample of third-stage units is selected from each second-stage unit in the sample, we have

$$\sigma^2_{x'_{hij}} = Q_{hij}^2 \frac{Q_{hij} - q_{hij}}{Q_{hij} q_{hij}} S^2_{3hijX} \tag{4.11}$$

and

$$S^2_{3hijX} = \frac{\sum^{Q_{hij}} (X_{hijk} - \overline{\overline{\overline{X}}}_{hij})^2}{Q_{hij} - 1} \tag{4.12}$$

By Theorem 6, Ch. 3 (p. 49), the contribution of the third stage of sampling in the hth stratum is

$$\sigma^2_{3x'_h} = E\sigma^2_{z_3|[1,2]} = \frac{M_h^2}{m_h} \frac{m_h}{M_h} \sum_{i}^{M_h} \frac{N_{hi}^2}{n_{hi}^2} \frac{n_{hi}}{N_{hi}} \sum_{j}^{N_{hi}} Q_{hij}^2 \frac{Q_{hij} - q_{hij}}{Q_{hij} q_{hij}} S^2_{3hijX}$$

$$= \frac{M_h}{m_h} \sum_{i}^{M_h} \frac{N_{hi}}{n_{hi}} \sum_{j}^{N_{hi}} Q_{hij}^2 \frac{Q_{hij} - q_{hij}}{Q_{hij} q_{hij}} S^2_{3hijX} \tag{4.13}$$

In the same way we can show that the third-stage contributions to $\sigma^2_{y'_h}$ and $\sigma_{x'_h y'_h}$ are, respectively,

$$\sigma^2_{3y'_h} = \frac{M_h}{m_h} \sum_{i}^{M_h} \frac{N_{hi}}{n_{hi}} \sum_{j}^{N_{hi}} Q_{hij}^2 \frac{Q_{hij} - q_{hij}}{Q_{hij} q_{hij}} S^2_{3hijY} \tag{4.14}$$

and

$$\sigma_{3x'_h y'_h} = \frac{M_h}{m_h} \sum_{i}^{M_h} \frac{N_{hi}}{n_{hi}} \sum_{j}^{N_{hi}} Q_{hij}^2 \frac{Q_{hij} - q_{hij}}{Q_{hij} q_{hij}} S_{3hijXY} \tag{4.15}$$

where

$$S_{3hijXY} = \frac{\sum_{k}^{Q_{hij}} (X_{hijk} - \overline{\overline{\overline{X}}}_{hij})(Y_{hijk} - \overline{\overline{\overline{Y}}}_{hij})}{Q_{hij} - 1} \tag{4.16}$$

and

$$S^2_{3hijY} = S_{3hijYY} \tag{4.17}$$

Now, the third-stage contribution to the variance of $x' = \sum^{L} x'_h$ is, by Eq. 4.3,

$$\sum_{h}^{L} \sigma^2_{3x'_h} = \sigma^2_{3x'} \tag{4.18}$$

Similarly,

$$\sum_{h}^{L} \sigma^2_{3y'_h} = \sigma^2_{3y'} \tag{4.19}$$

$$\sum_{h}^{L} \sigma_{3x'_h y'_h} = \sigma_{3x'y'} \tag{4.20}$$

We now have from Eq. 4.2 that the third-stage contribution to the rel-variance of x'/y' is

$$\frac{1}{X^2}(\sigma^2_{3x'} + R^2\sigma^2_{3y'} - 2R\sigma_{3x'y'})$$

$$= \frac{1}{X^2}\sum_{h}^{L}\frac{M_h}{m_h}\sum_{i}^{M_h}\frac{N_{hi}}{n_{hi}}\sum_{j}^{N_{hi}}\frac{Q^2_{hij}}{q_{hij}}\,\frac{Q_{hij} - q_{hij}}{Q_{hij}}\,S^2_{3hij} \tag{4.21}$$

where

$$S^2_{3hij} = S^2_{3hijX} + R^2S^2_{3hijY} - 2RS_{3hijXY} \tag{4.22}$$

In Sec. 1, the rel-variance of $r = x'/y'$ was developed for a two-stage design. Hence, the terms of Eq. 1.2 represent the contributions from the first and second stages of sampling. It follows that V_r^2 for a three-stage design (Eq. 4.1) is given by Eq. 1.2 + Eq. 4.21.

5. Gains due to stratification with cluster sampling and a comparison with gains due to stratification with simple random sampling of listing units (Vol. I, Ch. 7, Sec. 4). *To prove:* For a proportionate stratified random sample of equal-sized clusters from equal sized strata, the relative gain due to stratification is given approximately by

$$\frac{\sum^{L}(\bar{\bar{X}}_h - \bar{\bar{X}})^2\bar{N}}{L\sigma^2[1 + \delta(\bar{N} - 1)]} \tag{5.1 or I-7.4.1}$$

From Sec. 7 of Ch. 5, the relative gain due to stratification when a simple random sample of listing units is selected from the same strata is approximately

$$\frac{\sum^{L}(\bar{\bar{X}}_h - \bar{\bar{X}})^2}{L\sigma^2} \tag{5.2 or I-7.4.2}$$

The ratio of Eq. 5.1 to Eq. 5.2 is equal to

$$\frac{\bar{N}}{1 + \delta(\bar{N} - 1)} \tag{5.3}$$

and is the factor by which the relative gain due to stratification with simple random sampling of listing units must be multiplied in order to obtain the relative gain due to stratification with cluster sampling providing the same strata are used for both simple random sampling of listing units and for cluster sampling.

In Eq. 5.1 and 5.2,

$\bar{\bar{X}}_h = \dfrac{\sum^{\hat{M}} \sum^{\bar{N}} X_{hij}}{\hat{M}\bar{N}}$ is the average per listing unit in the hth stratum.

$\bar{N}$ = the number of listing units per cluster.

$\hat{M}$ = the number of clusters per stratum.

$\bar{\bar{X}} = \sum^{L} \bar{\bar{X}}_h / L$ is the average per listing unit over all strata.

$\sigma^2 = \dfrac{\sum^{L} \sum^{\hat{M}} \sum^{\bar{N}} (X_{hij} - \bar{\bar{X}})^2}{N}$ is the population variance for a simple random sample of listing units.

δ = the intraclass correlation among listing units given by Eq. 6.1, Ch. 6.

Proof. From Remark 1, Sec. 1, Ch. 5, the variance of a mean for a proportionate stratified sample with equal-sized clusters from equal-sized strata is

$$\sigma_1^2 = \frac{1-f}{m} \frac{\sum^{L} S_{hX}^2}{L} \tag{5.4}$$

where

$$f = \frac{\hat{m}}{\hat{M}} = \frac{m}{M}, \quad \hat{m} = \frac{m}{L}$$

$$S_{hX}^2 = \frac{\sum^{\hat{M}} (\bar{\bar{X}}_{hi} - \bar{\bar{X}}_h)^2}{\hat{M} - 1} \tag{5.5}$$

$$= \frac{\hat{M}}{\hat{M} - 1} \left[\frac{\sum^{\hat{M}} (\bar{\bar{X}}_{hi} - \bar{\bar{X}})^2}{\hat{M}} + (\bar{\bar{X}}_h - \bar{\bar{X}})^2 \right] \tag{5.6}$$

where $\bar{\bar{X}}_{hi} = X_{hi}/N_{hi}$

Thus,

$$\sigma_1^2 \doteq \frac{1-f}{m} \left[\frac{\sum^{L} \sum^{\hat{M}} (\bar{\bar{X}}_{hi} - \bar{\bar{X}})^2}{M} + \frac{\sum^{L} (\bar{\bar{X}}_h - \bar{\bar{X}})^2}{L} \right] \tag{5.7}$$

Now, the variance of a mean for a simple random sample of m clusters is (see Sec. 2, Ch. 4)

$$\sigma_2^2 = \frac{1-f}{m} \frac{\sum^L \sum^{\hat{M}} (\bar{X}_{hi} - \bar{\bar{X}})^2}{M} \tag{5.8}$$

which may also be written, from Sec. 5, Ch. 6,

$$\sigma_2^2 = \frac{1-f}{m} \frac{\sigma^2}{\bar{N}} [1 + \delta(\bar{N} - 1)] \tag{5.9}$$

Hence, the relative gain

$$\frac{\sigma_1^2 - \sigma_2^2}{\sigma_2^2} \doteq \frac{\text{Eq. 5.7} - \text{Eq. 5.8}}{\text{Eq. 5.9}}$$

is given by Eq. 5.1.

6. Optimum values for a two-stage stratified sampling design with variable sampling fractions and a simple cost function (Vol. I, Ch. 7, Sec. 9). *To prove:* For a two-stage stratified design with variable sampling fractions among strata and proportionate sampling within sample first-stage units in a stratum, the optimum expected size of ultimate cluster, $\bar{n}_h$, and the optimum number of sample primary units, m_h, for a stratum, when the expected cost of the survey is

$$C = \sum^L C_{1h} m_h + \sum^L C_{2h} m_h \bar{n}_h \tag{6.1 or I-7.9.1}$$

and the total expenditure is fixed, are

$$\text{opt. } \bar{n}_h = \bar{N}_h \sqrt{\frac{C_{1h}}{C_{2h}} \frac{S_{2h}^2}{S_{1h}^2 - \bar{N}_h S_{2h}^2}} = \sqrt{\frac{C_{1h}}{C_{2h}} \frac{W_h^2}{B_h^2 - W_h^2/\bar{N}_h}} \tag{6.2 or I-7.9.2}$$

$$\text{opt. } m_h = \frac{C N_h S_{2h}/\bar{n}_h \sqrt{C_{2h}}}{\sum_h^L [(C_{1h} + C_{2h}\bar{n}_h) N_h S_{2h}/\bar{n}_h \sqrt{C_{2h}}]} \tag{6.3 or I-7.9.3}$$

where S_{1h}^2, S_{2h}^2, B_h^2, and W_h^2 are defined in Sec. 1, and where C_{1h} is the cost per first-stage unit in the sample from the hth stratum, C_{2h} is the cost per second-stage unit in the sample from the hth stratum, and where $\bar{n}_h$ in Eq. 6.3 is the optimum value.

Proof. The rel-variance for a two-stage stratified design with uniform second-stage sampling fractions within strata is (see Eq. 1.9)

$$V_r^2 \doteq \frac{1}{X^2} \sum^L \frac{M_h^2}{m_h} \frac{M_h - m_h}{M_h} S_{1h}^2 + \frac{1}{X^2} \sum^L \frac{N_h^2}{m_h \bar{n}_h} \frac{\bar{N}_h - \bar{n}_h}{\bar{N}_h} S_{2h}^2 \tag{6.4}$$

where the terms are defined in Sec. 1. To determine the values of m_h

and $\bar{n}_h$ which minimize V_r^2 subject to a fixed total expenditure C, with the cost of the survey given by Eq. 6.1, we set up the Lagrangian F:

$$F = V_r^2 + \lambda(\sum^L C_{1h}m_h + \sum^L C_{2h}m_h\bar{n}_h - C)$$

Then the solution to the equations $\partial F/\partial m_h = 0$, $\partial F/\partial \bar{n}_h = 0$, and the cost equation will be the optimum. Now

$$\frac{\partial F}{\partial m_h} = -\frac{M_h^2 S_{1h}^2}{X^2 m_h^2} - \frac{N_h^2 S_{2h}^2}{X^2 m_h^2 \bar{n}_h} + \frac{N_h^2 S_{2h}^2}{X^2 m_h^2 \bar{N}_h} + \lambda(C_{1h} + C_{2h}\bar{n}_h) = 0 \quad (6.5)$$

$$\frac{\partial F}{\partial \bar{n}_h} = -\frac{N_h^2 S_{2h}^2}{X^2 m_h \bar{n}_h^2} + \lambda C_{2h} m_h = 0 \quad (6.6)$$

From Eq. 6.6 we obtain

$$\lambda X^2 m_h^2 = \frac{N_h^2 S_{2h}^2}{C_{2h}\bar{n}_h^2} \quad (6.7)$$

Multiplying Eq. 6.5 by m_h and subtracting Eq. 6.6 multiplied by $\bar{n}_h$, we obtain

$$\lambda X^2 m_h^2 = \frac{M_h^2 S_{1h}^2 - N_h^2 S_{2h}^2/\bar{N}_h}{C_{1h}} \quad (6.8)$$

Equating the right-hand members of Eq. 6.7 and 6.8 and solving for $\bar{n}_h$, substituting $M_h\bar{N}_h = N_h$, we obtain the optimum $\bar{n}_h$ as given in Eq. 6.2. The alternative form of $\bar{n}_h$ is obtained by making the substitutions $B_h^2 = S_{1h}^2/\bar{X}_h^2$ and $W_h^2 = S_{2h}^2/\bar{\bar{X}}_h^2$.

Now, substituting

$$m_h = \frac{1}{\sqrt{\lambda}} \frac{N_h S_{2h}}{X\bar{n}_h\sqrt{C_{2h}}}$$

from Eq. 6.7, into the cost equation and solving for $\sqrt{\lambda}$, we obtain

$$\sqrt{\lambda} = \sum^L (C_{1h} + C_{2h}\bar{n}_h) \frac{N_h S_{2h}}{X\bar{n}_h\sqrt{C_{2h}}} \Big/ C \quad (6.9)$$

Substituting Eq. 6.9 into Eq. 6.7 and simplifying, we obtain the optimum m_h as given in Eq. 6.3.

7. Optimum values for a two-stage stratified sampling design with variable sampling fractions and a more complicated cost function (Vol. I, Ch. 7, Sec. 11). *a.* Consider again the situation in Sec. 6 but with an added term $C_0\sqrt{m}$ in the cost function measuring the travel and perhaps other costs of the survey, i.e.

$$C = C_0\sqrt{m} + \sum^L C_{1h}m_h + \sum^L C_{2h}m_h\bar{n}_h \quad \text{(7.1 or I–7.11.1)}$$

with $m = \Sigma m_h$.

To prove: The optimum $\bar{n}_h$ is

$$\text{opt. } \bar{n}_h = \frac{N_h S_{2h}}{X a_h \sqrt{C_{2h}}} = \frac{X_h W_h}{X a_h \sqrt{C_{2h}}} \qquad \text{(7.2 or I–7.11.5)}$$

and the optimum m_h is

$$\text{opt. } m_h = \frac{C_0^2 a_h}{4d^2 \sum^L a_h} \qquad \text{(7.3 or I–7.11.6)}$$

where a_h and d are determined from the Eq. 7.4, 7.5, and 7.6 below, following a process of successive approximation similar to that described in Sec. 11, Ch. 6.

$$a_h = \frac{\frac{M_h}{X}\sqrt{S_{1h}^2 - \bar{N}_h S_{2h}^2}}{\sqrt{d + C_{1h}}} = \frac{\frac{X_h}{X}\sqrt{B_h^2 - W_h^2/\bar{N}_h}}{\sqrt{d + C_{1h}}} \qquad \text{(7.4 or I–7.11.2)}$$

$$\left.\begin{aligned} b &= \frac{\sum^L C_{1h} a_h + \sum^L \frac{N_h S_{2h}}{X}\sqrt{C_{2h}}}{\sum^L a_h} \\ &= \frac{\sum^L C_{1h} a_h + \sum^L \frac{X_h}{X} W_h \sqrt{C_{2h}}}{\sum^L a_h} \end{aligned}\right\} \qquad \text{(7.5 or I–7.11.3)}$$

$$d = \frac{C_0}{2\sqrt{m}} = \frac{b}{\sqrt{1 + \frac{4C}{C_0^2} b} - 1} \qquad \text{(7.6 or I–7.11.4)}$$

where S_{1h}^2, S_{2h}^2, B_h^2, and W_h^2 are defined in Sec. 1.

Proof. In this case the Lagrangian F is

$$F = V_r^2 + \lambda\left(C_0\sqrt{m} + \sum^L C_{1h} m_h + \sum^L C_{2h} m_h \bar{n}_h - C\right)$$

with V_r^2 given in Eq. 6.4. Then

$$\frac{\partial F}{\partial m_h} = -\frac{M_h^2 S_{1h}^2}{X^2 m_h^2} - \frac{N_h^2 S_{2h}^2}{X^2 m_h^2 \bar{n}_h} + \frac{N_h^2 S_{2h}^2}{X^2 m_h^2 \bar{N}_h} + \lambda\left(\frac{C_0}{2\sqrt{m}} + C_{1h} + C_{2h}\bar{n}_h\right) = 0 \qquad (7.7)$$

$$\frac{\partial F}{\partial \bar{n}_h} = -\frac{N_h^2 S_{2h}^2}{X^2 m_h \bar{n}_h^2} + \lambda C_{2h} m_h = 0 \qquad (7.8)$$

From Eq. 7.8 we obtain

$$\frac{\lambda X^2 m_h^2}{N_h^2} = \frac{S_{2h}^2}{C_{2h}\bar{n}_h^2} \tag{7.9}$$

Multiplying Eq. 7.7 by m_h and subtracting Eq. 7.8 multiplied by $\bar{n}_h$, we obtain

$$\frac{\lambda X^2 m_h^2}{N_h^2} = \frac{\dfrac{S_{1h}^2}{\bar{N}_h^2} - \dfrac{S_{2h}^2}{\bar{N}_h}}{\dfrac{C_0}{2\sqrt{m}} + C_{1h}} \tag{7.10}$$

Equating the right-hand members of Eq. 7.9 and 7.10 and solving for $\bar{n}_h$, we obtain

$$\bar{n}_h = \bar{N}_h \sqrt{\frac{\dfrac{C_0}{2\sqrt{m}} + C_{1h}}{C_{2h}} \frac{S_{2h}^2}{S_{1h}^2 - \bar{N}_h S_{2h}^2}} \tag{7.11}$$

Substituting Eq. 7.4 and 7.6 into Eq. 7.11, we obtain the optimum $\bar{n}_h$ as given in Eq. 7.2. Now, if we let

$$\frac{\sum^L C_{1h} m_h + \sum^L C_{2h} m_h \bar{n}_h}{m} = b \tag{7.12}$$

then the cost equation may be written

$$C = C_0\sqrt{m} + bm$$

Considering the cost equation as a quadratic in $\sqrt{m}$, we obtain

$$\sqrt{m} = \frac{-C_0 + \sqrt{C_0^2 + 4Cb}}{2b} \tag{7.13}$$

and

$$\frac{C_0}{2\sqrt{m}} = d = \frac{b}{\sqrt{1 + \dfrac{4C}{C_0^2} b} - 1} \tag{7.6}$$

Now, solving Eq. 7.8 for m_h, we obtain

$$m_h = \frac{N_h S_{2h}}{\sqrt{\lambda} X \bar{n}_h \sqrt{C_{2h}}} = \frac{a_h}{\sqrt{\lambda}} \tag{7.14}$$

Substituting Eq. 7.11 for $\bar{n}_h$, we obtain a_h as given in Eq. 7.4. Further,

$$\sum^L m_h = m = \frac{\sum^L a_h}{\sqrt{\lambda}} \tag{7.15}$$

Hence, substituting $1/\sqrt{\lambda} = m/\sum^{L} a_h$ into Eq. 7.14, the optimum m_h is

$$m_h = \frac{ma_h}{\sum^{L} a_h} = \frac{C_0^2 a_h}{4d^2 \sum^{L} a_h} \text{ (from Eq. 7.6)} \tag{7.3}$$

and substituting $a_h/\sqrt{\lambda}$ for m_h, $N_h S_{2h}/X\sqrt{\lambda}\sqrt{C_{2h}}$ for $m_h \bar{n}_h$ from Eq. 7.14, and $\sum^{L} a_h/\sqrt{\lambda}$ for m from Eq. 7.15, into Eq. 7.12, we have b as given in Eq. 7.5.

b. Suppose that the cost of travel between psu's varies from one stratum to another and thus the cost function is expressed as follows:

$$C = \sum^{L} C_{0h}\sqrt{m_h} + \sum^{L} C_{1h} m_h + \sum^{L} C_{2h} m_h \bar{n}_h \tag{7.16}$$

The rel-variance is again of the form given in Eq. 6.4. Then the optimum values of m_h and $\bar{n}_h$ for a fixed total expenditure are as follows:

$$\text{opt. } m_h = \frac{C_{0h}^2}{4d_h^2} \qquad \text{(7.17 or I–7.11.10)}$$

$$\left.\begin{aligned} \text{opt. } \bar{n}_h &= \frac{N_h S_{2h}}{X a_h \sqrt{C_{2h}}} = \bar{N}_h \sqrt{\frac{S_{2h}^2}{S_{1h}^2 - \bar{N}_h S_{2h}^2} \frac{(C_{0h}/2\sqrt{m_h}) + C_{1h}}{C_{2h}}} \\ &= \frac{X_h W_h}{X a_h \sqrt{C_{2h}}} = \sqrt{\frac{W_h^2}{B_h^2 - W_h^2/\bar{N}_h} \frac{(C_{0h}/2\sqrt{m_h}) + C_{1h}}{C_{2h}}} \end{aligned}\right\} \text{(7.18 or I–7.11.11)}$$

where a_h and d_h are determined from the following iterative equations:

$$a_h = \frac{\frac{M_h}{X}\sqrt{S_{1h}^2 - \bar{N}_h S_{2h}^2}}{\sqrt{d_h + C_{1h}}} = \frac{\frac{X_h}{X}\sqrt{B_h^2 - W_h^2/\bar{N}_h}}{\sqrt{d_h + C_{1h}}} \qquad \text{(7.19 or I–7.11.8)}$$

and

$$\left.\begin{aligned} d_h &= \frac{C_{0h}}{2\sqrt{m_h}} \\ &= \frac{\left(\sum^{L} C_{1h} a_h + \sum^{L} \frac{N_h S_{2h}}{X}\sqrt{C_{2h}}\right) C_{0h}/\sqrt{a_h}}{\sqrt{\left(\sum^{L} C_{0h}\sqrt{a_h}\right)^2 + 4C\left(\sum^{L} C_{1h} a_h + \sum^{L} \frac{N_h S_{2h}}{X}\sqrt{C_{2h}}\right)} - \sum^{L} C_{0h}\sqrt{a_h}} \\ &= \frac{\left(\sum^{L} C_{1h} a_h + \sum^{L} \frac{X_h}{X} W_h \sqrt{C_{2h}}\right) C_{0h}/\sqrt{a_h}}{\sqrt{\left(\sum^{L} C_{0h}\sqrt{a_h}\right)^2 + 4C\left(\sum^{L} C_{1h} a_h + \sum^{L} \frac{X_h}{X} W_h \sqrt{C_{2h}}\right)} - \sum^{L} C_{0h}\sqrt{a_h}} \end{aligned}\right\} \text{(7.20 or I–7.11.9)}$$

The derivation of these formulas follows the same general steps as the derivation of the optimum values in Part a above and is left to the reader as an exercise.

8. Optimum values for a stratified sampling design with joint use of one- and two-stage sampling, variable sampling fractions, and a simple cost function (Vol. I, Ch. 7, Sec. 10). *To prove:* For a stratified sampling design with two-stage sampling in L_2 strata and one-stage sampling in L_1 strata and with the sampling fraction within first-stage units, n_{hi}/N_{hi}, equal to f_{2h}, then the optimum m_k, m_h, and $\bar{n}_h$, when the expected cost of the survey is

$$C = \sum_h^{L_2} m_h C_{1h} + \sum_h^{L_2} m_h \bar{n}_h C_{2h} + \sum_k^{L_1} m_k C'_k \quad \text{(8.1 or I-7.10.1)}$$

and the total expenditure is fixed, are

$$\text{opt. } m_k = \frac{M_k S_k}{\sqrt{C'_k}} a \quad \text{(8.2 or I-7.10.2)}$$

$$\text{opt. } m_h = \frac{M_h \sqrt{S_{1h}^2 - \bar{N}_h S_{2h}^2}}{\sqrt{C_{1h}}} a \quad \text{(8.3 or I-7.10.3)}$$

$$\text{opt. } \bar{n}_h = \bar{N}_h \sqrt{\frac{S_{2h}^2}{S_{1h}^2 - \bar{N}_h S_{2h}^2} \frac{C_{1h}}{C_{2h}}} \quad \text{(8.4 or I-7.10.4)}$$

$$a = C \Big/ \left(\sum_h^{L_2} M_h \sqrt{C_{1h}} \sqrt{S_{1h}^2 - \bar{N}_h S_{2h}^2} + \sum_h^{L_2} N_h S_{2h} \sqrt{C_{2h}} + \sum_k^{L_1} M_k S_k \sqrt{C'_k} \right) \quad \text{(8.5 or I-7.10.5)}$$

Proof. The rel-variance of the design is

$$V_r^2 \doteq \frac{1}{X^2} \sum_h^{L_2} \frac{M_h^2}{m_h} \frac{M_h - m_h}{M_h} S_{1h}^2 + \frac{1}{X^2} \sum_h^{L_2} \frac{N_h^2}{m_h \bar{n}_h} \frac{\bar{N}_h - \bar{n}_h}{\bar{N}_h} S_{2h}^2 + \frac{1}{X^2} \sum_k^{L_1} \frac{M_k^2}{m_k} \frac{M_k - m_k}{M_k} S_k^2 \quad \text{(8.6)}$$

where S_{1h}^2 and S_{2h}^2 are defined in Sec. 1 and S_k^2 is defined as S_{1h}^2 but over the L_1 strata. The Lagrangian F is

$$F = V_r^2 + \lambda \left(\sum^{L_2} C_{1h} m_h + \sum^{L_2} C_{2h} m_h \bar{n}_h + \sum^{L_1} C'_k m_k - C \right)$$

Then

$$\frac{\partial F}{\partial m_h} = -\frac{M_h^2 S_{1h}^2}{X^2 m_h^2} - \frac{N_h^2 S_{2h}^2}{X^2 m_h^2 \bar{n}_h} + \frac{N_h^2 S_{2h}^2}{X^2 m_h^2 \bar{N}_h} + \lambda(C_{1h} + C_{2h}\bar{n}_h) = 0 \quad (8.7)$$

$$\frac{\partial F}{\partial \bar{n}_h} = -\frac{N_h^2 S_{2h}^2}{X^2 m_h \bar{n}_h^2} + \lambda(C_{2h} m_h) = 0 \quad (8.8)$$

$$\frac{\partial F}{\partial m_k} = -\frac{M_k^2 S_k^2}{X^2 m_k^2} + \lambda(C_k') = 0 \quad (8.9)$$

From Eq. 8.8, we obtain

$$\frac{\lambda m_h^2 X^2}{N_h^2} = \frac{S_{2h}^2}{C_{2h}\bar{n}_h^2} \quad (8.10)$$

Multiplying Eq. 8.7 by m_h and subtracting Eq. 8.8 multiplied by $\bar{n}_h$, we obtain

$$\frac{\lambda m_h^2 X^2}{N_h^2} = \frac{\bar{N}_h^2 S_{1h}^2 - S_{2h}^2/\bar{N}_h}{C_{1h}} \quad (8.11)$$

Equating the right-hand members of Eq. 8.10 and 8.11 and solving for $\bar{n}_h$, we obtain the optimum $\bar{n}_h$ as given in Eq. 8.4. From Eq. 8.9

$$m_k = \frac{M_k S_k}{\sqrt{C_k'}} \frac{1}{X\sqrt{\lambda}} \quad (8.12)$$

Let $1/(X\sqrt{\lambda}) = a$ to obtain m_k as given in Eq. 8.2. Now, solving Eq. 8.8 for m_h, we have

$$m_h = \frac{N_h S_{2h}}{\bar{n}_h \sqrt{C_{2h}}} \frac{1}{X\sqrt{\lambda}} \quad (8.13)$$

and, substituting $1/(X\sqrt{\lambda}) = a$ and $\bar{n}_h$ from Eq. 8.4, we obtain the optimum m_h as given in Eq. 8.3.

Finally, substituting Eq. 8.2, 8.3, and 8.4 into the cost function and solving for a, we obtain a as given in Eq. 8.5.

REFERENCES

(1) W. G. Cochran, "The Use of Analysis of Variance in Enumeration by Sampling," *J. Amer. Stat. Assn.*, **34** (1939), 492–510.

(2) R. J. Jessen, "Statistical Investigation of a Sample Survey for Obtaining Farm Facts," *Iowa Agr. Exp. Stat. Res. Bull.* 304 (1942).

CHAPTER 8

Control of Variation in Size of Cluster in Estimating Totals, Averages, or Ratios

DERIVATIONS, PROOFS, AND SOME EXTENSIONS OF THEORY FOR CH. 8 OF VOL. I*

NOTE. When primary sampling units vary in size, i.e., in the number of elementary units or listing units that they contain, some methods for control of variation in the size of cluster in the selection of a sample and in the estimation are sometimes useful. The derivations in this chapter relate to this problem. Ordinarily, the use of some method for controlling the variation in size of cluster is much more important in estimating totals than in estimating ratios, although many of the results for which proofs are needed and are given in this chapter deal with the problem of estimating ratios.

1. Sample estimates and their variances for a two-stage sampling design when first-stage units are selected with varying probabilities (Vol. I, Ch. 8, Sec. 14). *To prove:* If first-stage units are selected with varying probabilities, with replacement, and with any second-stage sampling fractions,

$$x' = \frac{1}{m}\sum^{m}\frac{x_i'}{P_i} \tag{1.1}$$

is an unbiased estimate of the population total X, where $x_i' = N_i x_i/n_i$ is an unbiased estimate of the psu total X_i, P_i is the probability of selecting the ith psu on a single draw, m is the number of psu's in the sample, N_i is the total number of second-stage units in the ith psu, and n_i is the number subsampled from the ith psu. The ratio $r = x'/y'$, where x' and y' are defined by Eq. 1.1, is an estimate of $R = X/Y$.

The rel-variance of r (and of x' as a special case) is

$$V_r^2 \doteq V_{x'}^2 + V_{y'}^2 - 2V_{x'y'} \tag{1.2}$$

---

* Appropriate references to Vol. I are shown in parentheses after section or subsection headings. The number following I– after some equations gives the chapter, section, and number of that particular equation in Vol. I.

where

$$V_{x'y'} = \frac{\sum^{M} P_i \left(\frac{X_i}{P_i} - X\right)\left(\frac{Y_i}{P_i} - Y\right)}{mXY} + \frac{\sum^{M} \frac{N_i^2}{P_i} \frac{N_i - n_i}{N_i n_i} S_{iXY}}{mXY} \quad \text{(1.3 or I–8.14.7)}$$

with

$$S_{iXY} = \frac{\sum^{N_i}(X_{ij} - \bar{\bar{X}}_i)(Y_{ij} - \bar{\bar{Y}}_i)}{N_i - 1} \quad (1.4)$$

$$V_{x'}^2 = V_{x'x'} \quad \text{and} \quad V_{y'}^2 = V_{y'y'}$$

Remark. The estimate given by Eq. 1.1 would be unbiased for samples drawn with varying probabilities, P_i for the ith psu, whether or not the psu's were selected with replacement. The variance, given by Eq. 1.2 and 1.3, holds only for sampling with replacement but may be a satisfactory approximation for sampling without replacement, especially if no P_i is large relative to 1, and if m/M is small.

Proof. By Theorem 6, Ch. 3 (p. 49),

$$Ex' = E\frac{1}{m}\sum_i^m \frac{N_i}{P_i n_i}\sum_j^{n_i} x_{ij}$$

$$= E\frac{1}{m}\sum_i^m \frac{N_i}{P_i n_i}\sum_j^{n_i} E_i x_{ij} \quad (1.5)$$

and since

$$E_i x_{ij} = \frac{1}{N_i}\sum_j^{N_i} X'_{ij} = \frac{X'_i}{N_i}$$

where X'_{ij} and X'_i are used to indicate the fact that X'_{ij} and X'_i are random variables depending on the result of the ith selection of first-stage unit,

$$Ex' = E\frac{1}{m}\sum_i^m \frac{X'_i}{P_i} = \frac{1}{m}\sum_i^m E\frac{X'_i}{P_i}$$

$$= \frac{1}{m}\sum_i^m \sum_i^M P_i \frac{X_i}{P_i} = X$$

Similarly, $Ey' = Y$.

Consider now $V_{x'y'} = \sigma_{x'y'}/XY$. By the corollary to Theorem 17, Ch. 3 (p. 68), with $K = 2$ and $u = x'$, $w = y'$, we may write

$$\sigma_{x'y'} = E\sigma_{x'y'|[1]} + \sigma_{E(x'|[1])E(y'|[1])} \quad (1.6)$$

where [1] refers to a fixed set of first-stage units in the sample. Consider the first term in the right-hand member of Eq. 1.6. Since for a fixed set of m first-stage units the sampling from one of the units is independent of the sampling from any other first-stage unit in the set, then, by the corollary to Theorem 12, Ch. 3 (p. 58),

$$\sigma_{x'y'|[1]} = \frac{1}{m^2} \sum_i^m \frac{1}{P_i^2} \sigma_{x_i'y_i'|[1]}$$

Since within the ith selected first-stage unit the n_i units are a simple random sample from the N_i units, then by Sec. 3, Ch. 4,

$$\sigma_{x_i'y_i'|[1]} = N_i^2 \frac{N_i - n_i}{N_i n_i} S_{iXY}$$

Hence,

$$E\sigma_{x'y'|[1]} = \frac{1}{m^2} \sum_i^m E \frac{1}{P_i^2} \sigma_{x_i'y_i'|[1]}$$

$$= \frac{1}{m^2} \sum_i^m \sum_i^M \frac{P_i}{P_i^2} N_i^2 \frac{N_i - n_i}{N_i n_i} S_{iXY} \tag{1.7}$$

which when divided by XY is equal to the second term in the right-hand member of Eq. 1.3.

Consider now the second term in the right-hand member of Eq. 1.6, namely, $\sigma_{E(x'|[1])E(y'|[1])}$.

By Theorem 6, Ch. 3 (p. 49),

$$E(x'|[1]) = \frac{1}{m} \sum_i^m E_i \frac{1}{P_i} \frac{N_i}{n_i} \sum_j^{n_i} x_{ij} = \frac{1}{m} \sum_i^m \frac{X_i'}{P_i}$$

Similarly,

$$E(y'|[1]) = \frac{1}{m} \sum_i^m \frac{Y_i'}{P_i}$$

Therefore,

$$\sigma_{E(x'|[1])E(y'|[1])} = \sigma_{\frac{1}{m}\sum_i^m \frac{X'_i}{P_i}, \frac{1}{m}\sum_i^m \frac{Y'_i}{P_i}} \tag{1.8}$$

and since the m first-stage units are selected with replacement, X_i'/P_i is independent of X_j'/P_j, and by the corollary to Theorem 12, Ch. 3 (p. 58),

$$\text{Eq. } 1.8 = \frac{1}{m^2} \sum_i^m \sigma_{\frac{X'_{(i)}}{P_{(i)}}, \frac{Y'_{(i)}}{P_{(i)}}} \tag{1.9}$$

Since the probability of selecting the ith first-stage unit is P_i,

$$E \frac{X_i'}{P_i} = \sum_i^M P_i \frac{X_i}{P_i} = X$$

Similarly,

$$E \frac{Y_i'}{P_i} = Y$$

Therefore,

$$\sigma_{\frac{X'_i}{P_i}, \frac{Y'_i}{P_i}} = E\left(\frac{X'_i}{P_i} - X\right)\left(\frac{Y'_i}{P_i} - Y\right)$$

$$= \sum_i^M P_i \left(\frac{X_i}{P_i} - X\right)\left(\frac{Y_i}{P_i} - Y\right) \tag{1.10}$$

Equation 1.10, substituted into Eq. 1.9 and divided by XY, is equal to the first term in the right-hand member of Eq. 1.3.

2. Determination of optimum probabilities for a two-stage sampling design in which the first-stage units are selected with replacement, the sample is self-weighting, and a simple cost function is used (Vol. I, Ch. 8, Sec. 14). *To prove:* Consider a subsampling design in which primary units are sampled with replacement and a subsample of listing units is selected within sample psu's. Suppose that the population consists of L size classes with M_h primary units in the hth size class and N_{hi} listing units in the hith psu, and that P_h is the probability of selection of a primary unit in the hth size class on a single draw. Then the estimate $r = x'/y'$, where x' and y' are given in Eq. 1.1 may be written

$$r = \frac{x'}{y'} = \frac{\dfrac{1}{m}\sum_h^L \sum_i^{m_h} \dfrac{N_{hi}}{P_h} \dfrac{\sum_j^{n_{hi}} x_{hij}}{n_{hi}}}{\dfrac{1}{m}\sum_h^L \sum_i^{m_h} \dfrac{N_{hi}}{P_h} \dfrac{\sum_j^{n_{hi}} y_{hij}}{n_{hi}}} \tag{2.1}$$

where $m = \sum_h^L m_h$ is the number of psu's in the sample, and n_{hi} is the number of elements included in the sample from the hith psu if it is drawn.

If also the sample is self-weighting, i.e., $P_h(n_{hi}/N_{hi}) = k$, then $f = mk$ is the over-all sampling ratio, $r = x/y$, and the rel-variance of r becomes

$$V_r^2 \doteq \frac{1}{mX^2}\left[\sum_h^L \frac{M_h}{P_h}(\Delta_h^2 - \bar{N}_h S_{2h}^2) + \frac{NS_2^2}{k}\right] \tag{2.2}$$

where

$$\Delta_h^2 = \frac{\sum_i^{M_h} Y_{hi}^2 \left(\dfrac{X_{hi}}{Y_{hi}} - \dfrac{X}{Y}\right)^2}{M_h} \tag{2.3}$$

$$S_{2h}^2 = \frac{\sum_i^{M_h} N_{hi} S_{2hi}^2}{M_h \bar{N}_h} \tag{2.4}$$

with S_{2hi}^2 defined by Eq. 1.5, Ch. 7, and where

$$S_2^2 = \frac{\sum_h^L M_h \bar{N}_h S_{2h}^2}{N} \quad \text{and} \quad \bar{N}_h = \frac{1}{M_h} \sum_i^{M_h} N_{hi} \tag{2.5}$$

The optimum values of P_h, m, and k subject to a fixed total cost and subject to the condition $\sum^L M_h P_h = 1$, where the cost function is

$$C = C_1'm + C_1''m\sum^L P_h M_h \bar{N}_h + C_2 mkN \quad \text{(2.6 or I–8.14.8)}$$

are given by

$$P_h = \frac{\sqrt{\dfrac{\Delta_h^2 - \bar{N}_h S_{2h}^2}{C_1' + C_1''\bar{N}_h}}}{\sum^L M_h \sqrt{\dfrac{\Delta_h^2 - \bar{N}_h S_{2h}^2}{C_1' + C_1''\bar{N}_h}}} \tag{2.7}$$

$$k = \frac{\sqrt{S_2^2/C_2}}{\sum^L M_h \sqrt{\dfrac{\Delta_h^2 - \bar{N}_h S_{2h}^2}{C_1' + C_1''\bar{N}_h}}} \tag{2.8}$$

$$m = C/(C_1' + C_1''\sum^L P_h M_h \bar{N}_h + C_2 kN) \tag{2.9}$$

For a particular fixed system of probabilities the optimum k is

$$k = \sqrt{\frac{S_2^2 \left(C_1' + C_1''\sum_h^L P_h M_h \bar{N}_h\right)}{C_1''\sum_h^L \dfrac{1}{P_h} M_h(\Delta_h^2 - \bar{N}_h S_{2h}^2)}} \tag{2.10}$$

and the optimum m is as given by Eq. 2.9.

Proof. If the estimate is given by Eq. 2.1, the rel-variance is (from Sec. 1) $V_r^2 \doteq V_{x'}^2 + V_{y'}^2 - 2V_{x'y'}$, with terms defined as follows:

$$V_{x'y'} = \frac{\sum_h^L \sum_i^{M_h} P_h \left(\dfrac{X_{hi}}{P_h} - X\right)\left(\dfrac{Y_{hi}}{P_h} - Y\right)}{mXY} + \frac{\sum_h^L \sum_i^{M_h} \dfrac{N_{hi}^2}{P_h} \dfrac{N_{hi} - n_{hi}}{N_{hi} n_{hi}} S_{2hiXY}}{mXY} \tag{2.11}$$

Substituting X for Y in Eq. 2.11, we obtain $V_{x'}^2$, and substituting Y for X, we obtain $V_{y'}^2$. V_r^2 may also be written in the following form:

$$V_r^2 \doteq \frac{1}{mX^2}\left[\sum_h^L \frac{M_h}{P_h}\Delta_h^2 + \sum_h^L\sum_i^{M_h}\frac{N_{hi}^2}{P_h}\frac{N_{hi}-n_{hi}}{N_{hi}n_{hi}}S_{2hi}^2\right] \tag{2.12}$$

Substituting $n_{hi} = kN_{hi}/P_h$, for a self-weighting sample, we obtain V_r^2 as given in Eq. 2.2.

To obtain the values of P_h, m, and k which make V_r^2 a minimum subject to a fixed total cost and $\sum_h^L P_hM_h = 1$, we set up the Lagrangian F:

$$F = V_r^2 + \lambda_1\left(C_1'm + C_1''m\sum_h^L P_hM_h\bar{N}_h + C_2mkN - C\right) + \lambda_2\left(\sum_h^L M_hP_h - 1\right)$$

Then the solution to the equations $\partial F/\partial P_h = 0$, $\partial F/\partial k = 0$, $\partial F/\partial m = 0$, $\partial F/\partial \lambda_1 = 0$, $\partial F/\partial \lambda_2 = 0$ will give the optimum values of P_h, m, and k.

$$\frac{\partial F}{\partial P_h} = -\frac{1}{P_h^2}\frac{M_h(\Delta_h^2 - \bar{N}_hS_{2h}^2)}{mX^2} + \lambda_1C_1''mM_h\bar{N}_h + \lambda_2M_h = 0 \tag{2.13}$$

$$\frac{\partial F}{\partial k} = -\frac{1}{k^2}\frac{NS_2^2}{mX^2} + \lambda_1C_2mN = 0 \tag{2.14}$$

$$\frac{\partial F}{\partial m} = -\frac{1}{m^2X^2}\left[\sum_h^L \frac{M_h(\Delta_h^2 - \bar{N}_hS_{2h}^2)}{P_h} + \frac{NS_2^2}{k}\right]$$

$$+ \lambda_1\left[C_1' + C_1''\sum_h^L P_hM_h\bar{N}_h + C_2kN\right] = 0 \tag{2.15}$$

$\partial F/\partial \lambda_1 = 0$ is the cost condition, and $\partial F/\partial \lambda_2 = 0$ is the condition $\sum^L P_hM_h = 1$. From Eq. 2.14,

$$\lambda_1 = \frac{S_2^2}{C_2m^2X^2k^2}$$

Multiplying Eq. 2.13 by P_h/m, summing to L, and subtracting Eq. 2.15, we obtain

$$\lambda_2 = mC_1'\lambda_1 = \frac{C_1'}{C_2}\frac{S_2^2}{mX^2k^2}$$

Substituting these values of λ_1 and λ_2 into Eq. 2.13 and solving for P_h^2, we obtain

$$P_h^2 = \frac{(\Delta_h^2 - \bar{N}_hS_{2h}^2)}{C_1' + C_1''\bar{N}_h}\frac{C_2k^2}{S_2^2} \tag{2.16}$$

$$\sum^{L} M_h P_h = 1 = \sum^{L} M_h \sqrt{\frac{\Delta_h^2 - \bar{N}_h S_{2h}^2}{C_1' + C_1'' \bar{N}_h}} \frac{\sqrt{C_2}}{S_2} k$$

and

$$k = \frac{1}{\sum^{L} M_h \sqrt{\frac{\Delta_h^2 - \bar{N}_h S_{2h}^2}{C_1' + C_1'' \bar{N}_h}} \frac{\sqrt{C_2}}{S_2}}$$

Hence,

$$P_h = \frac{\sqrt{\frac{\Delta_h^2 - \bar{N}_h S_{2h}^2}{C_1' + C_1'' \bar{N}_h}}}{\sum^{L} M_h \sqrt{\frac{\Delta_h^2 - \bar{N}_h S_{2h}^2}{C_1' + C_1'' \bar{N}_h}}} \tag{2.7}$$

Substituting Eq. 2.7 into Eq. 2.16 and solving for k, we obtain the optimum k as given in Eq. 2.8. And, solving the cost equation for m, we obtain the optimum m as given in Eq. 2.9.

For a particular fixed system of probabilities, we have the Lagrange equations 2.14 and 2.15, and the cost condition. Solving these equations, we obtain the optimum k as given in Eq. 2.10. This solution is straightforward and is left to the reader.

3. Optimum values for a two-stage stratified sampling design with a uniform over-all sampling fraction and a simple cost function, and a comparison of optimum probabilities with stratification by size as a control on variation in size of psu (Vol. I, Ch. 8, Sec. 12 and 14). *To prove:* For a two-stage stratified design with a uniform over-all sampling fraction the optimum $\bar{n}_h$ and m_h when the cost function is of the simple form

$$C = C_1' m + C_1'' \sum^{L} m_h \bar{N}_h + C_2 f N \qquad \text{(3.1 or I–8.12.1)}$$

and the total expenditure is fixed, are

$$\text{opt. } \bar{n}_h = \bar{N}_h S_2 \sqrt{\frac{C_1' + C_1'' \bar{N}_h}{C_2 (S_{1h}^2 - \bar{N}_h S_{2h}^2)}} \qquad \text{(3.2 or I–8.12.2)}$$

and

$$\text{opt. } m_h = M_h f \frac{\bar{N}_h}{\bar{n}_h} = \frac{M_h \sqrt{S_{1h}^2 - \bar{N}_h S_{2h}^2}}{\sqrt{C_1' + C_1'' \bar{N}_h}} \frac{f \sqrt{C_2}}{S_2} \qquad \text{(3.3 or I–8.12.5)}$$

where

$$S_2^2 = \frac{\sum^{L} N_h S_{2h}^2}{N} \qquad \text{(3.4 or I–8.12.3)}$$

and the optimum f is

$$\text{opt.} f = \frac{CS_2}{\sqrt{C_2}\left[\sum^L \sqrt{C_1' + C_1''\bar{N}_h}\, M_h \sqrt{S_{1h}^2 - \bar{N}_h S_{2h}^2} + \sqrt{C_2}\, NS_2\right]}$$

(3.5 or I–8.12.6)

Proof. When the sample is self-weighting, i.e., $f = f_{1h} f_{2h}$, the rel-variance given by Eq. 1.9, Ch. 7, becomes

$$V_r^2 = \frac{1}{X^2} \sum^L \left(\frac{f_{2h}}{f} - 1\right) M_h S_{1h}^2 + \frac{1}{X^2} \sum^L \frac{1}{f}(1 - f_{2h}) N_h S_{2h}^2 \qquad (3.6)$$

where S_{1h}^2 is defined by Eq. 1.3, Ch. 7, and S_{2h}^2 is defined by Eq. 1.10, Ch. 7. The cost function may be written

$$C = \sum^L (C_1' + C_1''\bar{N}_h) M_h \frac{f}{f_{2h}} + C_2 f N \qquad (3.7)$$

To obtain the values of f_{2h} and f which minimize V_r^2 subject to a fixed total expenditure set up the Lagrangian F:

$$F(f_{2h}, f, \lambda) = V_r^2 + \lambda \left\{\sum^L (C_1' + C_1''\bar{N}_h) M_h \frac{f}{f_{2h}} + C_2 f N - C\right\}$$

Then the solution of the equations $\partial F/\partial f_{2h} = 0$, $\partial F/\partial f = 0$, and the cost equation 3.7 will give the optimum values as shown below:

$$\frac{\partial F}{\partial f_{2h}} = \frac{M_h S_{1h}^2}{X^2 f} - \frac{N_h S_{2h}^2}{X^2 f} - \lambda \frac{(C_1' + C_1''\bar{N}_h) M_h f}{f_{2h}^2} = 0 \qquad (3.8)$$

$$\frac{\partial F}{\partial f} = -\frac{\sum^L M_h S_{1h}^2 f_{2h}}{X^2 f^2} - \frac{\sum^L N_h S_{2h}^2}{X^2 f^2} + \frac{\sum^L N_h S_{2h}^2 f_{2h}}{X^2 f^2} + \lambda \left\{\frac{\sum^L (C_1' + C_1''\bar{N}_h) M_h}{f_{2h}} + C_2 N\right\} = 0 \qquad (3.9)$$

Multiplying Eq. 3.8 by f_{2h}, summing to L, and adding Eq. 3.9 multiplied by f, we obtain

$$\lambda f^2 = \frac{\sum^L N_h S_{2h}^2}{C_2 N X^2} \qquad (3.10)$$

From Eq. 3.8

$$\lambda f^2 = \frac{(S_{1h}^2 - \bar{N}_h S_{2h}^2) f_{2h}^2}{X^2 (C_1' + C_1''\bar{N}_h)} \qquad (3.11)$$

Equate the right-hand members of Eq. 3.10 and 3.11 and solve for the optimum expected size of ultimate cluster in the hth stratum, $\bar{n}_h = \bar{N}_h f_{2h}$, to obtain Eq. 3.2.

Now, since $m_h = M_h f / f_{2h}$, substitute Eq. 3.2 for $\bar{N}_h f_{2h}$ to obtain the optimum m_h given in Eq. 3.3. Substitute Eq. 3.3 into the cost function for $m = \sum^{L} m_h$ and for m_h and solve for f to obtain the optimum f given in Eq. 3.5.

Comparison of optimum probabilities with stratification by size. If the stratification is by size and the strata are the size classes used in Sec. 2, then S_{2h}^2 and S_2^2 in Sec. 2 are the same as S_{2h}^2 and S_2^2 in this section. We saw in Sec. 2 that the over-all sampling fraction is $f = mP_h\bar{n}_h/\bar{N}_h$. Hence mP_h is comparable to m_h/M_h. Note from Eq. 2.7 and Eq. 3.3 that mP_h and m_h/M_h are proportionate to quantities that differ only in that Δ_h^2 in Eq. 2.7 replaces S_{1h}^2 in Eq. 3.3. The difference between Δ_h^2 and S_{1h}^2 is approximately

$$\bar{Y}_h^2 \left(\frac{X_h}{Y_h} - \frac{X}{Y}\right)^2$$

so that the two expressions are approximately the same whenever the X_h/Y_h do not vary a great deal from stratum to stratum.

Proof. From Eq. 2.3,

$$\Delta_h^2 = \frac{1}{M_h}\sum_i^{M_h} Y_{hi}^2 \left(\frac{X_{hi}}{Y_{hi}} - \frac{X}{Y}\right)^2 = \frac{1}{M_h}\sum_i^{M_h}(X_{hi} - RY_{hi})^2$$

$$= \frac{1}{M_h}\sum_i^{M_h}(X_{hi}^2 + R^2Y_{hi}^2 - 2RX_{hi}Y_{hi}) \tag{3.12}$$

with $R = X/Y$. From Eq. 1.3, Ch. 7,

$$S_{1h}^2 = S_{1hX}^2 + R^2S_{1hY}^2 - 2RS_{1hXY}$$

$$= \frac{1}{(M_h - 1)}\left(\sum_i^{M_h} X_{hi}^2 - M_h\bar{X}_h^2 + R^2\sum_i^{M_h} Y_{hi}^2 - R^2M_h\bar{Y}_h^2 - 2R\sum_i^{M_h} X_{hi}Y_{hi} + 2RM_h\bar{X}_h\bar{Y}_h\right) \tag{3.13}$$

For M_h large, so that the assumption $M_h \doteq (M_h - 1)$ is valid,

$$\Delta_h^2 - S_{1h}^2 = (\bar{X}_h - R\bar{Y}_h)^2 = \bar{Y}_h^2\left(\frac{X_h}{Y_h} - \frac{X}{Y}\right)^2$$

4. Comparison of $\hat{V}^2$ (Eq. 5.4 of Ch. 6) and V^2 (Eq. 5.11 of Ch. 6) (Vol. I, Ch. 8, Sec. 1 and 11, also Vol. I, Ch. 6, Sec. 8). *To prove:* For the class of populations described in Sec. 8, Ch. 6, we may expect that $(\hat{V}^2/V^2) > 1$, where $\hat{V}^2$ is given by Eq. 5.4 of Ch. 6 and V^2 is given by Eq. 5.11 of Ch. 6.

Proof. $\hat{V}^2$ and V^2 can be restated in the forms

$$\hat{V}^2 = \frac{\sum_i^M N_i^2 \bar{\bar{Z}}_i^2}{M\bar{N}^2\bar{\bar{X}}^2} + \frac{\frac{\bar{N}-1}{\bar{N}} \sum_i^M \frac{N_i}{N_i - 1} \sum_j^{N_i} (Z_{ij} - \bar{\bar{Z}}_i)^2}{N\bar{\bar{X}}^2} \tag{4.1}$$

$$V^2 = \frac{\sum_i^M N_i \bar{\bar{Z}}_i^2}{M\bar{N}\bar{\bar{X}}^2} + \frac{\sum_i^M \sum_j^{N_i} (Z_{ij} - \bar{\bar{Z}}_i)^2}{N\bar{\bar{X}}^2} \tag{4.2}$$

where for V^2 we have assumed that N is large so that $N/(N-1)$ is very close to 1, and

$$Z_{ij} = X_{ij} - RY_{ij} \tag{4.3}$$

$$Z_i = \sum_j^{N_i} Z_{ij} = X_i - RY_i \tag{4.4}$$

and

$$\bar{\bar{Z}}_i = \frac{Z_i}{N_i} \tag{4.5}$$

The last term in Eq. 4.1 will be nearly equivalent to the last term in Eq. 4.2 provided the N_i are moderately large (it may be about equal under less stringent conditions). When these last terms are about equal, then

$$\hat{V}^2 - V^2 \doteq \frac{\sum^M N_i^2 \bar{\bar{Z}}_i^2 - \bar{N} \sum^M N_i \bar{\bar{Z}}_i^2}{M\bar{N}^2\bar{\bar{X}}^2}$$

$$= \frac{\sigma_{N_i, N_i \bar{\bar{Z}}_i^2}}{M\bar{N}^2\bar{\bar{X}}^2} \tag{4.6}$$

where $\sigma_{N_i, N_i \bar{\bar{Z}}_i^2}$ is the covariance of N_i and $N_i\bar{\bar{Z}}_i^2$. For the class of populations described in Sec. 8, Ch. 6, this covariance is positive, and thus for many common sampling problems

$$\frac{\hat{V}^2}{V^2} > 1$$

Illustrations are given in Case Study D of Ch. 12, Vol. I. (Compare, also, Sec. 2, Ch. 9, where it is shown that sampling with probability proportionate to size, under the same conditions, gains over sampling with equal probability.)

5. Effect of variation in size of cluster in estimating totals (Vol. I, Ch. 8, Sec. 4). Consider a one- or more stage cluster sample design with a uniform over-all sampling fraction, f, and with m primary units included in the sample; it is desired to estimate X. If a simple unbiased estimate, x', is used, where

$$x' = \frac{1}{f}x \tag{5.1}$$

and $x = \sum^{m} x_i$ is the aggregate value of the X-characteristic for the units in the sample, then,

$$V_{x'}^2 \doteq V_r^2 + V_n^2 + 2\rho_{rn}V_rV_n \qquad \text{(5.2 or I–8.4.1)}$$

where $n = \sum^{m} n_i$ is the number of elementary units in the sample, or some other aggregate measure of size associated with the units in the sample, and where V_r^2 is the rel-variance of x/n, V_n^2 is the rel-variance of n, and ρ_{rn} is the coefficient of correlation of r and n.

Exercises

5.1. Show that Eq. 5.2 holds approximately and that the last two terms become zero if the sample selection is made in such a manner that, if two or more stages of sampling are used, the first-stage units are selected with probability proportionate to N_i, the size of the ith unit.

5.2. Let $n_i = f_2N_i$, where N_i is the size of the ith first-stage unit, and where the first-stage units are a simple random sample. Show that the last two terms of Eq. 5.2 vanish if the estimate $(x/n)N$ is used.

Remark. The importance of the relationship given by Eq. 5.2 is that it separates the rel-variance of an estimated total into a component due to the variation in size of cluster and a component representing the variance that would arise if the problem were to estimate the ratio x/n. The variation in size of cluster usually has much less effect on the variance of a ratio than on an estimated total based on an expansion of the sample by the reciprocal of the sampling fraction, and therefore the latter two terms in Eq. 5.2 represent the principal contribution of variation in size of cluster. The last term will often be small, in which case the contribution of the variation in size of cluster is given approximately by V_n^2.

REFERENCES

(1) U.S. Bureau of the Census, Sampling Staff, *A Chapter in Population Sampling*, U.S. Government Printing Office, Washington, D.C., 1947.

(2) W. G. Cochran, "Sampling Theory When the Sampling Units Are of Unequal Size," *J. Amer. Stat. Assn.*, **37** (1942), 199–212.

(3) M. H. Hansen and W. N. Hurwitz, "On the Determination of Optimum Probabilities in Sampling," *Annals Math. Stat.*, **20** (1949), 426–432.

(4) D. G. Horvitz and D. J. Thompson, "A Generalization of Sampling without Replacement from a Finite Universe," *J. Amer. Stat. Assn.*, **47** (1952), 663–685.

(5) Hiroshi Midzuno, "An Outline of the Theory of Sampling Systems," *Annals Inst. Stat. Math. (Japan)*, **1** (1950), 149–156.

CHAPTER 9

Multi-stage Sampling with Large Primary Sampling Units

DERIVATIONS, PROOFS, AND SOME EXTENSIONS OF THEORY FOR CH. 9 OF VOL. I*

NOTE. So far as the theory is concerned, there is no distinction in multi-stage sampling whether the primary sampling units are large or small. However, some principles and methods become more important with large psu's, and these are emphasized in this chapter. They include substratification in sampling second-stage units, extensive use of varying probabilities in the selection of primary units, inclusion of a small number of primary units in the sample per primary stratum, the determination of optimum sizes of strata, and allowance for travel within psu's by a separate term in the cost function.

Of course, the theory introduced here for large psu's is applicable to any problem in which the principles are applied, whatever the size of the psu's.

Some notation used in this chapter.

X_{hiajk} = value of the X-characteristic for the *hiajk*th third-stage unit *in the population*; i.e., X_{hiajk} is the value for the *k*th third-stage unit in the *j*th second-stage unit in the *a*th substratum in the *i*th primary unit in the *h*th primary stratum.

x_{hiajk} = value of the X-characteristic for the *hiajk*th third-stage unit *in the sample.*

Y_{hiajk} and y_{hiajk} represent similar values for a Y-characteristic.

Numbers of sampling units and strata are as follows:

	IN THE SAMPLE	IN THE POPULATION
Number of third-stage units in *hiaj*th second-stage unit	q_{hiaj}	Q_{hiaj}
Number of second-stage units in *a*th substratum of *hi*th primary unit	n_{hia}	N_{hia}

* Appropriate references to Vol. I are shown in parentheses after section or subsection headings. The number following I– after some equations gives the chapter, section, and number of that particular equation in Vol. I.

	In the sample	In the population
Number of substrata in *hi*th primary unit	D_{hi}	D_{hi}
Number of primary units in *h*th stratum	m_h	M_h
Number of primary strata	L	L

With the notation above the following additional notation is defined:

In the sample	In the population
$x_{hiaj} = \sum_k^{q_{hiaj}} x_{hiajk}$	$X_{hiaj} = \sum_k^{Q_{hiaj}} X_{hiajk}$
$x_{hia} = \sum_j^{n_{hia}} x_{hiaj}$	$X_{hia} = \sum_j^{N_{hia}} X_{hiaj}$
$\cdot$	$\cdot$
$\cdot$	$\cdot$
$\cdot$	$\cdot$
$x = \sum_h^L x_h$	$X = \sum_h^L X_h$
$q_{hia} = \sum_j^{n_{hia}} q_{hiaj}$	$Q_{hia} = \sum_j^{N_{hia}} Q_{hiaj}$
$\cdot$	$\cdot$
$\cdot$	$\cdot$
$\cdot$	$\cdot$
$q = \sum_h^L q_h$	$Q = \sum_h^L Q_h$
$m = \sum_h^L m_h$	$M = \sum_h^L M_h$

Average values per third-stage unit

$\bar{\bar{\bar{x}}}_{hiaj} = x_{hiaj}/q_{hiaj}$	$\bar{\bar{\bar{X}}}_{hiaj} = X_{hiaj}/Q_{hiaj}$
$\cdot$	$\cdot$
$\cdot$	$\cdot$
$\cdot$	$\cdot$
$\bar{\bar{\bar{x}}} = x/q$	$\bar{\bar{\bar{X}}} = X/Q$

Average values per second-stage unit

$\bar{\bar{x}}_{hia} = x_{hia}/n_{hia}$	$\bar{\bar{X}}_{hia} = X_{hia}/N_{hia}$
$\cdot$	$\cdot$
$\cdot$	$\cdot$
$\cdot$	$\cdot$
$\bar{\bar{x}} = x/n$	$\bar{\bar{X}} = X/N$

IN THE SAMPLE	IN THE POPULATION
$\bar{\bar{q}}_{hia} = q_{hia}/n_{hia}$	$\bar{\bar{Q}}_{hia} = Q_{hia}/N_{hia}$
.	.
.	.
.	.
$\bar{\bar{q}} = q/n$	$\bar{\bar{Q}} = Q/N$

Average values per primary unit

$\bar{x}_h = x_h/m_h$	$\bar{X}_h = X_h/M_h$
$\bar{x} = x/m$	$\bar{X} = X/M$
$\bar{n} = n/m$	$\bar{N} = N/M$

Average values per primary stratum

	$\hat{X} = X/L$
$\hat{m} = m/L$	$\hat{M} = M/L$
	$\hat{N} = N/L$

Ratios

$r = x/y$	$R = X/Y$
$r_h = x_h/y_h$	$R_h = X_h/Y_h$
$r_{hi} = x_{hi}/y_{hi}$	$R_{hi} = X_{hi}/Y_{hi}$

(If no substrata are used the a is omitted, and if no primary strata are used the h is omitted.)

Simple unbiased estimates of totals are

x' is estimate of X,
x'_h is estimate of X_h,
x'_{hi} is estimate of X_{hi}, etc.

See Sec. 1 for further definitions of these estimates.

P_{hi} = probability of drawing ith psu from hth stratum in a single draw (i.e., for a sample of 1 psu).

Sampling fractions are

$f_{hiaj} = q_{hiaj}/Q_{hiaj}$ is the third-stage sampling fraction.

$f_{hia} = n_{hia}/N_{hia}$ is the second-stage sampling fraction in the ath substratum.

f is a uniform over-all sampling fraction.

$f = m_h P_{hi} f_{hia} f_{hiaj}$ for three-stage sampling with a uniform over-all sampling fraction.

***1. The rel-variance of a ratio estimate for a multi-stage stratified sampling design** (Vol. I, Ch. 9, Sec. 14 and 26). *a. General case.* Assume that we have a population consisting of L primary strata, M_h psu's in the hth primary stratum, and D_{hi} substrata in the ith psu of the hth stratum. Then an estimate of $R = X/Y$ is given by

$$r = \frac{x'}{y'} = \frac{\sum_h^L \frac{1}{m_h} \sum_i^{m_h} \frac{1}{P_{hi}} x'_{hi}}{\sum_h^L \frac{1}{m_h} \sum_i^{m_h} \frac{1}{P_{hi}} y'_{hi}} \tag{1.1}$$

where m_h is the number of first-stage units included in the sample from the hth stratum, P_{hi} is the probability of selection of the ith psu of the hth stratum on a single draw, and x'_{hi} and y'_{hi} are unbiased estimates of X_{hi} and Y_{hi}. Assume further that the first-stage units are selected with replacement but that the second- and subsequent stage units are selected without replacement.

To prove: We shall show that

$$V_r^2 \doteq V_{x'}^2 + V_{y'}^2 - 2V_{x'y'} \tag{1.2}$$

with

$$V_{x'y'} = \frac{\sum_h^L \frac{1}{m_h} S_{hXY}}{XY} + \frac{\sum_h^L \frac{1}{m_h} \sum_i^{M_h} \frac{1}{P_{hi}} \sigma_{x'_{hi}y'_{hi}}}{XY} \tag{1.3}$$

$$V_{x'}^2 = V_{x'x'} \quad \text{and} \quad V_{y'}^2 = V_{y'y'}$$

where

$$S_{hXY} = \sum_i^{M_h} P_{hi} \left(\frac{X_{hi}}{P_{hi}} - X_h\right)\left(\frac{Y_{hi}}{P_{hi}} - Y_h\right) \tag{1.4}$$

and $\sigma_{x'_{hi}y'_{hi}}$ is the covariance between x'_{hi} and y'_{hi}, the estimated totals for the ith psu in the hth stratum.

Proof. From the general theorem on the rel-variance of a ratio (see Ch. 4, Sec. 11), the rel-variance of r may be written

$$V_r^2 = V_{(x'/y')}^2 \doteq V_{x'}^2 + V_{y'}^2 - 2V_{x'y'} \tag{1.5}$$

We shall show that $V_{x'y'}$ is given by Eq. 1.3. By definition,

$$V_{x'y'} = \frac{\sigma_{x'y'}}{XY} \tag{1.6}$$

Since the sample is drawn independently from stratum to stratum,

$$\sigma_{x'y'} = \sum_h^L \sigma_{x'_h y'_h} \tag{1.7}$$

* May be deferred.

where

$$x'_h = \frac{1}{m_h} \sum_i^{m_h} \frac{1}{P_{hi}} x'_{hi}, \quad y'_h = \frac{1}{m_h} \sum_i^{m_h} \frac{1}{P_{hi}} y'_{hi} \tag{1.8}$$

and

$$Ex'_{hi} = X_{hi}, \quad Ey'_{hi} = Y_{hi} \tag{1.9}$$

To evaluate $\sigma_{x'_h y'_h}$, we make use of Theorem 15, Ch. 3 (p. 65), which expresses the covariance in terms of a contribution from sampling first-stage units and a contribution from sampling within first-stage units. Thus, from Theorem 15, Ch. 3, with $u = x'_h$, $v = y'_h$, and the random event $b^\star$ being the selection of a set of m_h first-stage units, we have

$$\sigma_{x'_h y'_h} = E\sigma_{x'_h y'_h|[1]} + \sigma_{E(x'_h|[1])E(y'_h|[1])} \tag{1.10}$$

where $\sigma_{x'_h y'_h|[1]}$ represents the covariance between x'_h and y'_h within fixed first-stage units, and $E\sigma_{x'_h y'_h|[1]}$ is the expected value of this conditional covariance. In the second term of Eq. 1.10, $E(x'_h|[1])$, which we will also write as $E_{[1]}x'_h$, represents the conditional expected value of x'_h for a fixed set of first-stage units, $E(y'_h|[1])$ is similarly defined, and $\sigma_{E_{[1]}x'_h,\, E_{[1]}y'_h}$ is the covariance between first-stage units within strata.

Consider the evaluation of $\sigma_{E_{[1]}x'_h,\, E_{[1]}y'_h}$. By Theorem 14, Ch. 3 (p. 61),

$$E_{[1]}x'_h = \frac{1}{m_h} \sum_i^{m_h} \frac{1}{P_{hi}} E_{hi} x'_{hi}$$

where E with a subscript represents the conditional expected value of the expression following it for the fixed values represented by the subscripts. Since x'_{hi} is an unbiased estimate of X_{hi},

$$E_{[1]}x'_h = \frac{1}{m_h} \sum_i^{m_h} \frac{1}{P_{hi}} X_{hi} \tag{1.11}$$

Similarly,

$$E_{[1]}y'_h = \frac{1}{m_h} \sum_i^{m_h} \frac{1}{P_{hi}} Y_{hi}$$

Therefore, since the psu's are selected with replacement,

$$\sigma_{E_{[1]}x'_h,\, E_{[1]}y'_h} = \frac{1}{m_h^2} \sum_i^{m_h} \sigma_{\frac{X_{hi}}{P_{hi}},\, \frac{Y_{hi}}{P_{hi}}} \tag{1.12}$$

Now,

$$E \frac{X_{hi}}{P_{hi}} = \sum_i^{M_h} P_{hi} \frac{X_{hi}}{P_{hi}} = X_h$$

since the probability of selecting the *hi*th primary unit is P_{hi}. Similarly, $E(Y_{hi}/P_{hi}) = Y_h$.

Hence,

$$\sigma_{E_{[1]}x'_h,\, E_{[1]}y'_h} = \frac{1}{m_h^2} \sum_i^{m_h} E\left(\frac{X_{hi}}{P_{hi}} - X_h\right)\left(\frac{Y_{hi}}{P_{hi}} - Y_h\right)$$

$$= \frac{1}{m_h} \sum_i^{M_h} P_{hi}\left(\frac{X_{hi}}{P_{hi}} - X_h\right)\left(\frac{Y_{hi}}{P_{hi}} - Y_h\right)$$

$$= \frac{1}{m_h} S_{hXY} \tag{1.13}$$

where S_{hXY} is given by Eq. 1.4.

Consider now the evaluation of $E\sigma_{x'_h, y'_h|[1]}$. By Eq. 6.2, Ch. 3,

$$\sigma_{x'_h, y'_h|[1]} = E\left[(x'_h - E_{[1]}x'_h)(y'_h - E_{[1]}y'_h)|[1]\right] \tag{1.14}$$

Since the sampling is carried out independently within the selected primary units, the conditional covariance, Eq. 1.14, becomes

$$\sigma_{x'_h y'_h|[1]} = E\left[\frac{1}{m_h^2} \sum_i^{m_h} \frac{1}{P_{hi}^2}(x'_{hi} - X_{hi})(y'_{hi} - Y_{hi})|[1]\right]$$

$$= \frac{1}{m_h^2} \sum_i^{m_h} E_{hi} \frac{1}{P_{hi}^2}(x'_{hi} - X_{hi})(y'_{hi} - Y_{hi})$$

$$= \frac{1}{m_h^2} \sum_i^{m_h} \frac{1}{P_{hi}^2} \sigma_{x'_{hi} y'_{hi}}$$

and

$$E\sigma_{x'_h y'_h|[1]} = \frac{1}{m_h} \sum_i^{M_h} \frac{1}{P_{hi}} \sigma_{x'_{hi} y'_{hi}} \tag{1.15}$$

Substitute Eq. 1.13 and 1.15 into 1.10 and then substitute Eq. 1.10 into Eq. 1.6 to obtain $V_{x'y'}$ as given by Eq. 1.3.

b. The rel-variance of r for a three-stage design with substratification. For this design r is given by Eq. 1.1 with

$$x'_{hi} = \sum_a^{D_{hi}} \frac{1}{f_{hia}} \sum_j^{n_{hia}} \frac{1}{f_{hiaj}} x_{hiaj} = \sum_a^{D_{hi}} x'_{hia} \tag{1.16}$$

where $f_{hia} = n_{hia}/N_{hia}$, $f_{hiaj} = q_{hiaj}/Q_{hiaj}$, N_{hia} is the number of second-stage units in the population in the ath substratum in the hith psu, Q_{hiaj} is the number of third-stage units in the population in the $hiaj$th second-stage unit, n_{hia} and q_{hiaj} are the corresponding numbers in the sample, and $x_{hiaj} = \sum_k^{q_{hiaj}} x_{hiajk}$. The y'_{hi} are similarly defined.

V_r^2 is given by Eq. 1.2 and 1.3 with

$$\sigma_{x'_{hi}y'_{hi}} = \sum_a^{D_{hi}} N_{hia}^2 \frac{N_{hia} - n_{hia}}{N_{hia}} \frac{S_{hiaXY}}{n_{hia}} + \sum_a^{D_{hi}} \frac{N_{hia}}{n_{hia}} \sum_j^{N_{hia}} Q_{hiaj}^2 \frac{Q_{hiaj} - q_{hiaj}}{Q_{hiaj}} \frac{S_{hiajXY}}{q_{hiaj}} \tag{1.17}$$

where

$$S_{hiaXY} = \frac{\sum_j^{N_{hia}} (X_{hiaj} - \bar{\bar{X}}_{hia})(Y_{hiaj} - \bar{\bar{Y}}_{hia})}{N_{hia} - 1} \tag{1.18}$$

$$\bar{\bar{X}}_{hia} = \frac{\sum_j^{N_{hia}} X_{hiaj}}{N_{hia}}, \quad \bar{\bar{Y}}_{hia} = \frac{\sum_j^{N_{hia}} Y_{hiaj}}{N_{hia}}$$

and

$$S_{hiajXY} = \frac{\sum_k^{Q_{hiaj}} (X_{hiajk} - \bar{\bar{\bar{X}}}_{hiaj})(Y_{hiajk} - \bar{\bar{\bar{Y}}}_{hiaj})}{Q_{hiaj} - 1} \tag{1.19}$$

$$\bar{\bar{\bar{X}}}_{hiaj} = \frac{\sum_k^{Q_{hiaj}} X_{hiajk}}{Q_{hiaj}}, \quad \bar{\bar{\bar{Y}}}_{hiaj} = \frac{\sum_k^{Q_{hiaj}} Y_{hiajk}}{Q_{hiaj}}$$

Equation 1.17 may be obtained by applying Theorem 15, Ch. 3 (p. 65), to $\sigma_{x'_{hi}y'_{hi}}$. The details are the same as written out in Sec. 1 of Ch. 7 and Sec. 1 of Ch. 6 and are left to the reader.

Note that by combining corresponding terms of $V_{x'}^2$, $V_{y'}^2$, and $V_{x'y'}$ we may write the rel-variance as follows:

$$V_r^2 \doteq \frac{1}{X^2} \sum_h^L \frac{1}{m_h} S_h^2 + \frac{1}{X^2} \sum_h^L \frac{1}{m_h} \sum_i^{M_h} \frac{1}{P_{hi}} \sum_a^{D_{hi}} N_{hia}^2 \frac{N_{hia} - n_{hia}}{N_{hia}} \frac{S_{hia}^2}{n_{hia}} + \frac{1}{X^2} \sum_h^L \frac{1}{m_h} \sum_i^{M_h} \frac{1}{P_{hi}} \sum_a^{D_{hi}} \frac{N_{hia}}{n_{hia}} \sum_j^{N_{hia}} Q_{hiaj}^2 \frac{Q_{hiaj} - q_{hiaj}}{Q_{hiaj}} \frac{S_{hiaj}^2}{q_{hiaj}} \tag{1.20}$$

where

$$S_h^2 = S_{hX}^2 + R^2 S_{hY}^2 - 2RS_{hXY} \qquad \text{(1.21 or I–9.30.1)}$$

$$S_{hia}^2 = S_{hiaX}^2 + R^2 S_{hiaY}^2 - 2RS_{hiaXY} \qquad \text{(1.22 or I–9.30.2)}$$

$$S_{hiaj}^2 = S_{hiajX}^2 + R^2 S_{hiajY}^2 - 2RS_{hiajXY} \qquad \text{(1.23 or I–9.30.3)}$$

c. Special case when $P_{hi} = 1/M_h$ (i.e., equal probability of selection of psu's within a stratum) and the sampling of primary units is made without replacement. The evaluation of the covariance proceeds exactly as above for the contributions from the second and third stages of sampling. However, the evaluation of the first-stage contribution is made under the assumption that the psu's were drawn without replacement, and with equal probability of selection within a stratum, whereas in Part *a* it was assumed that the psu's were drawn independently (with replacement), and with varying probabilities of selection within a stratum. Hence, by substituting $1/M_h$ for P_{hi} in Eq. 1.11 and then recognizing that m_h primary units are selected without replacement, we have for the first-stage contribution

$$E\left(\frac{M_h}{m_h}\sum_i^{m_h} X_{hi} - X_h\right)\left(\frac{M_h}{m_h}\sum_i^{m_h} Y_{hi} - Y_h\right) \tag{1.24}$$

and, recognizing again that the covariance of $\sum_i^{m_h} X_{hi}/m_h$ and $\sum_i^{m_h} Y_{hi}/m_h$ is given by the procedure of Sec. 3, Ch. 4, we have

$$\text{Eq. 1.24} = M_h^2 \frac{M_h - m_h}{m_h M_h} \frac{\sum_i^{M_h}(X_{hi} - \bar{X}_h)(Y_{hi} - \bar{Y}_h)}{M_h - 1} \tag{1.25}$$

Thus, $\sigma_{x'_h y'_h}$ is given by Eq. 1.25 plus Eq. 1.15 with $1/M_h$ substituted for P_{hi}.

d. V_r^2 for a three-stage design when we have a self-weighting sample with $m_h = \hat{m}$ for all h, with the second-stage sampling fraction small relative to 1 and with constant third-stage sampling fractions. In this case the rel-variance may be written

$$V_r^2 \doteq \frac{B^2}{m} + \frac{W_b^2}{m\bar{n}} + \frac{\bar{\bar{Q}} - \bar{\bar{q}}}{\bar{\bar{Q}}} \frac{W_w^2}{m\bar{n}\bar{\bar{q}}} \tag{1.26}$$

where

$$B^2 = B_X^2 + B_Y^2 - 2B_{XY}$$

$$B_{XY} = \frac{\sum_h^L \sum_i^{M_h} P_{hi}\left(\frac{X_{hi}}{P_{hi}} - X_h\right)\left(\frac{Y_{hi}}{P_{hi}} - Y_h\right)}{L\hat{X}\hat{Y}} = \frac{\sum_h^L S_{hXY}}{L\hat{X}\hat{Y}}$$

$$B_X^2 = B_{XX} \quad \text{and} \quad B_Y^2 = B_{YY}$$

$$\hat{X} = \frac{\sum^L X_h}{L} \quad \text{and} \quad \hat{Y} = \frac{\sum^L Y_h}{L}$$

and

$$W_b^2 = W_{bX}^2 + W_{bY}^2 - 2W_{bXY}$$

$$W_{bXY} = \frac{\sum_h^L \sum_i^{M_h} \sum_a^{D_{hi}} N_{hia} S_{hiaXY}}{N\bar{\bar{X}}\bar{\bar{Y}}}$$

$$W_{bX}^2 = W_{bXX} \quad \text{and} \quad W_{bY}^2 = W_{bYY}$$

$$W_w^2 = W_{wX}^2 + W_{wY}^2 - 2W_{wXY}$$

and where in Eq. 1.26 and throughout this subsection, $\bar{\bar{q}} = f_3\bar{\bar{Q}}$ is the expected number of third-stage units in the sample per second-stage unit in the sample, and $\bar{n} = fN/f_3m$ is the expected number of second-stage units in the sample per primary unit in the sample.

$$W_{wXY} = \frac{\sum_h^L \sum_i^{M_h} \sum_a^{D_{hi}} \sum_j^{N_{hia}} Q_{hiaj} S_{hiajXY}}{Q\bar{\bar{\bar{X}}}\bar{\bar{\bar{Y}}}}$$

$$W_{wX}^2 = W_{wXX} \quad \text{and} \quad W_{wY}^2 = W_{wYY}$$

These results follow from Eq. 1.20 with the appropriate substitutions, namely,

$$m_h = \hat{m} = \frac{m}{L}$$

$$1 - f_{hia} = 1$$

$$f = \frac{q}{Q} = \hat{m}P_{hi}f_{hia}f_{hiaj}$$

$$f_{hiaj} = f_3 = \frac{\bar{\bar{q}}}{\bar{\bar{Q}}}$$

$$X = L\hat{X} = N\bar{\bar{X}} = Q\bar{\bar{\bar{X}}}$$

$$Y - L\hat{Y} = N\bar{\bar{Y}} = Q\bar{\bar{\bar{Y}}}$$

$$n = m\bar{n}, \quad q = m\bar{n}\bar{\bar{q}}$$

2. Conditions under which a gain is to be expected by using probability proportionate to size (Vol. I, Ch. 9, Sec. 11). *To prove:* For many common populations, and with a fixed number, m, of primary units in the sample, a smaller variance will be obtained by sampling with probability proportionate to size than by sampling with equal probability, but the relative reduction in variance will not exceed $V_N^2/(1 + V_N^2)$.

Proof. Consider drawing a sample, with replacement, of m psu's from a population of M psu's, and making the estimate $r = x'/y'$ (Eq. 1.1). Then, from Eq. 1.2 and 1.3, the between-psu contribution to the rel-variance of r can be stated in the form:*

---

* See also Sec. 4, Ch. 8, for a related comparison.

(a) If the psu's are sampled with equal probability:

$$V_a^2 = \frac{1}{mM\bar{X}^2}\sum^{M} N_i^2(\bar{\bar{X}}_i - R\bar{\bar{Y}}_i)^2$$

$$= \frac{1}{mM\bar{N}^2\bar{\bar{X}}^2}\sum^{M} N_i^2 d_i^2 \tag{2.1}$$

where

$$d_i = \bar{\bar{X}}_i - R\bar{\bar{Y}}_i \tag{2.2}$$

(b) If the psu's are sampled with probability proportionate to size (pps):

$$V_b^2 = \frac{1}{mM\bar{N}\bar{\bar{X}}^2}\sum^{M} N_i(\bar{\bar{X}}_i - R\bar{\bar{Y}}_i)^2$$

$$= \frac{1}{mM\bar{N}\bar{\bar{X}}^2}\sum^{M} N_i d_i^2 \tag{2.3}$$

Then we gain by using pps whenever $V_a^2 - V_b^2 > 0$, or whenever $m\bar{N}^2\bar{\bar{X}}^2(V_a^2 - V_b^2) > 0$. Notice that

$$m\bar{N}^2\bar{\bar{X}}^2(V_a^2 - V_b^2) = \frac{1}{M}\left(\sum^{M} N_i^2 d_i^2 - \bar{N}\sum^{M} N_i d_i^2\right) \tag{2.4}$$

is the covariance of N_i and $N_i d_i^2$ and will be positive whenever N_i and $N_i d_i^2$ are positively correlated, which is the case for the commonly encountered class of populations described in Sec. 8 of Ch. 6.

Also, for this class of populations, i.e., when N_i and $N_i d_i^2$ are positively correlated but N_i and d_i^2 are negatively correlated, the intercept of the least-squares regression line of $N_i d_i^2$ on N_i is positive, and consequently

$$\frac{V_a^2 - V_b^2}{V_b^2} < V_N^2 \tag{2.5}$$

where V_N^2 is the rel-variance of the sizes of the psu's. This follows since, if $w = c + du$ is the regression of w on u, the least-squares value of the intercept, c, is $\bar{W} - (\sigma_{wu}/\bar{U}V_u^2)$, where $Ew = \bar{W}$ and $Eu = \bar{U}$.

In our case $U_i = N_i$ and $W_i = N_i d_i^2$. Therefore,

$$c = \frac{\sum^{M} N_i d_i^2}{M} - \frac{\sum^{M} N_i^2 d_i^2 - \bar{N}\sum^{M} N_i d_i^2}{M\bar{N}V_N^2} \tag{2.6}$$

Then the condition of a positive intercept, $c > 0$, is

$$\frac{\sum^{M} N_i d_i^2}{M} > \frac{\sum^{M} N_i^2 d_i^2 - \bar{N}\sum^{M} N_i d_i^2}{M\bar{N}V_N^2} \tag{2.7}$$

from which it follows that

$$\frac{V_a^2 - V_b^2}{V_b^2} < V_N^2 \tag{2.8}$$

It also follows that

$$\frac{V_a^2 - V_b^2}{V_a^2} < \frac{V_N^2}{1 + V_N^2} \tag{2.9}$$

3. When to equalize the sizes of the strata (Vol. I, Ch. 9, Sec. 8 and 24). *To prove:* A rule of thumb which will provide a rough guide to the optimum sizes of strata when a constant number of psu's are selected from each stratum is to make the strata equal in terms of X_h when the psu rel-variances are about the same and remain about the same on adjusting the sizes of strata.

Proof. Assume that the population is divided into L strata. Assume further that the strata are grouped into G classes so that the strata within a class have about the same rel-variances between psu's within strata. If there is no subsampling, an estimate of X can be written

$$x' = \sum_g^G \frac{1}{\bar{m}_g} \sum_h^{L_g} \sum_i^{\bar{m}_g} \frac{X_{ghi}}{P_{ghi}} \tag{3.1}$$

where X_{ghi} is the total for the ith psu in the hth stratum of the gth group, and $\bar{m}_g = m_{gh}$ is the number of psu's in the sample from each stratum in the gth group.

$$\sigma_{x'}^2 = \sum_g^G \frac{1}{\bar{m}_g} \sum_h^{L_g} X_{gh}^2 B_{gh}^2 \tag{3.2}$$

where

$$B_{gh}^2 = \frac{\sum_i^{M_{gh}} P_{ghi} \left(\frac{X_{ghi}}{P_{ghi}} - X_{gh} \right)^2}{M_{gh} X_{gh}^2} \tag{3.3}$$

If, by a shift of some of the psu's from one stratum to another within the gth group, the B_{gh}^2 remain about the same for all L_g strata, then the values of X_{gh} which will minimize $\sigma_{x'}^2$ are determined by finding the values of X_{gh} which will minimize

$$\sum_h^{L_g} X_{gh}^2 B_{gh}^2 \text{ subject to the condition that } \sum_h^{L_g} X_{gh} = X_g$$

This minimum is given by the solution to

$$\frac{\partial F}{\partial X_{gh}} = 0 \quad \text{and} \quad \frac{\partial F}{\partial \lambda} = 0$$

where

$$F = \sum_h^{L_g} X_{gh}^2 B_{gh}^2 + \lambda \left(\sum_h^{L_g} X_{gh} - X_g \right)$$

and where λ is the usual Lagrangian multiplier.

Hence, the minimum is obtained when X_{gh} is proportionate to $1/B_{gh}^2$, and if the B_{gh}^2 are constant, X_{gh} should be made the same for all strata in a group. It follows that if the B_{gh} are the same for all g and h, and $\bar{m}_g$ is the same for all strata, then X_{gh} should be made the same for all strata.

The psu contribution to the variance with multi-stage sampling is the same as Eq. 3.2 above, and consequently the same values for X_{gh} minimize the between-psu variance with multi-stage sampling.

Remark. Similar results can be shown for ratio estimates. Note, also, for ratio estimates that if the variances are about equal in the different strata, instead of the rel-variances, then the Y_{gh} should be equalized instead of the X_{gh}. Often the Y_{gh} are known, and the X_{gh} are not, and the X_{gh} and Y_{gh} are highly correlated. Then the measures of size to be equalized in practice would be the Y_{gh}.

4. The estimate of the rel-variance of ratio estimates when more than one primary sampling unit is taken from each stratum (Vol. I, Ch. 9, Sec. 27). As was pointed out in Sec. 2, Ch. 6, when it is not of interest to estimate the components of the variance for a multi-stage sampling design, an unbiased estimate of the total variance and covariance of $\bar{u}$ and $\bar{w}$ as estimates of $\bar{U}$ and $\bar{W}$ is given by Eq. 4.1 whenever the sampling at the first-stage level is carried out with replacement and at least two first-stage units are included in the sample. Thus, an estimate of $\sigma_{\bar{u}\bar{w}}$ is

$$s_{\bar{u}\bar{w}} = \frac{\sum^m (u_i - \bar{u})(w_i - \bar{w})}{m(m-1)} \tag{4.1}$$

and

$$Es_{\bar{u}\bar{w}} = \sigma_{\bar{u}\bar{w}} \tag{4.2}$$

where m is the number of primary units in the sample.

u_i is an unbiased estimate of $\bar{U}$.

$\bar{u} = \frac{\sum_i^m u_i}{m}$ is an unbiased estimate of $\bar{U}$.

w_i is an unbiased estimate of $\bar{W}$.

$\bar{w} = \frac{\sum_i^m w_i}{m}$ is an unbiased estimate of $\bar{W}$.

a. Multi-stage stratified design with variable fractions. In a multi-stage stratified sampling design with either fixed or variable sampling fractions, we have within each stratum

$$u_i = \frac{x'_{hi}}{P_{hi}}, \quad w_i = \frac{y'_{hi}}{P_{hi}}$$

$$\bar{u} = \frac{\sum_i^{m_h} \frac{x'_{hi}}{P_{hi}}}{m_h} = x'_h, \quad \bar{w} = \frac{\sum_i^{m_h} \frac{y'_{hi}}{P_{hi}}}{m_h} = y'_h$$

$$\bar{U} = X_h, \qquad \bar{W} = Y_h$$

and

$$s_{\bar{u}\bar{w}} = s_{x'_h y'_h} = \frac{\sum_i^{m_h} \left(\frac{x'_{hi}}{P_{hi}} - x'_h\right)\left(\frac{y'_{hi}}{P_{hi}} - y'_h\right)}{m_h(m_h - 1)} \tag{4.3}$$

Also,

$$Es_{x'_h y'_h} = \sigma_{x'_h y'_h} \tag{4.4}$$

Hence,

$$v_{x'y'} = \frac{\sum_h^L s_{x'_h y'_h}}{x'y'} \tag{4.5}$$

is a consistent estimate of $V_{x'y'}$. Further,

$$v_{x'}^2 = v_{x'x'} \quad \text{and} \quad v_{y'}^2 = v_{y'y'} \tag{4.6 or I–9.27.3}$$

are consistent estimates of $V_{x'}^2$ and $V_{y'}^2$, respectively, and

$$v_r^2 = v_{x'}^2 + v_{y'}^2 - 2v_{x'y'} \tag{4.7}$$

is a consistent estimate of V_r^2 given in Eq. 1.2.

b. Fixed over-all sampling fraction. In a multi-stage stratified sampling design with a fixed over-all sampling fraction the estimate $v_{x'y'}$ in Eq. 4.5 above becomes

$$\frac{v_{xy}}{m} = \frac{1}{m} \frac{\sum_h^L m_h \frac{\sum_i^{m_h}(x_{hi} - \bar{x}_h)(y_{hi} - \bar{y}_h)}{m_h - 1}}{m\bar{x}\bar{y}} \tag{4.8}$$

This follows readily by making the appropriate substitutions as follows: $f = m_h P_{hi} f_{hia} f_{hiaj}$

$$\frac{x'_{hi}}{P_{hi}} = \frac{m_h x_{hi}}{f} \text{ with } x_{hi} = \sum_a^{D_{hi}} \sum_j^{n_{hia}} \sum_k^{q_{hiaj}} x_{hiajk}$$

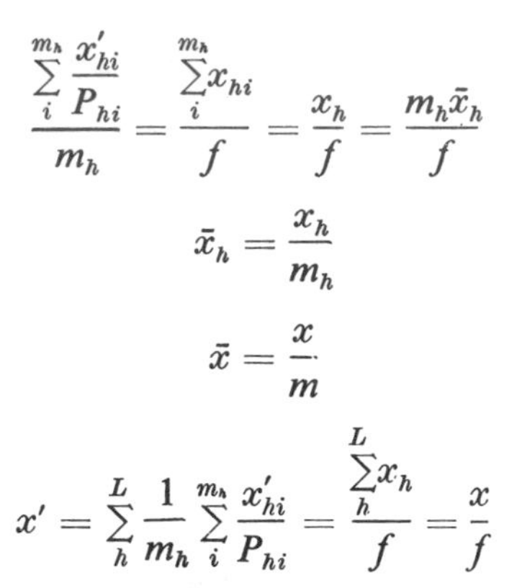

$$\frac{\sum_{i}^{m_h} \frac{x'_{hi}}{P_{hi}}}{m_h} = \frac{\sum_{i}^{m_h} x_{hi}}{f} = \frac{x_h}{f} = \frac{m_h \bar{x}_h}{f}$$

$$\bar{x}_h = \frac{x_h}{m_h}$$

$$\bar{x} = \frac{x}{m}$$

$$x' = \sum_{h}^{L} \frac{1}{m_h} \sum_{i}^{m_h} \frac{x'_{hi}}{P_{hi}} = \frac{\sum_{h}^{L} x_h}{f} = \frac{x}{f}$$

and similarly for the Y-characteristic. Further,

$$v_x^2 = v_{xx} \quad \text{and} \quad v_y^2 = v_{yy}$$

and

$$v_r^2 = v_{x'}^2 + v_{y'}^2 - 2v_{x'y'}$$

becomes

$$\frac{v_x^2 + v_y^2 - 2v_{xy}}{m} \tag{4.9}$$

Equation 2.7 of Ch. 7 becomes identical in form with Eq. 4.9 above if, in Eq. 2.7, the finite multiplier is close to 1 and if a uniform over-all sampling fraction is assumed.

***5. An estimate of the rel-variance of ratio estimates when only one primary unit is taken from each stratum** (Vol. I, Ch. 9, Sec. 15, Eq. 15.1, and Sec. 28). *a.* When only one primary unit is selected from each stratum, it is not usually possible to obtain a consistent estimate of the variance. However, it is possible to obtain an estimate which tends to be an overstatement, and, for many practical problems, the overstatement will not be serious. The procedure is to combine the strata into G groups with L_g strata in the gth group. Assume that we are estimating the rel-variance of $r = x'/y'$, where r is defined by Eq. 1.1.

To prove: We shall show that an estimate of the rel-variance of r that tends to be an overestimate is

$$v_r^2 = v_{x'}^2 + v_{y'}^2 - 2v_{x'y'} \tag{5.1}$$

where

$$v_{x'}^2 = \frac{s_{x'}^2}{x'^2} = \frac{1}{x'^2} \sum_{g}^{G} \frac{L_g}{L_g - 1} \sum_{h}^{L_g} \left(x'_{gh} - \frac{A_{gh}}{A_g} x'_g \right)^2 \tag{5.2}$$

---

* May be deferred.

$$v_{x'y'} = \frac{s_{x'y'}}{x'y'} = \frac{1}{x'y'} \sum_g^G \frac{L_g}{L_g - 1} \sum_h^{L_g} \left(x'_{gh} - \frac{A_{gh}}{A_g} x'_g \right) \left(y'_{gh} - \frac{A_{gh}}{A_g} y'_g \right) \quad \text{(5.3 or I-9.28.2)}$$

$$x'_{gh} = \frac{x'_{ghi}}{P_{ghi}}$$

and where x'_{ghi} is an unbiased estimate of a total for the ith psu in the sample from the hth stratum of the gth group, P_{ghi} is the corresponding probability of selecting the psu, and $Ex'_{gh} = X_{gh}$; A_{gh} is some measure associated with the ghth stratum that tends to be highly correlated with X_{gh}; $x'_g = \sum_h^{L_g} x'_{gh}$, and $A_g = \sum_h^{L} A_{gh}$; the y'_{gh} and y'_g are similarly defined; and $v^2_{y'} = v_{y'y'}$ is given by Eq. 5.2 or 5.3.

Proof. (1) In order to prove that v^2_r tends to be an overstatement of V^2_r and to indicate the mathematical conditions for the bias to be small we assume for now, and prove in (2), that

$$\frac{Es^2_{x'}}{(Ex')^2} = \frac{1}{X^2} \sum_g^G \frac{L_g - 1 + V^2_{A_{g(h)}} - 2V_{A_{g(h)}, \sigma^2_{x'_{g(h)}}}}{L_g - 1} \sum_h^{L_g} \sigma^2_{x'_{gh}}$$

$$+ \frac{1}{X^2} \sum_g^G \frac{L_g}{L_g - 1} \sum_h^{L_g} \left(X_{gh} - \frac{A_{gh}}{A_g} X_g \right)^2 \quad (5.4)$$

where

$$V_{A_{g(h)}, \sigma^2_{x'_{g(h)}}} = \frac{\sum_h^{L_g} A_{gh} \sigma^2_{x'_{gh}}}{\bar{A}_g \sigma^2_{x'_g}} - 1$$

with

$$\bar{A}_g = \frac{\sum_h^{L_g} A_{gh}}{L_g}$$

$$\sigma^2_{x'_g} = \sum_h^{L_g} \sigma^2_{x'_{gh}}$$

and where

$$V^2_{A_{g(h)}} = \frac{\sum_h^{L_g} A^2_{gh}}{L_g \bar{A}^2_g} - 1$$

Now, for sufficiently large samples (from Sec. 15, Ch. 4), $E(s^2_{x'}/x'^2)$ is approximately equal to $Es^2_{x'}/(Ex')^2$, and consequently $Ev^2_{x'}$ is approximately equal to Eq. 5.4. We shall assume that the term $V^2_{A_{g(h)}} - 2V_{A_{g(h)}, \sigma^2_{x'_{g(h)}}}$

is small relative to $L_g - 1$ (as will be the case if the strata do not vary widely in size within groups). It follows that

$$Ev_{x'}^2 \doteq \frac{1}{X^2}\sum_g^G \sum_h^{L_g} \sigma_{x'_{gh}}^2 + \frac{1}{X^2}\sum_g^G \frac{L_g}{L_g - 1}\sum_h^{L_g}\left(X_{gh} - \frac{A_{gh}}{A_g} X_g\right)^2 \tag{5.5}$$

and since the last term of Eq. 5.5 is positive, $v_{x'}^2$ tends to be an overstatement of $\sigma_{x'}^2/X^2 = \sum_g^G \sum_h^{L_g} \sigma_{x'_{gh}}^2/X^2$.

Similarly, we can obtain

$$Ev_{x'y'} = E\frac{s_{x'y'}}{x'y'} \doteq \frac{1}{XY}\sum_g^G \sum_h^{L_g} \sigma_{x'_{gh}y'_{gh}}$$

$$+ \frac{1}{XY}\sum_g^G \frac{L_g}{L_g - 1}\sum_h^{L_g}\left(X_{gh} - \frac{A_{gh}}{A_g} X_g\right)\left(Y_{gh} - \frac{A_{gh}}{A_g} Y_g\right) \tag{5.6}$$

and finally

$$Ev_{y'}^2 = E\frac{s_{y'}^2}{y'^2} \doteq \frac{Es_{y'}^2}{(Ey')^2} \tag{5.7}$$

From Eq. 5.5, 5.6, and 5.7 we have

$$Ev_r^2 = Ev_{x'}^2 + Ev_{y'}^2 - 2Ev_{x'y'}$$

$$\doteq \frac{1}{X^2}\sum_g^G \sum_h^{L_g} \sigma_{x'_{gh}}^2 + \frac{1}{Y^2}\sum_g^G \sum_h^{L_g} \sigma_{y'_{gh}}^2 - 2\frac{1}{XY}\sum_g^G \sum_h^{L_g} \sigma_{x'_{gh}y'_{gh}}$$

$$+ \sum_g^G \frac{L_g}{L_g - 1}\sum_h^{L_g}\left[\frac{1}{X}\left(X_{gh} - \frac{A_{gh}}{A_g} X_g\right) - \frac{1}{Y}\left(Y_{gh} - \frac{A_{gh}}{A_g} Y_g\right)\right]^2 \tag{5.8}$$

Since the sum of the first three terms of Eq. 5.8 is $\sigma_{(x'/y')}^2$, and the last term is positive, Eq. 5.1 tends to be an overstatement of the rel-variance of x'/y'.

(2) It remains to show that Eq. 5.4 holds. We consider the variance rather than the covariance for simplicity in notation, and the reader can follow the same steps to develop an analogous relationship for the covariance.

$$Es_{x'}^2 = \sum_g^G \frac{L_g}{L_g - 1} E\left(\sum_h^{L_g} x_{gh}'^2 - 2\sum_h^{L_g} x'_{gh}\frac{A_{gh}}{A_g} x'_g + \sum_h^{L_g}\frac{A_{gh}^2}{A_g^2} x_g'^2\right) \tag{5.9}$$

Consider the first term in the right-hand member of Eq. 5.9. Since, in general, $Eu^2 = \sigma_u^2 + (Eu)^2$, it follows that

$$E\sum_h^{L_g} x_{gh}'^2 = \sum_h^{L_g}(\sigma_{x'_{gh}}^2 + X_{gh}^2) \tag{5.10}$$

Consider next the second term in the right-hand member of Eq. 5.9:

$$E\sum_{h}^{L_g} x'_{gh} \frac{A_{gh}}{A_g} x'_g = E\left(\sum_{h}^{L_g} x'^2_{gh} \frac{A_{gh}}{A_g} + \sum_{h \neq l}^{L_g} x'_{gh} \frac{A_{gh}}{A_g} x'_{gl}\right)$$

$$= \sum_{h}^{L_g} \frac{A_{gh}}{A_g} (\sigma^2_{x'_{gh}} + X^2_{gh}) + \sum_{h}^{L_g} \frac{A_{gh}}{A_g} X_{gh} \sum_{h}^{L_g} X_{gh} - \sum_{h}^{L_g} \frac{A_{gh}}{A_g} X^2_{gh} \tag{5.11}$$

and, finally, the third term in the right-hand member of Eq. 5.9 is given as follows:

$$E\sum_{h}^{L_g} \frac{A^2_{gh}}{A^2_g} x'^2_g = \left(\sum_{h}^{L_g} \frac{A^2_{gh}}{A^2_g}\right)(\sigma^2_{x'_g} + X^2_g) \tag{5.12}$$

Substitute Eq. 5.10, 5.11, and 5.12 into 5.9 and simplify to obtain 5.13.

$$Es^2_{x'} = \sum_{g}^{G} \frac{L_g}{L_g - 1}\left[\sum_{h}^{L_g}\left(X_{gh} - \frac{A_{gh}}{A_g} X_g\right)^2 + \sigma^2_{x'_g}\left(1 + \sum_{h}^{L_g} \frac{A^2_{gh}}{A^2_g}\right) - 2\sum_{h}^{L_g} \frac{A_{gh}}{A_g} \sigma^2_{x'_{gh}}\right] \tag{5.13}$$

where

$$\sigma^2_{x'_g} = \sum_{h}^{L_g} \sigma^2_{x'_{gh}}$$

Note that

$$\sum_{h}^{L_g} \frac{A^2_{gh}}{A^2_g} = \frac{1 + V^2_{A_{g(h)}}}{L_g} \tag{5.14}$$

and

$$\frac{\sum_{h}^{L_g} A_{gh} \sigma^2_{x'_{gh}}}{A_g} = \frac{\sigma^2_{x'_g}}{L_g} (1 + V_{A_{g(h)}, \sigma^2_{x'_{g(h)}}}) \tag{5.15}$$

Substituting Eq. 5.14 and 5.15 into 5.13 and collecting terms, we obtain Eq. 5.4.

b. When the sample is self-weighting, the following relationships hold:

$$x'_{gh} = \frac{x'_{gh(i)}}{P_{gh(i)}} = \frac{x_{gh}}{f}$$

$$x'_g = \frac{x_g}{f}$$

$$x' = \frac{x}{f} = \frac{m\bar{x}}{f}$$

and similarly for the y's. Substituting these relationships into Eq. 5.3 above, we obtain

$$v_{x'y'} = \frac{\sum_{g}^{G} \frac{L_g}{L_g - 1} \sum_{h}^{L_g} \left(x_{gh} - \frac{A_{gh}}{A_g} x_g\right)\left(y_{gh} - \frac{A_{gh}}{A_g} y_g\right)}{m^2 \bar{x}\bar{y}}$$

$$= \frac{1}{m} v_{cxy}$$

***6. Rel-variance for a self-weighting sample in terms of measures of homogeneity** (Vol. I, Ch. 9, Eq. 17.1). *To prove:* The rel-variance as given by Eq. 1.26 can be restated as follows:

$$V_r^2 \doteq \frac{\hat{V}_1^2}{m\bar{Q}} [1 + \delta_1(\bar{Q} - 1)] + \frac{\hat{V}_2^2}{m\bar{n}\bar{\bar{q}}} [1 + \delta_2(\bar{\bar{q}} - 1)] \quad \text{(6.1 or I-9.17.1)}$$

where δ_1 and δ_2 are measures of homogeneity and are defined by Eq. 6.3 and 6.6, and where $\hat{V}_1^2$ and $\hat{V}_2^2$ are defined by Eq. 6.4 and 6.7, and where, in this section $\bar{\bar{q}} = E\bar{q}$ is the expected number of third-stage units in the sample per second-stage unit in the sample, and $\bar{n} = E\bar{n}$ is the expected number of second-stage units in the sample per first-stage unit in the sample.

Proof.

$$V_r^2 \doteq \frac{B^2}{m} + \frac{W_b^2}{m\bar{n}} + \frac{\bar{\bar{Q}} - \bar{\bar{q}}}{\bar{\bar{Q}}} \frac{W_w^2}{m\bar{n}\bar{\bar{q}}} \quad \text{(6.2 or I-9.14.1)}$$

where the terms in Eq. 6.2 are as defined in Sec. 1, *d*.

Consider first B^2. By definition, for M large,

$$\delta_1 = \frac{B^2 - \frac{W^2}{\bar{Q}}}{\hat{V}_1^2} \quad \text{(6.3 or I-9.17.6)}$$

$$\hat{V}_1^2 = B^2 + \frac{\bar{Q} - 1}{\bar{Q}} W^2 \quad \text{(6.4 or I-9.17.2)}$$

From Eq. 6.3 and 6.4,

$$B^2 = \frac{\hat{V}_1^2}{\bar{Q}} [1 + \delta_1(\bar{Q} - 1)] \quad (6.5)$$

Consider next W_b^2 and W_w^2. By definition,

$$\delta_2 = \frac{W_b^2 - \frac{W_w^2}{\bar{\bar{Q}}}}{\hat{V}_2^2} \quad \text{(6.6 or I-9.17.8)}$$

* May be deferred.

$$\hat{V}_2^2 = W_b^2 + \frac{\bar{\bar{Q}} - 1}{\bar{\bar{Q}}} W_w^2 \qquad \text{(6.7 or I–9.17.7)}$$

From Eq. 6.6 and 6.7,

$$W_b^2 = \frac{\hat{V}_2^2}{\bar{\bar{Q}}} [1 + \delta_2(\bar{\bar{Q}} - 1)] \qquad (6.8)$$

and

$$W_w^2 = \hat{V}_2^2(1 - \delta_2) \qquad (6.9)$$

Substitute Eq. 6.5, 6.8, and 6.9 into Eq. 6.2 and simplify to obtain Eq. 6.1.

***7. Optimum allocation for a fixed total expenditure for a self-weighting, three-stage stratified design** (Vol. I, Ch. 9, Sec. 19 and 20). *To prove:* For a three-stage stratified sampling design which is self-weighting and for which also the sampling within second-stage units is constant, the optimum m, $\bar{n}$, and $\bar{\bar{q}}$ subject to a fixed total expenditure are given by the iterative formulas 7.3, 7.4, and 7.5 below. The rel-variance for the design is given by

$$V_r^2 \doteq \frac{B^2}{m} + \frac{W_b^2}{m\bar{n}} + \frac{\bar{\bar{Q}} - \bar{\bar{q}}}{\bar{\bar{Q}}} \frac{W_w^2}{m\bar{n}\bar{\bar{q}}} \qquad \text{(7.1 or I–9.14.1)}$$

where the terms are as defined in Sec. 1, *d*. The expected cost of the survey is given by

$$C = C_0\sqrt{m} + C_1 m + C_2 m\bar{n} + C_3 m\sqrt{\bar{n}} + m\bar{n}\bar{\bar{q}} \qquad (7.2)$$

The optimum values are given by

$$\bar{\bar{q}} = \frac{W_w}{\sqrt{W_b^2 - W_w^2/\bar{\bar{Q}}}} \sqrt{\frac{C_3}{2\sqrt{\bar{n}}} + C_2} \qquad \text{(7.3 or I–9.19.1)}$$

$$a = \tfrac{1}{4}k + \sqrt{\tfrac{1}{4}k(C_3\sqrt{\bar{n}} + C_2\bar{n} + \bar{n}\bar{\bar{q}} + C_1 + \tfrac{1}{4}k)} \qquad \text{(7.4 or I–9.19.2)}$$

where $k = C_0^2/C$ and $a = C_0/2\sqrt{m}$,

$$\bar{n} = \frac{1}{\bar{\bar{q}}} \frac{W_w}{B} \sqrt{\frac{C_3\sqrt{\bar{n}}}{2} + C_1 + a} \qquad \text{(7.5 or I–9.19.3)}$$

Proof. To obtain the values of m, $\bar{n}$, and $\bar{\bar{q}}$ which minimize the error subject to a fixed cost, set up the Lagrangian F:

$$F = V_r^2 + \lambda(C_0\sqrt{m} + C_1 m + C_2 m\bar{n} + C_3 m\sqrt{\bar{n}} + m\bar{n}\bar{\bar{q}} - C)$$

* May be deferred.

Then the solution to the equations $\partial F/\partial m = 0$, $\partial F/\partial \bar{n} = 0$, $\partial F/\partial \bar{\bar{q}} = 0$, and the cost equation gives the optimum values.

$$\frac{\partial F}{\partial m} = -\frac{B^2}{m^2} - \frac{W_b^2}{m^2\bar{n}} - \frac{\bar{\bar{Q}} - \bar{\bar{q}}}{\bar{\bar{Q}}}\frac{W_w^2}{m^2\bar{n}\bar{\bar{q}}} + \lambda\left(\frac{C_0}{2\sqrt{m}} + C_1 + C_2\bar{n} + C_3\sqrt{\bar{n}} + \bar{n}\bar{\bar{q}}\right) = 0 \tag{7.6}$$

$$\frac{\partial F}{\partial \bar{n}} = -\frac{W_b^2}{m\bar{n}^2} - \frac{\bar{\bar{Q}} - \bar{\bar{q}}}{\bar{\bar{Q}}}\frac{W_w^2}{m\bar{n}^2\bar{\bar{q}}} + \lambda\left(C_2 m + \frac{C_3 m}{2\sqrt{\bar{n}}} + m\bar{\bar{q}}\right) = 0 \tag{7.7}$$

$$\frac{\partial F}{\partial \bar{\bar{q}}} = -\frac{W_w^2}{m\bar{n}\bar{\bar{q}}^2} + \lambda m\bar{n} = 0 \tag{7.8}$$

From Eq. 7.8:

$$\lambda m^2\bar{n}^2 = \frac{W_w^2}{\bar{\bar{q}}^2} \tag{7.9}$$

From Eq. 7.7:

$$\lambda m^2\bar{n}^2 = \frac{W_b^2 + \dfrac{W_w^2}{\bar{\bar{q}}} - \dfrac{W_w^2}{\bar{\bar{Q}}}}{C_2 + \dfrac{C_3}{2\sqrt{\bar{n}}} + \bar{\bar{q}}} \tag{7.10}$$

From Eq. 7.9 = 7.10:

$$\bar{\bar{q}} = \frac{W_w}{\sqrt{W_b^2 - W_w^2/\bar{\bar{Q}}}}\sqrt{\frac{C_3}{2\sqrt{\bar{n}}} + C_2} \tag{7.3}$$

Multiplying Eq. 7.6 by m and subtracting Eq. 7.7 multiplied by $\bar{n}$, we obtain

$$\frac{1}{\lambda m^2} = \frac{1}{B^2}\left(\frac{C_0}{2\sqrt{m}} + C_1 + \frac{C_3\sqrt{\bar{n}}}{2}\right) \tag{7.11}$$

Substitute for λm^2 from Eq. 7.9 and let $C_0/2\sqrt{m} = a$ to obtain $\bar{n}$ as given by Eq. 7.5. Treating the cost function as a quadratic equation in the $\sqrt{m}$, we obtain

$$\sqrt{m} = \frac{-C_0 + \sqrt{C_0^2 + 4C(C_1 + C_2\bar{n} + C_3\sqrt{\bar{n}} + \bar{n}\bar{\bar{q}})}}{2(C_1 + C_2\bar{n} + C_3\sqrt{\bar{n}} + \bar{n}\bar{\bar{q}})}$$

and $\dfrac{C_0}{2\sqrt{m}} = a$ is given by Eq. 7.4

Remark 1. Note that if we use the approximation to V_r^2 given by Eq. 6.1 with

$$\frac{\hat{V}_1^2}{V^2} = 1, \quad \frac{\hat{V}_2^2}{V^2} = 1, \quad \text{and} \quad \frac{1 + \delta_1(\bar{Q} - 1)}{\bar{Q}} = \delta_1$$

the optimum values are given by

$$\bar{\bar{q}} = \sqrt{\frac{1-\delta_2}{\delta_2}\left(\frac{C_3}{2\sqrt{\bar{n}}} + C_2\right)}$$

which is identical with Eq. 7.3 above, where δ_2 is defined by Eq. 6.6, and

$$\bar{n} = \frac{1}{\bar{\bar{q}}}\sqrt{\frac{1-\delta_2}{\delta_1}\left(\frac{C_3\sqrt{\bar{n}}}{2} + C_1 + a\right)}$$

which is approximately equal to Eq. 7.5. above.

Of course, a, given by Eq. 7.4, retains the same form since the cost function is unchanged.

Remark 2. If the cost can be expressed by the simplified cost function

$$C = C_1 m + C_2 m\bar{n} + C_4 m\bar{n}\bar{\bar{q}} \qquad \text{(7.12 or I–9.19.5)}$$

The optimum values are given explicitly by

$$\bar{\bar{q}} = \frac{W_w}{\sqrt{W_b^2 - W_w^2/\bar{\bar{Q}}}}\sqrt{\frac{C_2}{C_4}} = \sqrt{\frac{1-\delta_2}{\delta_2}\frac{C_2}{C_4}} \qquad \text{(7.13 or I–9.19.6)}$$

$$\bar{n} = \frac{1}{\bar{\bar{q}}}\frac{W_w}{B}\sqrt{\frac{C_1}{C_4}} \doteq \frac{1}{\bar{\bar{q}}}\sqrt{\frac{1-\delta_2}{\delta_1}\frac{C_1}{C_4}} \qquad \text{(7.14 or I–9.19.7)}$$

$$m = \frac{C}{C_1 + C_2\bar{n} + C_4\bar{n}\bar{\bar{q}}} \qquad \text{(7.15 or I–9.17.8)}$$

Remark 3. If the precision of the survey is fixed at $\varepsilon^2 = V_r^2$, the numbers of units in the sample yielding a minimum cost are given by Eq. 7.3 and 7.5 with a now given by Eq. 7.16 below in place of Eq. 7.4:

$$a = \frac{\varepsilon C_0}{2\sqrt{B^2 + \frac{1}{\bar{n}}\left(W_b^2 - \frac{W_w^2}{\bar{\bar{Q}}}\right) + \frac{1}{\bar{n}\bar{\bar{q}}}W_w^2}}$$

$$\doteq \frac{\varepsilon C_0}{V\sqrt{\delta_1 + \frac{\delta_2}{\bar{n}} + \frac{1-\delta_2}{\bar{n}\bar{\bar{q}}}}} \qquad \text{(7.16 or I–9.20.1)}$$

with $V^2 = V_X^2 + V_Y^2 - 2V_{XY}$, and V_{XY} is the within-strata rel-covariance for a simple random sample of listing units, $V_X^2 = V_{XX}$, and $V_Y^2 = V_{YY}$. The development follows the same lines as given above and is left to the reader.

***8. The variance of ratio estimates by specific subclasses that can make use of both current and past information** (Vol. I, Ch. 9, Sec. 23, and Vol. I, Ch. 12, Case Study B). Let x'_{ac} and y'_{ac} be unbiased estimates of X_{ac} and Y_{ac}, which are aggregate values for the acth subclass of a population,

* May be deferred.

$a = 1, 2, \cdots, k$, and $c = 1, 2, \cdots, l$, with

$$X_c = \sum_a^k X_{ac}, \quad X_a = \sum_c^l X_{ac}, \quad Y_c = \sum_a^k Y_{ac}, \quad Y_a = \sum_c^l Y_{ac}$$

Assume that z'_c is an unbiased estimate of Z_c, an aggregate value for the cth class of some related characteristic, perhaps for a past date, and assume that the Z_c and the Y_a are known from sources other than the sample.

An estimate which makes use of both sets of known data for the specific subclasses is

$$\sum_a^k \frac{\sum_c^l \dfrac{x'_{ac}}{z'_c} Z_c}{\sum_c^l \dfrac{y'_{ac}}{z'_c} Z_c} Y_a = x'''$$

The variance of x''' can be approximated by the method of Sec. 11, Ch. 4, and is

$$\begin{aligned}
\sigma^2_{x'''} \doteq \sum_c^l [& X_c^2(V^2_{x'_c} + 3V^2_{z'_c} - 4V_{x'_c z'_c} + 1) + \hat{Y}_c^2(V^2_{\hat{y}'_c} + 3V^2_{z'_c} - 4V_{\hat{y}'_c z'_c} + 1) \\
& - 2X_c\hat{Y}_c(1 + 3V^2_{z'_c} - 2V_{\hat{y}'_c z'_c} - 2V_{x'_c z'_c} + V_{x'_c \hat{y}'_c})] \\
+ \sum_{c \neq d}^{l} [& X_c X_d(1 + V^2_{z'_c} + V^2_{z'_d} + V_{z'_c z'_d} + V_{x'_c x'_d} \\
& - V_{x'_c z'_c} - V_{x'_c z'_d} - V_{x'_d z'_c} - V_{x'_d z'_d}) \\
& + \hat{Y}_c \hat{Y}_d(1 + V^2_{z'_c} + V^2_{z'_d} + V_{z'_c z'_d} + V_{\hat{y}'_c \hat{y}'_d} \\
& - V_{\hat{y}'_c z'_c} - V_{\hat{y}'_c z'_d} - V_{\hat{y}'_d z'_c} - V_{\hat{y}'_d z'_d}) \\
& - 2X_c\hat{Y}_d(1 + V^2_{z'_c} + V^2_{z'_d} + V_{z'_c z'_d} + V_{x'_c \hat{y}'_d} \\
& - V_{x'_c z'_c} - V_{x'_c z'_d} - V_{\hat{y}'_d z'_c} - V_{\hat{y}'_d z'_d})] \qquad (8.1)
\end{aligned}$$

where

$$\hat{Y}_c = \sum_a^k \frac{X_a}{Y_a} Y_{ac}$$

$$\hat{y}'_c = \sum_a^k \frac{X_a}{Y_a} y'_{ac}$$

and where $V_u^2 = E(u - Eu)^2/(Eu)^2$ is the rel-variance of u, and $V_{uw} = E(u - Eu)(w - Ew)/(Eu)(Ew)$ is the rel-covariance of u and w, where u and w are the various random variables indicated by the subscripts in Eq. 8.1.

The estimate of the variance is obtained by substituting sample estimates for each of the terms in Eq. 8.1 above.

The danger of *increasing* the variance if an estimate of the type given by x''' is carried into too many subclasses is discussed in Ch. 5, Sec. 13.

***9. Reduction in variance due to stratification, when psu's are large** (Vol. I, Ch. 9, Sec. 8). *To prove:* Stratification will usually introduce a relatively larger reduction in the primary unit contribution to the variance when the psu's are large than when they are small. It is assumed that the same strata are used for both large and small psu's.

Proof. Assume that we have L strata. Let us denote the average size of the large clusters within a stratum by $\bar{N}_{hL}$, and assume that $a\bar{N}_{hL} = \bar{N}_{hS}$, the average size of the small cluster, where $a < 1$.

If we let

$$x' = \sum_h^L \frac{M_h}{m_h} \sum_i^{m_h} x_{hi}$$

then

$$V_{x'}^2 = \frac{\hat{M} - \hat{m}}{\hat{M}m} \frac{\sum_h^L M_h \dfrac{\sum_i^{M_h}(X_{hi} - \bar{X}_h)^2}{M_h - 1}}{\sum_h^L M_h \bar{X}^2} \tag{9.1}$$

when $m_h = kM_h$ and m_h = number of psu's in the sample from the hth stratum, M_h = number of psu's in the population from the hth stratum, and

$$\hat{M} = \frac{\sum_h^L M_h}{L}, \quad \bar{X} = \frac{X}{M}, \quad \hat{m} = \frac{\sum_h^L m_h}{L}$$

If the M_h are large, then

$$V_{x'}^2 \doteq \frac{\hat{M} - \hat{m}}{\hat{M}m} \frac{\Sigma\Sigma(X_{hi} - \bar{X}_h)^2}{M\bar{X}^2} = \frac{\hat{M} - \hat{m}}{\hat{M}m} \left[\frac{\Sigma\Sigma(X_{hi} - \bar{X})^2}{M\bar{X}^2} - \frac{\Sigma M_h(\bar{X}_h - \bar{X})^2}{M\bar{X}^2}\right] \tag{9.2}$$

But the rel-variance of a simple random sample of m psu's is approximately

$$\frac{\hat{M} - \hat{m}}{\hat{M}m} \frac{\Sigma\Sigma(X_{hi} - \bar{X})^2}{M\bar{X}^2}$$

and the relative gain due to stratification is

$$\frac{\Sigma M_h(\bar{X}_h - \bar{X})^2}{M\bar{X}^2} \Bigg/ \frac{\Sigma\Sigma(X_{hi} - \bar{X})^2}{M\bar{X}^2} \tag{9.3}$$

* May be deferred.

Assume that in the equations given above we let $M_h = M_{hL}$ for the number of large clusters in the stratum and $= M_{hS}$ for the number of small clusters in the hth stratum. Moreover, we let $M = M_L$ for the number of large clusters in the population, and $M = M_S$ for the number of small clusters in the population. Then, since $M_{hL} = N_h/\bar{N}_{hL}$, and $M_{hS} = N_h/a\bar{N}_{hL}$, it follows that $M_{hL}/\Sigma M_{hL} = M_{hS}/\Sigma M_{hS}$. Similarly, $\bar{X}_{hS} = a\bar{X}_{hL}$ and $\bar{X}_S = a\bar{X}_L$. Therefore,

$$\frac{\Sigma M_{hL}(\bar{X}_{hL} - \bar{X}_L)^2}{M_L\bar{X}_L^2} = \frac{\Sigma M_{hS}(\bar{X}_{hS} - \bar{X}_S)^2}{M_S\bar{X}_S^2}$$

or the numerator of Eq. 9.3 remains the same for the change in the average size of cluster. From Ch. 6, Eq. 5.6, the denominator of Eq. 9.3 can be written as

$$\frac{\hat{V}_L^2}{\bar{N}_L}[1 + \delta_L(\bar{N}_L - 1)] \tag{9.4}$$

for large clusters, and as

$$\frac{\hat{V}_S^2}{\bar{N}_S}[1 + \delta_S(\bar{N}_S - 1)] \tag{9.5}$$

for small sizes of clusters, and the ratio of Eq. 9.4 to Eq. 9.5 is

$$\frac{\bar{N}_S\hat{V}_L^2\,[1 + \delta_L(\bar{N}_L - 1)]}{\bar{N}_L\hat{V}_S^2\,[1 + \delta_S(\bar{N}_S - 1)]} \tag{9.6}$$

It follows, if $\hat{V}_L^2 \doteq \hat{V}_S^2$, $\delta_L > 0$, $\delta_L/\delta_S < 1$, and $(\bar{N}_S/\bar{N}_L) < \dfrac{(1 - \delta_S)}{(1 - \delta_L)}$, that Eq. 9.6 is less than unity and hence the relative gain from the stratification is greater for large clusters than for small clusters.

The reader can show that in the case where the estimate is $x'/y' = r$, the relative gain is given by

$$\frac{\dfrac{\Sigma M_h(\bar{X}_h - \bar{X})^2}{M\bar{X}^2} + \dfrac{\Sigma M_h(\bar{Y}_h - \bar{Y})^2}{M\bar{Y}^2} - \dfrac{2\Sigma M_h(\bar{X}_h - \bar{X})(\bar{Y}_h - \bar{Y})}{M\bar{X}\bar{Y}}}{\dfrac{\Sigma\Sigma(X_{hi} - \bar{X})^2}{M\bar{X}^2} + \dfrac{\Sigma\Sigma(Y_{hi} - \bar{Y})^2}{M\bar{Y}^2} - \dfrac{2\Sigma\Sigma(X_{hi} - \bar{X})(Y_{hi} - \bar{Y})}{M\bar{X}\bar{Y}}} \tag{9.7}$$

and when δ is defined by Eq. 5.5 of Ch. 6, the relationship in Eq. 9.6 holds. It is for the ratio estimate that the assumption that $\hat{V}_L^2 \doteq \hat{V}_S^2$ will usually hold.

***10. Consistent estimates of the components of the rel-variance when more than one psu is selected with replacement from each stratum** (Vol. I, Ch. 9, Sec. 30, 31, and 32). We will consider a three-stage stratified

* May be deferred.

sampling design for which the estimate is given by Eq. 1.1 and 1.16 and the rel-variance is given by Eq. 1.20. Assume that we have a sample of m'_h psu's from the hth stratum, n'_{hia} second-stage units from the ath substratum in the hith psu, and q'_{hiaj} third-stage units in the $hiaj$th second-stage unit, for estimating the components of the rel-variance.

a. To prove: A consistent estimate of the contribution to the rel-variance due to sampling third-stage units, i.e., the third term of Eq. 1.20, is given by

$$\frac{1}{x'^2}\sum_h^L \frac{1}{m_h}\frac{1}{m'_h}\sum_i^{m'_h}\frac{1}{P_{hi}^2}\sum_a^{D_{hi}}\frac{N_{hia}}{n_{hia}}\frac{N_{hia}}{n'_{hia}}\sum_j^{n'_{hia}} Q_{hiaj}^2 \frac{Q_{hiaj}-q_{hiaj}}{Q_{hiaj}}\frac{s_{hiaj}^2}{q_{hiaj}} \tag{10.1}$$

where

$$s_{hiaj}^2 = s_{hiajX}^2 + r^2 s_{hiajY}^2 - 2r s_{hiajXY} \tag{10.2}$$

with

$$s_{hiajXY} = \frac{\sum_k^{q'_{hiaj}} (x_{hiajk} - \bar{\bar{\bar{x}}}_{hiaj})(y_{hiajk} - \bar{\bar{\bar{y}}}_{hiaj})}{q'_{hiaj} - 1} \tag{10.3}$$

$$\bar{\bar{\bar{x}}}_{hiaj} = \sum_k^{q'_{hiaj}} \frac{x_{hiajk}}{q'_{hiaj}}; \quad \bar{\bar{\bar{y}}}_{hiaj} = \sum_k^{q'_{hiaj}} \frac{y_{hiajk}}{q'_{hiaj}}$$

$$s_{hiajX}^2 = s_{hiajXX} \quad \text{and} \quad s_{hiajY}^2 = s_{hiajYY}$$

We will show that Eq. 10.1 with x' replaced by X and s_{hiaj}^2 replaced by S_{hiaj}^2 (defined in Eq. 1.23) is an unbiased estimate of the third-stage contribution to the rel-variance. Then we need to show that x'^2 is a consistent estimate of X^2, and s_{hiaj}^2 is a consistent estimate of S_{hiaj}^2, for Eq. 10.1 to be a consistent estimate of the third term of Eq. 1.20.

Proof. The expected value of Eq. 10.1 with x' substituted for X and s_{hiaj}^2 substituted for S_{hiaj}^2 is, by Theorems 6 and 14, Ch. 3 (pp. 49 and 61), equal to

$$\frac{1}{X^2}\sum_h^L \frac{1}{m_h}\frac{1}{m'_h}\sum_i^{m'_h} E_{hi}\frac{1}{P_{hi}^2}\sum_a^{D_{hi}} E_{hia}\frac{N_{hia}}{n_{hia}}\frac{N_{hia}}{n'_{hia}}\sum_j^{n'_{hia}} E_{hiaj} Q_{hiaj}^2 \frac{Q_{hiaj}-q_{hiaj}}{Q_{hiaj}}\frac{S_{hiaj}^2}{q_{hiaj}}$$

$$= \frac{1}{X^2}\sum_h^L \frac{1}{m_h}\frac{1}{m'_h}\sum_i^{m'_h}\sum_i^{M_h} P_{hi}\frac{1}{P_{hi}^2}\sum_a^{D_{hi}}\frac{N_{hia}}{n_{hia}}\frac{N_{hia}}{n'_{hia}}\sum_j^{n'_{hia}}\frac{1}{N_{hia}}\sum_j^{N_{hia}} Q_{hiaj}^2 \frac{Q_{hiaj}-q_{hiaj}}{Q_{hiaj}}\frac{S_{hiaj}^2}{q_{hiaj}}$$

$$= \text{the third term of Eq. 1.20} \tag{10.4}$$

Now, by Sec. 21 of Ch. 4, r is a consistent estimate of R. Since the q'_{hiaj} units are a simple random sample from the Q_{hiaj} units in the population, s_{hiajXY} is an unbiased estimate of S_{hiajXY} (by Sec. 3, Ch. 4).

Similarly, s^2_{hiajX} is an unbiased estimate of S^2_{hiajX}, and s^2_{hiajY} is an unbiased estimate of S^2_{hiajY}. Since also x' is an unbiased estimate of X, then, by the corollary to Theorem 19 and Corollary 2 to Theorem 20, Ch. 3 (p. 75), Eq. 10.1 is a consistent estimate of the third term of Eq. 1.20.

b. To prove: A consistent estimate of the contribution to the rel-variance due to sampling second-stage units, i.e., the second term of Eq. 1.20, is given by

$$\frac{1}{x'^2}\sum_h^L \frac{1}{m_h}\frac{1}{m'_h}\sum_i^{m'_h}\frac{1}{P^2_{hi}}\sum_a^{D_{hi}} N^2_{hia}\frac{N_{hia}-n_{hia}}{N_{hia}}\frac{s^2_{hia}}{n_{hia}} \tag{10.5}$$

where

$$s^2_{hia} = \hat{s}^2_{hia} - \tilde{s}^2_{hia} \tag{10.6}$$

and

$$\hat{s}^2_{hia} = \hat{s}^2_{hiaX} + r^2\hat{s}^2_{hiaY} - 2r\hat{s}_{hiaXY} \tag{10.7}$$

with

$$\hat{s}_{hiaXY} = \frac{\sum_j^{n'_{hia}} (x'_{hiaj} - \bar{x}'_{hia})(y'_{hiaj} - \bar{y}'_{hia})}{n'_{hia} - 1} \tag{10.8}$$

$$x'_{hiaj} = \frac{Q_{hiaj}}{q'_{hiaj}}\sum_k^{q'_{hiaj}} x_{hiajk} \tag{10.9}$$

$$\bar{x}'_{hia} = \frac{\sum_j^{n'_{hia}} x'_{hiaj}}{n'_{hia}} \tag{10.10}$$

y'_{hiaj} and $\bar{y}'_{hia}$ similarly defined, and

$$\hat{s}^2_{hiaX} = \hat{s}_{hiaXX} \quad \text{and} \quad \hat{s}^2_{hiaY} = \hat{s}_{hiaYY}$$

and where

$$\tilde{s}^2_{hia} = \frac{\sum_j^{n'_{hia}} Q^2_{hiaj}\frac{Q_{hiaj}-q'_{hiaj}}{Q_{hiaj}}\frac{s^2_{hiaj}}{q'_{hiaj}}}{n'_{hia}} \tag{10.11}$$

Proof. It follows from the same considerations made in Eq. 10.4 that the expected value of Eq. 10.5 with x' replaced by X and s^2_{hia} replaced by S^2_{hia} (defined in Eq. 1.22) is the second term of Eq. 1.20. We wish now to show that s^2_{hia} is a consistent estimate of S^2_{hia}. By Sec. 3, Ch. 6, we have

$$E_{hia}\hat{s}_{hiaXY} = S_{hiaXY} + \frac{\sum_j^{N_{hia}} Q^2_{hiaj}\frac{Q_{hiaj}-q'_{hiaj}}{Q_{hiaj}}\frac{S_{hiajXY}}{q'_{hiaj}}}{N_{hia}} \tag{10.12}$$

By Theorems 6 and 14, Ch. 3 (pp. 49 and 61), and Sec. 3, Ch. 4,

$$E_{hia}\tilde{s}_{hiaXY} = \frac{\sum_{j}^{N_{hia}} Q_{hiaj}^2 \dfrac{Q_{hiaj} - q'_{hiaj}}{Q_{hiaj}} \dfrac{S_{hiajXY}}{q'_{hiaj}}}{N_{hia}} \tag{10.13}$$

Hence, subtracting Eq. 10.13 from Eq. 10.12, it follows that

$$E_{hia}s_{hiaXY} = E_{hia}\hat{s}_{hiaXY} - E_{hia}\tilde{s}_{hiaXY} = S_{hiaXY} \tag{10.14}$$

Similarly, s_{hiaX}^2 is an unbiased estimate of S_{hiaX}^2, and s_{hiaY}^2 is an unbiased estimate of S_{hiaY}^2. Since also r is a consistent estimate of R, s_{hia}^2 is a consistent estimate of S_{hia}^2.

c. To prove: A consistent estimate of the contribution to the rel-variance due to sampling first-stage units, i.e., the first term of Eq. 1.20, is given by

$$\frac{1}{x'^2} \sum_{h}^{L} \frac{1}{m_h} s_h^2 \tag{10.15}$$

where

$$s_h^2 = \hat{s}_h^2 - \tilde{s}_h^2 \tag{10.16}$$

$$\hat{s}_h^2 = \hat{s}_{hX}^2 + r^2\hat{s}_{hY}^2 - 2r\hat{s}_{hXY} \tag{10.17}$$

with

$$\hat{s}_{hXY} = \frac{1}{m'_h - 1} \sum_{i}^{m'_h} \left(\frac{x'_{hi}}{P_{hi}} - \frac{\sum_{i}^{m'_h} \dfrac{x'_{hi}}{P_{hi}}}{m'_h} \right) \left(\frac{y'_{hi}}{P_{hi}} - \frac{\sum_{i}^{m'_h} \dfrac{y'_{hi}}{P_{hi}}}{m'_h} \right) \tag{10.18}$$

$$x'_{hi} = \sum_{a}^{D_{hi}} \frac{N_{hia}}{n'_{hia}} \sum_{j}^{n'_{hia}} x'_{hiaj} \tag{10.19}$$

and x'_{hiaj} given by Eq. 10.9. The definition of y'_{hi} is similar to that of x'_{hi}. Finally, $\tilde{s}_h^2$ in Eq. 10.16 is

$$\tilde{s}_h^2 = \frac{1}{m'_h} \sum_{i}^{m'_h} \frac{1}{P_{hi}^2} \sum_{a}^{D_{hi}} \frac{N_{hia}^2}{n'_{hia}} \left[\frac{N_{hia} - n'_{hia}}{N_{hia}} \hat{s}_{hia}^2 + \frac{n'_{hia}}{N_{hia}} \tilde{s}_{hia}^2 \right] \tag{10.20}$$

Proof. From Sec. 4 it follows that $\dfrac{1}{x'^2} \sum_{h}^{L} \dfrac{1}{m_h} \hat{s}_h^2$ is a consistent estimate of $V_r'^2$, the rel-variance of r for a sample of m_h, n'_{hia}, and q'_{hiaj} sampling units, i.e., $V_r'^2$ is given by Eq. 1.20 with n_{hia} replaced by n'_{hia} and q_{hiaj} replaced by q'_{hiaj}.

From Part b above it follows that $\frac{1}{x'^2}\sum_h^L \frac{1}{m_h}\tilde{s}_h^2$ is a consistent estimate of:

$$\frac{1}{X^2}\sum_h^L \frac{1}{m_h}\sum_i^{M_h}\frac{1}{P_{hi}}\sum_a^{D_{hi}} N_{hia}^2 \frac{N_{hia}-n'_{hia}}{N_{hia}}\frac{S_{hia}^2}{n'_{hia}}$$

$$+\frac{1}{X^2}\sum_h^L \frac{1}{m_h}\sum_i^{M_h}\frac{1}{P_{hi}}\sum_a^{D_{hi}}\frac{N_{hia}}{n'_{hia}}\frac{N_{hia}-n'_{hia}}{N_{hia}}\sum_j^{N_{hia}} Q_{hiaj}^2 \frac{Q_{hiaj}-q'_{hiaj}}{Q_{hiaj}}\frac{S_{hiaj}^2}{q'_{hiaj}}$$

$$+\frac{1}{X^2}\sum_h^L \frac{1}{m_h}\sum_i^{M_h}\frac{1}{P_{hi}}\sum_a^{D_{hi}}\sum^{N_{hia}} Q_{hiaj}^2 \frac{Q_{hiaj}-q'_{hiaj}}{Q_{hiaj}}\frac{S_{hiaj}^2}{q'_{hiaj}} \tag{10.21}$$

Equation 10.21 can be shown to be equal to the sum of the second and third terms of Eq. 1.20 but with n_{hia} replaced by n'_{hia} and q_{hiaj} replaced by q'_{hiaj}. Therefore, a consistent estimate of the total rel-variance minus a consistent estimate of the second and third terms combined will be a consistent estimate of the contribution to the rel-variance due to sampling first-stage units, i.e.,

$$\frac{1}{x'^2}\sum_h^L \frac{1}{m_h}s_h^2 = \frac{1}{x'^2}\sum_h^L \frac{1}{m_h}\hat{s}_h^2 - \frac{1}{x'^2}\sum_h^L \frac{1}{m_h}\tilde{s}_h^2$$

is a consistent estimate of the first term of $V_r'^2$, which is the same as the first term of V_r^2.

***11. Optimum values for a three-stage stratified design** (Vol. I, Ch. 9, Sec. 6 and 26). *To prove:* The values of m_h, the number of psu's in the sample from the hth stratum, of $f_{hia} = n_{hia}/N_{hia}$, the sampling fractions for second-stage units, and of $f_{hia(j)}$, the sampling fractions for third-stage units, with

$$f_{hia(j)} = \frac{E\bar{\bar{q}}_{hia}}{\bar{\bar{Q}}_{hia}}, \text{ where } E\bar{\bar{q}}_{hia} = \frac{\sum^{N_{hia}} q_{hiaj}}{N_{hia}}$$

(i.e., the sampling fractions for third-stage units are assumed to be constant for all second-stage units in the hiath substratum), which minimize the variance subject to a fixed total cost, are given in Eq. 11.7 to 11.10 below. The cost condition is given in Eq. 11.5 below. It will also be shown that a uniform sampling fraction is optimum when $\tilde{S}_{whia}/\sqrt{C_{3hia}}$ is a constant for all h, i, and a.

The rel-variance of x'/y' for a general three-stage stratified design with varying probabilities of selection of primary units is given by Eq. 1.20 in Sec. 1. If we specify that a uniform third-stage sampling fraction be

* May be deferred.

applied to all second-stage units within a substratum of a selected psu, i.e., if we let

$$f_{hia(j)} = \frac{E\bar{\bar{q}}_{hia}}{\bar{\bar{Q}}_{hia}} = f_{hiaj}$$

the rel-variance may be written

$$V^2_{(x'/y')} = \frac{1}{X^2}\left[\sum_h^L \frac{1}{m_h}\Delta_h + \sum_h^L \frac{1}{m_h}\sum_i^{M_h}\frac{1}{P_{hi}}\sum_a^{D_{hi}}\frac{N_{hia}}{f_{hia}}(S^2_{hia} - \bar{\bar{Q}}_{hia}\tilde{S}^2_{whia}) + \sum_h^L \frac{1}{m_h}\sum_i^{M_h}\frac{1}{P_{hi}}\sum_a^{D_{hi}}\frac{1}{f_{hia}}\frac{1}{f_{hia(j)}}Q_{hia}\tilde{S}^2_{whia}\right] \tag{11.1}$$

where

$$\tilde{S}^2_{whia} = \frac{\sum_j^{N_{hia}} Q_{hiaj}S^2_{hiaj}}{Q_{hia}} \tag{11.2}$$

with S^2_{hiaj} given by Eq. 1.23, and where

$$S^2_{hia} = S^2_{hiaX} + R^2S^2_{hiaY} - 2RS_{hiaXY} \tag{11.3}$$

which is also given by Eq. 1.22,

$$\Delta_h = S^2_h - \sum_i^{M_h}\frac{1}{P_{hi}}\sum_a^{D_{hi}} N_{hia}S^2_{hia} \tag{11.4}$$

where S^2_h is given by Eq. 1.21, and other terms are as defined in Sec. 1. Assume the following cost function:

$$C = \sum_h^L C_{1h}m_h + \sum_h^L m_h \sum_i^{M_h} P_{hi}\sum_a^{D_{hi}} n_{hia}C_{2hia} + \sum_h^L m_h \sum_i^{M_h} P_{hi}\sum_a^{D_{hi}} n_{hia}\bar{\bar{q}}_{hia}C_{3hia} \tag{11.5}$$

where C_{1h} is the cost per primary unit for the hth primary stratum, C_{2hia} is the cost per second-stage unit in the ath substratum of the ith primary unit in the hth primary stratum, and C_{3hia} is the cost per listing unit for the hiath substratum, and $\bar{\bar{q}}_{hia}$ is the expected number of listing units in the sample per second-stage unit in the sample in the ath substratum of the hith psu.

The cost function may also be written as follows:

$$C = \sum_h^L C_{1h}m_h + \sum_h^L m_h \sum_i^{M_h} P_{hi}\sum_a^{D_{hi}} N_{hia}f_{hia}C_{2hia} + \sum_h^L m_h \sum_i^{M_h} P_{hi}\sum_a^{D_{hi}} N_{hia}f_{hia}\bar{\bar{Q}}_{hia}f_{hia(j)}C_{3hia} \tag{11.6}$$

The optimum values subject to a fixed total cost are given as follows:

$$m_h = \frac{1}{\sqrt{\lambda}} \sqrt{\frac{\Delta_h}{C_{1h}}} \tag{11.7}$$

$$f_{hia} = \frac{1}{P_{hi}} \sqrt{\frac{C_{1h}}{C_{2hia}} \frac{S^2_{hia} - \bar{\bar{Q}}_{hia}\tilde{S}^2_{whia}}{\Delta_h}} \tag{11.8}$$

$$f_{hia(j)} = \sqrt{\frac{C_{2hia}}{C_{3hia}} \frac{\tilde{S}^2_{whia}}{S^2_{hia} - \bar{\bar{Q}}_{hia}\tilde{S}^2_{whia}}} \tag{11.9}$$

where

$$\sqrt{\lambda} = \frac{1}{C}\left(\sum_h^L \sqrt{C_{1h}\Delta_h} + \sum_h^L \sum_i^{M_h} \sum_a^{D_{hi}} N_{hia} \sqrt{S^2_{hia} - \bar{\bar{Q}}_{hia}\tilde{S}^2_{whia}} \sqrt{C_{2hia}} + \sum_h^L \sum_i^{M_h} \sum_a^{D_{hi}} Q_{hia}\tilde{S}_{whia} \sqrt{C_{3hia}}\right) \tag{11.10}$$

Proof. To obtain the optimum values we set up the Lagrangian F:

$$F = X^2 V^2_{(x'/y')} + \lambda \text{ (Cost function} - C)$$

Then the solution of the equations $\partial F/\partial m_h = 0$, $\partial F/\partial f_{hia} = 0$, $\partial F/\partial f_{hia(j)} = 0$, and the cost condition, Eq. 11.6, yields the optimum values given in Eq. 11.7 to 11.10. The process of obtaining the solution, using the Lagrange multipliers, is illustrated in Sec. 7 above and in Ch. 5, Sec. 9, and is left to the reader.

Note that the optimum over-all sampling fraction is a uniform over-all sampling fraction when $\tilde{S}^2_{whia}/C_{3hia}$ is constant, since for a uniform over-all sampling fraction

$$f = m_h P_{hi} f_{hia} f_{hia(j)}$$

and, substituting the optimum values for m_h, f_{hia}, and $f_{hia(j)}$, we obtain

$$\sqrt{\lambda} f = \frac{\tilde{S}_{whia}}{\sqrt{C_{3hia}}}$$

12. Adjustment for changes in probabilities when initial sample is selected with varying probabilities.* Given a population classified into strata; one unit is selected from each stratum with a specified probability. It is desired to determine a method for drawing a sample of units with probabilities differing from the original probabilities of selection, but still retaining a maximum number of the originally selected units in the sample. The method is applicable, for example, when probability proportionate to a

* This result is due to Keyfitz (3).

measure of size has been used in the initial selection of a sample, the measures of size have since been brought up to date, and it is desired to redraw the sample, using the more recent measures of size, and still retain as much as possible of the original sample.

Illustration of method: Consider one of the strata, and assume that the population consists of units A, B, C, and D with original probabilities of selection equal to α, β, γ, and δ and new probabilities equal to a, b, c, and d. Assume that $a > \alpha$, $b > \beta$, $c < \gamma$, and $d < \delta$. In this case, if either A or B was chosen originally, it would be retained in the sample. However, if C or D was originally selected, some chance of rejecting it must be introduced. The appropriate probability of rejecting C or D would be $(\gamma - c)/\gamma$, or $(\delta - d)/\delta$, respectively.

Suppose that C is the one which was originally selected. We may then determine from a table of random numbers whether or not to reject it, by selecting a random number between 0 and 1. C is rejected from the sample if the random number is between 0 and $(\gamma - c)/\gamma$. If the random number is greater than $(\gamma - c)/\gamma$ the original sample selection is retained.

If we have determined that C (or D) is to be rejected, our next problem is to choose between A and B. The choice is made by selecting another random number between 0 and 1 and determining whether or not it is between 0 and $(a - \alpha)/(a - \alpha + b - \beta)$. If so, A is selected; if not, B is selected.

The proof that this method yields a sample with probabilities of selection proportionate to a, b, c, d, and that it results in the minimum probability of change from the original sample, is left to the reader. The extension to any number of units in the stratum is immediate.

REFERENCES

(1) M. H. Hansen and W. N. Hurwitz, "On the Theory of Sampling from Finite Populations," *Annals Math. Stat.*, **14** (1943), 332–362.

(2) Emil H. Jebe, "Estimation for Subsampling Designs Employing the County as a Primary Sampling Unit," *J. Amer. Stat. Assn.*, **47** (1952), 49–70.

(3) Nathan Keyfitz, "Sampling with Probabilities Proportional to Size: Adjustment for Changes in the Probabilities," *J. Amer. Stat. Assn.*, **46** (1951), 105–109.

CHAPTER 10

Estimating Variances

DERIVATIONS, PROOFS, AND SOME EXTENSIONS OF THEORY FOR CH. 10 OF VOL. I*

NOTE. This chapter contains some relationships (in addition to those given in Ch. 4) useful for evaluating the precision of variance estimates. Some methods are also given for simplifying the estimation of variances from sample returns. Additional methods are given in Ch. 10 of Vol. I.

1. The rel-variance of the estimate of the rel-variance and of the coefficient of variation with simple random sampling (Vol. I, Ch. 10, Sec. 5). *a.* The rel-variance of the estimated rel-variance based on a simple random sample of n units drawn with replacement from a population of N units is

$$V_{v^2}^2 \doteq \frac{\beta - 1}{n} + \frac{4V_X^2}{n} - \frac{4\mu_3/\bar{X}^3 V_X^2}{n} \tag{1.1}$$

where

$$v^2 = \frac{s^2}{\bar{x}^2}, \qquad \bar{x} = \frac{\sum^n x_i}{n}$$

$$s^2 = \frac{\sum^n (x_i - \bar{x})^2}{n - 1}, \qquad \beta = \frac{\mu_4}{\sigma^4}$$

$$\sigma^2 = \frac{\sum^N (X_i - \bar{X})^2}{N}, \quad \mu_4 = \frac{\sum^N (X_i - \bar{X})^4}{N}$$

$$V_X^2 = \frac{\sigma^2[N/(N-1)]}{\bar{X}^2}, \quad \bar{X} = \frac{\sum^N X_i}{N}$$

$$\mu_3 = \frac{\sum^N (X_i - \bar{X})^3}{N}$$

---

* Appropriate references to Vol. I are shown in parentheses after section or subsection headings. The number following I– after some equations gives the chapter, section, and number of that particular equation in Vol. I.

and where x_i is the value for the ith unit in the sample, and X_i is the value for the ith unit in the population.

The proof follows the same steps outlined in Sec. 18 of Ch. 4 for deriving the rel variance of the estimated variance of a ratio, and is left to the reader.

b. The rel-variance of the estimated coefficient of variation based on a simple random sample drawn with replacement is

$$V_v^2 \doteq \frac{\beta - 1}{4n} + \frac{V_X^2}{n} - \frac{\mu_3/\bar{X}^3 V_X^2}{n} \doteq \frac{V_{v^2}^2}{4} \qquad (1.2 \text{ or I–10.5.1})$$

This follows immediately from Part *a* and from Sec. 6, Ch. 4.

Sampling with replacement is assumed here for simplicity and will be a good approximation for sampling without replacement whenever the number of sampling units in the population is large relative to the number in the sample.

2. The rel-variance of the estimated variance for a stratified random sample (Vol. I, Ch. 10, Sec. 7 and 9). *To prove:* The rel-variance, $V_{s^2_{x'}}^2$, of the estimated variance, $s_{x'}^2$, with a stratified random sample of $n = \sum_h^L n_h$ units when N_h is large relative to n_h, is

$$V_{s^2_{x'}}^2 \doteq \frac{1}{S_{x'}^4} \sum_h^L N_h^4 \frac{1}{n_h^3} \left(\beta_h - \frac{n_h - 3}{n_h - 1}\right) S_h^4 \qquad (2.1 \text{ or I–10.7.3})$$

where

$$x' = \sum_h^L \frac{N_h}{n_h} x_h$$

$$s_{x'}^2 = \sum_h^L N_h^2 \frac{s_h^2}{n_h} \qquad (2.2)$$

with

$$s_h^2 = \frac{\sum_i^{n_h} (x_{hi} - \bar{x}_h)^2}{n_h - 1} \qquad (2.3 \text{ or I–5.14.1})$$

and

$$S_{x'}^2 = \sum_h^L N_h^2 \frac{S_h^2}{n_h} \qquad (2.4 \text{ or I–10.7.2})$$

with

$$S_h^2 = \frac{\sum_i^{N_h} (X_{hi} - \bar{X}_h)^2}{N_h - 1} \qquad (2.5 \text{ or I–5.1.2})$$

and

$$\beta_h = \frac{\mu_{4h}}{S_h^4}$$

$$\mu_{4h} = \frac{\sum_i^{N_h}(X_{hi} - \bar{X}_h)^4}{N_h}$$

and x_{hi} is the value of the ith unit in the sample from the hth stratum, and X_{hi} is the value of the hith unit in the population. The n_h represent the number of units in the sample from the N_h units in the hth stratum.

Proof. By definition,

$$V_{s^2_{x'}}^2 = \frac{\sigma_{s^2_{x'}}^2}{(Es_{x'}^2)^2} \tag{2.6}$$

By Theorem 6, Ch. 3 (p. 49), and Sec. 4, Ch. 4,

$$Es_{x'}^2 = \sum_h^L N_h^2 \frac{1}{n_h} S_h^2 = S_{x'}^2 \tag{2.7}$$

By Corollary 1 to Theorem 11, Ch. 3 (p. 56),

$$\sigma_{s^2_{x'}}^2 = \sum_h^L N_h^4 \frac{1}{n_h^2} \sigma_{s^2_h}^2 \tag{2.8}$$

and by Sec. 5, Ch. 4, with N_h large relative to n_h,

$$\sigma_{s^2_{x'}}^2 = \sum_h^L N_h^4 \frac{1}{n_h^2} \frac{\beta_h - \dfrac{n_h - 3}{n_h - 1}}{n_h} S_h^4 \tag{2.9}$$

Substituting Eq. 2.7 and 2.9 into 2.6, we obtain Eq. 2.1. It follows that

$$V_{s^2_{\bar{x}}}^2 = V_{s^2_{x'}}^2, \text{ where } \bar{x} = x'/N$$

Remark 1. If proportionate stratified sampling is used, i.e., $f_h = n_h/N_h = n/N = f$, $s_{x'}^2$ becomes

$$s_{x'}^2 = \frac{N}{f} s_w^2, \text{ where } s_w^2 = \frac{\Sigma N_h s_h^2}{N} \tag{2.10}$$

and $V_{s^2_{x'}}^2 = V_{s^2_w}^2$, and with the substitution $f_h = f$, we have

$$V_{s^2_w}^2 = \frac{\Sigma N_h \left(\beta_h - \dfrac{n_h - 3}{n_h - 1}\right) S_h^4}{N n S_w^4} \tag{2.11 or I–10.7.6}$$

where

$$S_w^2 = \frac{\Sigma N_h S_h^2}{N}$$

Remark 2. If proportionate stratified sampling is used and also the strata are of equal size, i.e., $N_h = \bar{N}$, and hence also $n_h = \bar{n}$, and the within-stratum variances are all the same, i.e., $S_h^2 = \bar{S}^2$ for all h, then $S_w^2 = \bar{S}^2$, $\bar{N} = N/L$, and Eq. 2.11 above becomes

$$V_{s^2_w}^2 = \frac{1}{n}\left(\bar{\beta} - \frac{\bar{n}-3}{\bar{n}-1}\right) \qquad (2.12 \text{ or I–10.9.1})$$

where

$$\bar{\beta} = \frac{\sum^L \mu_{4h}}{L\bar{S}^4}$$

Remark 3. If proportionate stratified sampling is used and also the β_h's are the same for all strata, say $\beta_h = \beta_w$ for all h, then Eq. 2.11 above becomes

$$V_{s^2_w}^2 \doteq \frac{\beta_w - 1}{n}\frac{\Sigma N_h S_h^4}{N S_w^4} \qquad (2.13)$$

and since

$$V_{S^2_h}^2 = \frac{\Sigma N_h\left(S_h^2 - \frac{\Sigma N_h S_h^2}{N}\right)^2}{N\left(\frac{\Sigma N_h S_h^2}{N}\right)^2} = \frac{\Sigma N_h S_h^4}{N S_w^4} - 1$$

we may write

$$V_{s^2_w}^2 \doteq \frac{\beta_w - 1}{n}(1 + V_{S^2_h}^2) \qquad (2.14 \text{ or I–10.9.2})$$

★3. Optimum allocation to strata of a subsample for estimating the variance of a stratified random sample (Vol. I, Ch. 10, Sec. 9). Suppose that the original sample consists of a stratified random sample of $n = \sum_h^L n_h$ sampling units with variance $S_{x'}^2 = \sum_h^L N_h^2(S_h^2/n_h)$, where S_h^2 is given by Eq. 2.5. Suppose also that we wish to estimate the variance $S_{x'}^2$, from a subsample, $n' = \sum_h^L n'_h$, of the original sample.

To prove: The variance of the estimated variance, $\sigma_{s^2_{x'}}^2$, will be close to the minimum when the subsample is allocated as follows:

$$n'_h = \frac{N_h^2 S_h^2 \sqrt{\beta_h - 1}/n_h}{\sum_h^L (N_h^2 S_h^2 \sqrt{\beta_h - 1}/n_h)} n' \qquad (3.1 \text{ or I–10.9.3})$$

★ May be deferred.

where

$$s_{x'}^2 = \sum_h^L \frac{N_h^2}{n_h} \frac{\sum_i^{n'_h}(x_{hi} - \bar{x}_h)^2}{n'_h - 1} \tag{3.2}$$

and from Eq .2.9

$$\sigma_{s^2_{x'}}^2 \doteq \sum_h^L \frac{N_h^4}{n_h^2} \left(\frac{\beta_h - \dfrac{n'_h - 3}{n'_h - 1}}{n'_h} \right) S_h^4 \tag{3.3}$$

$$\doteq \sum_h^L \frac{N_h^4}{n_h^2} \left(\frac{\beta_h - 1}{n'_h} \right) S_h^4, \text{ for } n'_h \text{ reasonably large} \tag{3.4}$$

and

$$\beta_h = \frac{\mu_{4h}}{S_h^4} \tag{3.5}$$

Proof. To obtain the values of n'_h which minimize the variance (Eq. 3.4) subject to the condition that $\sum_h^L n'_h = n'$, we set up the Lagrangian F:

$$F = \sum_h^L \frac{N_h^4}{n_h^2} \left(\frac{\beta_h - 1}{n'_h} \right) S_h^4 + \lambda \left(\sum_h^L n'_h - n' \right)$$

Then the solution to the $L + 1$ equations, $\partial F / \partial n'_h = 0$ and $\sum_h^L n'_h = n'$, in $L + 1$ unknowns (the n'_h and λ) will yield n'_h as given in Eq. 3.1. The solution is straightforward and follows the same steps used in Sec. 9, Ch. 5.

4. The rel-variance of an estimated variance based on random group totals (Vol. I, Ch. 10, Eq. 16.3). *To prove: a.* An unbiased estimate of the population variance based on a simple random sample of n units drawn without replacement from a population of N units may be obtained by subdividing the $n = tk$ units in the sample into t random groups of k units each and computing s_k^2, based on the variance in the t random group totals, i.e.,

$$Es_k^2 = S^2$$

where

$$s_k^2 = \sum_g (x_g - \bar{x}')^2 / k(t - 1) \tag{4.1}$$

x_{gi} = the value of the ith observation in the sample in the gth group

$x_g = \sum_i^k x_{gi}$ is the total value of the characteristic for the k observations in the gth group

$$\bar{x}' = \sum_{g}^{t} x_g / t$$

$$S^2 = \frac{\sum^{N}(X_i - \bar{X})^2}{N - 1}$$

and where X_i is the value for the ith unit in the population and $\bar{X} = \Sigma X_i / N$.

b. The rel-variance of s_k^2 is given by Eq. 4.2 whenever the sampling is with replacement or the sampling is without replacement and the number of units in the population is large relative to the number in the sample.

$$V_{s^2_k}^2 = \frac{1}{t}\left(\beta_k - \frac{t-3}{t-1}\right) \qquad \text{(4.2 or I–10.16.3)}$$

where

$$\beta_k = \frac{\beta}{k} + 3\,\frac{k-1}{k} \qquad (4.3)$$

$$\beta = \frac{\mu_4}{\sigma^4}$$

$$\mu_4 = \frac{\sum^{N}(X_i - \bar{X})^4}{N}, \quad \text{and} \quad \sigma^2 = \frac{\sum^{N}(X_i - \bar{X})^2}{N}$$

Proof. *a*. Consider first Es_k^2:

$$Es_k^2 = \frac{1}{k(t-1)}\left(\sum_{g}^{t} Ex_g^2 - tE\bar{x}'^2\right) \qquad (4.4)$$

Since x_g is a total from a simple random sample of k units (from Sec. 2, Ch. 4),

$$Ex_g^2 = k^2\bar{X}^2 + k\,\frac{N-k}{N}\,S^2$$

and since $\bar{x}' = k\bar{x}$, where $\bar{x} = \Sigma x_i / n$,

$$E\bar{x}'^2 = k^2\bar{X}^2 + k^2\,\frac{N-n}{N}\,\frac{S^2}{n}$$

It follows, by substituting these results in 4.4, that

$$Es_k^2 = S^2 \qquad (4.5)$$

Similarly, if the sampling had been with replacement, we would have

$$Es_k^2 = \sigma^2 \qquad (4.6)$$

b. Consider next $V^2_{s^2_k}$ when the sampling has been done with replacement. By definition,

$$V^2_{s^2_k} = \frac{\sigma^2_{s^2_k}}{(Es^2_k)^2} = \frac{Es^4_k - (Es^2_k)^2}{(Es^2_k)^2} \tag{4.7}$$

We may also write s^2_k as

$$s^2_k = k\,\frac{\sum_g^t(\bar{x}_g - \bar{x})^2}{t-1} \tag{4.8}$$

where

$$\frac{x_g}{k} = \bar{x}_g$$

$$\frac{\bar{x}'}{k} = \bar{x} = \sum_i^n \frac{x_i}{n}$$

Make the following transformation:

$$x_{gi} - \bar{X} = z_{gi}$$

$$\bar{x}_g - \bar{X} = \bar{z}_g = \sum_i^k \frac{z_{gi}}{k}$$

$$\bar{x} - \bar{X} = \bar{z} = \sum_g^t \frac{\bar{z}_g}{t}$$

Then

$$s^2_k = \frac{k}{t-1}\sum_g^t(\bar{z}_g - \bar{z})^2 \tag{4.9}$$

and

$$Es^4_k = \frac{k^2}{(t-1)^2}\,E\left[\sum_g^t(\bar{z}_g - \bar{z})^2\right]^2$$

$$= \frac{k^2}{(t-1)^2}\,E\left[\left(\sum_g^t \bar{z}^2_g\right)^2 - 2t\bar{z}^2\sum_g^t \bar{z}^2_g + t^2\bar{z}^4\right] \tag{4.10}$$

The first term in Eq. 4.10 becomes

$$E\left(\sum_g^t \bar{z}^2_g\right)^2 = E\sum_g^t \bar{z}^4_g + E\sum_{g\neq h}^t \bar{z}^2_g\bar{z}^2_h$$

$$= E\sum_g^t\left(\frac{\sum_i^k z_{gi}}{k}\right)^4 + E\sum_{g\neq h}^t\left(\frac{\sum_i^k z_{gi}}{k}\right)^2\left(\frac{\sum_i^k z_{hi}}{k}\right)^2$$

which by Sec. 5, Ch. 4,

$$= t\left[\frac{\mu_4}{k^3} + \frac{3(k-1)\sigma^4}{k^3}\right] + t(t-1)\,\frac{\sigma^4}{k^2} \tag{4.11}$$

The second term in Eq. 4.10 becomes

$$E\left(2t\bar{z}^2\sum_g^t \bar{z}_g^2\right) = 2tE\left[\left(\frac{\sum_g^t \bar{z}_g}{t}\right)^2 \sum_g^t \bar{z}_g^2\right]$$

$$= \frac{2}{t} E\left(\sum_g^t \bar{z}_g^2\right)^2$$

which by Eq. 4.11 above

$$= 2\left[\frac{\mu^4}{k^3} + \frac{3(k-1)\sigma^4}{k^3}\right] + 2(t-1)\frac{\sigma^4}{k^2} \tag{4.12}$$

The third term in Eq. 4.10 becomes

$$Et^2\bar{z}^4 = \frac{1}{t^2} E\left(\sum_g^t \bar{z}_g\right)^4 = \frac{1}{t^2} E\left(\sum_g^t \bar{z}_g^4 + 3\sum_{g\neq h}^t \bar{z}_g^2\bar{z}_h^2\right)$$

which by Sec. 5, Ch. 4,

$$= \frac{1}{t^2}\left\{t\left[\frac{\mu_4}{k^3} + \frac{3(k-1)\sigma^4}{k^3}\right] + 3t(t-1)\frac{\sigma^4}{k^2}\right\} \tag{4.13}$$

Combining Eq. 4.11, 4.12, and 4.13 into 4.10 and simplifying, we obtain

$$Es_k^4 = \frac{\mu_4}{kt} + \frac{3(k-1)}{kt}\sigma^4 + \frac{t^2 - 2t + 3}{(t-1)t}\sigma^4 \tag{4.14}$$

Substituting Eq. 4.6 and 4.14 into 4.7, we obtain

$$V_{s^2_k}^2 = \frac{1}{t}\left[\frac{1}{k}\frac{\mu_4}{\sigma^4} + \frac{3(k-1)}{k} - \frac{t-3}{t-1}\right]$$

$$= \frac{1}{t}\left[\frac{\beta}{k} + \frac{3(k-1)}{k} - \frac{t-3}{t-1}\right]$$

$$= \frac{1}{t}\left[\beta_k - \frac{t-3}{t-1}\right] \tag{4.2}$$

Exercise 4.1. By the procedure indicated in Sec. 2, extend $V_{s^2_k}^2$ to a stratified sampling case where random groups are set up in each stratum.

***5. The variance of the estimates of the components of the variance in a two-stage sample design with $\bar{n}$ elementary units in the sample from each first-stage unit in the sample, $\bar{N}$ elementary units in the population in each first-stage unit, and the sampling carried out with replacement at each stage of sampling (or small sampling fractions)** (Vol. I, Ch. 10, Sec. 6).

---

* May be deferred.

a. To prove: The variance of an estimate of the between component of the variance is

$$\sigma^2_{s^2_b} = \frac{\mu_{4b} - \sigma_b^4 \dfrac{m-3}{m-1}}{m} + \frac{4\sigma_b^2\sigma_w^2}{\bar{n}(m-1)} + \frac{2\sigma_w^4}{m\bar{n}^2} \frac{m\bar{n}-1}{(\bar{n}-1)(m-1)} + \frac{4\sigma_{\delta^2\sigma^2}}{m\bar{n}} \tag{5.1}$$

where $s_b^2 = \dfrac{\sum^m(\bar{x}_i - \bar{x})^2}{m-1} - \dfrac{s_w^2}{\bar{n}}$ is an unbiased estimate of

$$S_b^2 = \frac{M}{M-1}\sigma_b^2 = \frac{\sum^M(\bar{X}_i - \bar{X})^2}{M-1}, \quad s_w^2 = \frac{\sum^m\sum^{\bar{n}}(x_{ij} - \bar{x}_i)^2}{m(\bar{n}-1)}$$

$$\mu_{4b} = \frac{\sum^M(\bar{X}_i - \bar{X})^4}{M}; \quad \sigma_{\delta^2\sigma^2} = E[(\bar{X}_i - \bar{X})^2\sigma_i^2]$$

$$\sigma_i^2 = \frac{\sum^{\bar{N}}(X_{ij} - \bar{X}_i)^2}{\bar{N}}; \quad \text{and} \quad \sigma_w^2 = \frac{\sum^M\sum^{\bar{N}}(X_{ij} - \bar{X}_i)^2}{M\bar{N}}$$

The proof is left to the reader.

Hint: Let

$$x_{ij} = \bar{X} + \delta_i + \Delta_{ij}$$

where

$$\delta_i = \bar{X}_i - \bar{X}, \quad \text{and} \quad \Delta_{ij} = x_{ij} - \bar{X}_i$$

Then

$$s_b^2 = \frac{\sum^m(\bar{x}_i - \bar{x})^2}{m-1} - \frac{\sum^m\sum^{\bar{n}}(x_{ij} - \bar{x}_i)^2}{\bar{n}m(\bar{n}-1)}$$

$$= \frac{\sum^m(\delta_i - \bar{\delta})^2}{m-1} + \frac{2\sum^m(\delta_i - \bar{\delta})(\overline{\Delta}_i - \overline{\Delta})}{m-1} + \frac{\sum_i^m\sum_{j\neq l}^{\bar{n}}\Delta_{ij}\,\Delta_{il}}{m\bar{n}(\bar{n}-1)} - \frac{\sum_{i\neq j}^m\overline{\Delta}_i\overline{\Delta}_j}{m(m-1)}$$

where

$$\bar{\delta} = \frac{\sum_i^m \delta_i}{m}, \quad \overline{\Delta}_i = \frac{\sum_j^{\bar{n}}\Delta_{ij}}{\bar{n}}, \quad \text{and} \quad \overline{\Delta} = \frac{\sum_i^m\overline{\Delta}_i}{m}$$

Now

$$E\left[\frac{\sum^m(\delta_i - \bar{\delta})^2}{m-1} - \sigma_b^2\right]^2 = \frac{\mu_{4b} - \sigma_b^4\dfrac{m-3}{m-1}}{m}$$

Also

$$\frac{\sum^{m}(\delta_i - \bar{\delta})(\overline{\Delta}_i - \overline{\Delta})}{m-1} = \frac{\sum^{m}\delta_i\overline{\Delta}_i}{m} - \frac{\sum\limits_{i\neq j}^{m}\delta_i\overline{\Delta}_j}{m(m-1)}$$

and

$$E\left[\frac{\sum^{m}\delta_i\overline{\Delta}_i}{m} - \frac{\sum\limits_{i\neq j}^{m}\delta_i\overline{\Delta}_j}{m(m-1)}\right]^2 = E\,\frac{\delta_i^2\sigma_i^2}{\bar{n}} + \frac{\sigma_b^2\sigma_w^2}{m\bar{n}(m-1)}$$

$$E\left[\frac{\sum\limits_{i}^{m}\sum\limits_{j\neq l}^{\bar{n}}\Delta_{ij}\Delta_{il}}{m\bar{n}(\bar{n}-1)}\right]^2 = \frac{2\sigma_w^4}{m\bar{n}(\bar{n}-1)}$$

and

$$E\left[\frac{\sum\limits_{i\neq j}^{m}\overline{\Delta}_i\overline{\Delta}_j}{m(m-1)}\right]^2 = \frac{2\sigma_w^4}{m(m-1)\bar{n}^2}$$

b. The variance of an estimate of the within component of the variance is

$$\sigma^2_{s^2_w} = \frac{1}{m\bar{n}}\left(\mu_{4w} - \sigma_w^4\,\frac{\bar{n}-3}{\bar{n}-1}\right) \tag{5.2}$$

Exercises

5.1. Prove that

$$\sigma^2_{s^2_c} = \frac{\mu_{4b} - \sigma_b^4\,\dfrac{m-3}{m-1}}{m} + \frac{\mu_{4w}}{m\bar{n}^3} - \frac{\sigma_w^4}{m\bar{n}^3}\left[\frac{m(3-2\bar{n})-3}{m-1}\right]$$
$$+ \frac{4\sigma_b^2\sigma_w^2}{\bar{n}(m-1)} + \frac{6\sigma_{\delta^2\sigma^2}}{m\bar{n}} + \frac{4\sigma_{\delta\mu_3}}{m\bar{n}^2} \tag{5.3}$$

where

$$s_c^2 = \frac{\sum^{m}(\bar{x}_i - \bar{x})^2}{m-1}$$

Show that, when $\bar{n} = 1$, $V^2_{s^2_c} = \sigma^2_{s^2_c}/(Es_c^2)^2$ reduces to the rel-variance of s^2, where $s^2 = \sum^{n}(x_i - \bar{x})^2/n$, and its rel-variance is given by Eq. 5.1 of Ch. 4.

5.2. Show that the covariance of s_c^2 and $s_w^2/\bar{n}$ is

$$\sigma_{s^2_c,\, s^2_w/\bar{n}} = \frac{\mu_{4w}}{m\bar{n}^3} - \frac{3\sigma_w^4}{m\bar{n}^3} + \frac{2\sigma_{\delta\mu_3}}{m\bar{n}^2} + \frac{\sigma_{\delta^2\sigma^2}}{m\bar{n}} \tag{5.4}$$

5.3. For a simple random sample of n elements show that

$$\mu_4 = \mu_{4w} + 4E\delta_i\mu_{i3} + 6E\delta_i^2\sigma_i^2 + \mu_{4b} \tag{5.5}$$

and

$$\sigma^4 = \sigma_b^4 + 2\sigma_b^2\sigma_w^2 + \sigma_w^4 \tag{5.6}$$

***6. Use of known stratum means and totals in estimating variances** (Vol. I, Ch. 10, Sec. 19). *a. To prove:* When the stratum means are known, an unbiased estimate of S_w^2, the average within-stratum variance for a proportionate stratified random sample of $n = \sum_h^L n_h$ units, is

$$s_w^2 = \frac{1}{N}\sum_h^L \frac{N_h^2}{N_h - 1} \frac{\sum_i^{n_h}(x_{hi} - \bar{X}_h)^2}{n_h} \qquad \text{(6.1 or I–10.19.1)}$$

where n_h is the number of units selected from the hth stratum, and

$$S_w^2 = \frac{\sum_h^L N_h S_h^2}{N} \qquad \text{(6.2 or I–5.3.8)}$$

where S_h^2 is defined by Eq. 2.5.

Proof. By Theorem 6, Ch. 3 (p. 49),

$$Es_w^2 = \frac{1}{N}\sum_h^L \frac{N_h^2}{N_h - 1}\frac{1}{n_h}\sum_i^{n_h} E(x_{hi} - \bar{X}_h)^2$$

Since $\bar{X}_h$ is known, the probability of obtaining $(x_{hi} - \bar{X}_h)^2$ is the probability of obtaining x_{hi}, namely, $1/N_h$. Therefore,

$$E(x_{hi} - \bar{X}_h)^2 = \frac{1}{N_h}\sum_i^{N_h}(X_{hi} - \bar{X}_h)^2$$

and

$$Es_w^2 = \frac{1}{N}\sum_h^L \frac{N_h}{N_h - 1}\sum_i^{N_h}(X_{hi} - \bar{X}_h)^2$$

$$= \frac{1}{N}\sum_h^L N_h S_h^2 = S_w^2$$

b. To prove: When the stratum totals are known, an unbiased estimate of B_X^2, the between-psu contribution to the variance for a multi-stage stratified sampling design in which m_h primary units are selected with replacement from the hth stratum, is:

$$b_X^2 = \frac{1}{L\hat{X}^2}\sum_h^L \frac{1}{m_h}\sum_i^{m_h}\left(\frac{X'_{hi}}{P_{hi}} - X_h\right)^2 \qquad \text{(6.3 or I–10.19.2)}$$

where P_{hi} is the probability of selecting the ith primary unit from the hth stratum,

$$\hat{X} = \frac{\sum_h^L X_h}{L}$$

* May be deferred.

X'_{hi} is the total of the X-characteristic in the ith psu of the hth stratum,

$$B_X^2 = \frac{\sum_h^L \sum_i^{M_h} P_{hi} \left(\frac{X_{hi}}{P_{hi}} - X_h \right)^2}{L\hat{X}^2} \tag{6.4}$$

Proof. By Theorem 6, Ch. 3 (p. 49),

$$Eb_X^2 = \frac{1}{L\hat{X}^2} \sum_h^L \frac{1}{m_h} \sum_i^{m_h} E \left(\frac{X'_{hi}}{P_{hi}} - X_h \right)^2$$

Since X_h is known, the probability of obtaining $\left(\frac{X'_{hi}}{P_{hi}} - X_h \right)^2$ is the probability of obtaining X'_{hi}, namely, P_{hi}. Therefore,

$$E \left(\frac{X'_{hi}}{P_{hi}} - X_h \right)^2 = \sum_i^{M_h} P_{hi} \left(\frac{X_{hi}}{P_{hi}} - X_h \right)^2$$

and

$$Eb_X^2 = B_X^2$$

7. Confidence limits for the median and other position measures* (Vol. I, Ch. 10, Sec. 18). Let x_A and x_B be the estimated Ath and Bth percentiles of a distribution derived from a sample where A is less than 50 per cent and B is greater than 50 per cent. Let p be the percentage of items in the sample which are less than θ, the true median.

To prove: The x_A and x_B constitute confidence limits for θ with the following probability:

$$Pr(x_A < \theta < x_B) = Pr(A < p < B)$$

Proof. Let us assume that a given population is arrayed by size. The point θ (the true median) divides the population into two equal parts.† Let us assume further that a sample is drawn from this population, and by proper weighting of each sample item a reflection of the original population is obtained and ordered by size. This is illustrated graphically by the cumulative frequency curves in Fig. 1. The solid line is the cumulative frequency curve of the population; the dashed line represents the cumulative frequency curve derived from a sample.

Now let us draw two arbitrary horizontal lines across the graph to represent the limits A and B. Then, since these curves are nondecreasing, θ is within the limits x_A and x_B if and only if p is within the limits A and B.

* Proof developed by Ralph S. Woodruff (4).

† Note that this condition is not met where more than a negligible proportion of the population has a value exactly equal to θ.

Symbolically, this can be expressed as

$$Pr(x_A < \theta < x_B) = Pr(A < p < B)^*$$

Since it is often possible to make meaningful statements about probabilities on the right-hand side of this equation, the limits on the left-hand side constitute usable confidence limits for the median. For example, if simple random sampling is used, the probability of p falling within the limits A and B can be calculated by summing appropriate terms of the binomial distribution (either directly or by means of tables of the incomplete β function). More generally, if large samples are used (with

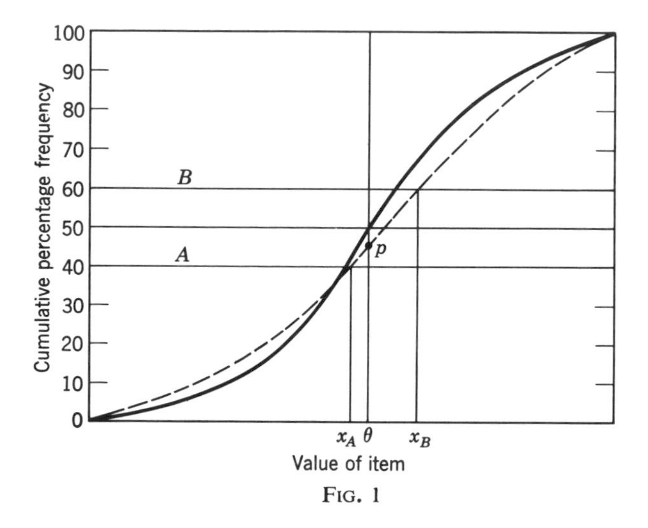

FIG. 1

any type of probability sample design), the distribution of p will often be near normal and a usable estimate of σ_p^2 can frequently be derived from the sample.† In this case we can obtain meaningful confidence limits for

* For the application of this principle with simple random sampling, see, for example, "On Confidence Ranges for the Median and Other Expectation Distributions for Population of Unknown Distribution Form," by W. R. Thompson, *Annals Math. Stat.*, **7** (1936), 122–128; "Order Statistics," by S. S. Wilks, *Bull. Amer. Math. Soc.*, **54** (1948), No. 1, p. 14; and A. M. Mood, *Introduction to the Theory of Statistics*, McGraw-Hill, 1950, pp. 388–389.

† Theoretically σ_p^2 should be the variance of the percentage of items less than θ, the true median. Since in practical problems θ is not known, the value of σ_p^2 must be estimated as the variance of the percentage of sample items less than θ', the median derived from the sample distribution. Where large samples are used, the substitution of θ' for θ has little effect on the estimate of σ_p^2.

the median by choosing our arbitrary limits A and B so that $A = 50$ per cent minus $K\sigma_p$ and $B = 50$ per cent plus $K\sigma_p$ (K being any positive number).

The method can be applied directly to other position measures as well as the median.

REFERENCES

(1) U.S. Bureau of the Census, Sampling Staff, *A Chapter in Population Sampling*, U.S. Government Printing Office, Washington, D.C., 1947.

(2) U.S. Bureau of the Census, *Notes on Precision of Sample Estimates* (Reprint of Appendix Section of the Special Study: Value of Farm Products by Color and Tenure of Farm Operators), U.S. Government Printing Office, Washington, D.C., 1945.

(3) M. H. Hansen and W. N. Hurwitz, "Relative Efficiencies of Various Sampling Units in Population Inquiries," *J. Amer. Stat. Assn.*, **37** (1942), 89–94.

(4) Ralph S. Woodruff, "Confidence Intervals for Medians and Other Position Measures," *J. Amer. Stat. Assn.*, **47** (1952), 635–646.

CHAPTER 11

Regression Estimates, Double Sampling, Sampling for Time Series, and Other Sampling Methods

DERIVATIONS, PROOFS, AND SOME EXTENSIONS OF THEORY FOR CH. 11 OF VOL. I*

NOTE. Many of the developments presented in this chapter can be applied as variants or extensions of the methods already introduced. Included among the topics covered are alternative methods of estimation; double sampling (often referred to as two-phase or multi-phase sampling); techniques for selection of sample units that introduce negative correlations in the selections between strata, and thus have the effect of extending the depth of stratification; an approach for estimation of characteristics of each unit in a population from a sample of such units; and sampling for time series.

1. The difference and regression estimates (Vol. I, Ch. 11, Eq. 2.1, 2.7, 2.2, 2.10). *a. The difference estimate. To prove:* The difference estimate, Eq. 1.1, is an unbiased estimate of X, and its variance is given by Eq. 1.2.

Proof. Let x' and y' be random variables that are unbiased estimates of X and Y, and let k be any arbitrary constant. Then the "difference estimate" of X is

$$x_1'' = x' + k(Y - y') \qquad (1.1 \text{ or I–11.2.1})$$

Since x' and y' are unbiased estimates of X and Y,

$$Ex_1'' = Ex' + k(Y - Ey') = X$$

and hence the difference estimate is an unbiased estimate of X.

* Appropriate references to Vol. I are shown in parentheses after section or subsection headings. The number following I– after some equations gives the chapter, section, and number of that particular equation in Vol. I.

The variance of x_1'' is

$$\sigma_1^2 = E\{x'^2 + 2kx'(Y - y') + k^2(Y - y')^2\} - X^2$$

$$= \sigma_{x'}^2 + k^2\sigma_{y'}^2 - 2k\rho_{x'y'}\sigma_{x'}\sigma_{y'} \qquad (1.2 \text{ or I-11.2.7})$$

b. The regression estimate. To prove: The regression estimate, Eq. 1.4, is a consistent estimate, and its variance is

$$\sigma_2^2 \doteq \sigma_{x'}^2(1 - \rho_{x'y'}^2) \qquad (1.3)$$

when x', y', and b are consistent estimates of X, Y, and β, respectively. The values of X, Y, and β, defined below, can be estimated from sample designs described in this book, such as stratified sampling, cluster sampling, and multi-stage sampling. In each case, one should investigate the precision of the approximation of Eq. 1.3.

Proof. Let

$$x_2'' = x' + b(Y - y') \qquad (1.4 \text{ or I-11.2.2})$$

and let

$$\hat{x}_2'' = x' + \beta(Y - y')$$

Since x', y', and b are consistent estimates of X, Y, and β, it follows from Corollary 2, Theorem 20, Ch. 3, that x_2'' is a consistent estimate of X. To obtain the variance of x_2'' we can write

$$x_2'' = \hat{x}_2'' + (b - \beta)(Y - y')$$

so that

$$E(x_2'' - X)^2 = \sigma_{\hat{x}_2''}^2 + 2E(\hat{x}_2'' - X)(Y - y')(b - \beta)$$

$$+ E(b - \beta)^2(Y - y')^2 \qquad (1.5)$$

The variance of $\hat{x}_2''$ is given by substituting $\beta = \rho_{x'y'}\sigma_{x'}/\sigma_{y'}$ for k in Eq. 1.2 and is equal to

$$\sigma_{\hat{x}_2''}^2 = \sigma_{x'}^2(1 - \rho_{x'y'}^2) \qquad (1.6)$$

where

$$\rho_{x'y'} = \frac{\sigma_{x'y'}}{\sigma_{x'}\sigma_{y'}}$$

If $E(\hat{x}_2'' - X)^2$ is to be approximated by $\sigma_{\hat{x}_2''}^2 \doteq \sigma_2^2$, the remainder term is

$$2E(\hat{x}_2'' - X)(Y - y')(b - \beta) + E(b - \beta)^2(Y - y')^2 \qquad (1.7)$$

and it remains to show that, for a sufficiently large sample size, these terms are negligible compared to σ_2^2. Thus, if x' is an estimate of the total X from a simple random sample of size n, σ_2^2 is of the order $1/n$.

By the inequality given on p. 56, Ch. 3,

$$E(\hat{x}_2'' - X)(Y - y')(b - \beta)$$
$$\leq [E(b-\beta)^2 E(\hat{x}_2'' - X)^2 (Y - y')^2]^{1/2}$$
$$\leq [E(b-\beta)^2]^{1/2} [E(\hat{x}_2'' - X)^4]^{1/4} [E(Y - y')^4]^{1/4}$$

Since $\hat{x}_2''$ and y' are arithmetic means, it follows that

$$E(\hat{x}_2'' - X)^4 = 0(n^{-2})$$
$$E(Y - y')^4 = 0(n^{-2})$$

where $0(n^{-2})$ indicates that the term is of the order of $1/n^2$ in n so that

$$[E(\hat{x}_2'' - X)^4]^{1/4} = 0(n^{-1/2})$$
$$[E(Y - y')^4]^{1/4} = 0(n^{-1/2})$$

Also, since b is a ratio estimate, we have

$$\frac{E(b-\beta)^2}{\beta^2} = \frac{\sigma^2_{s_{xy}}}{\sigma^2_{XY}} + \frac{\sigma^2_{s^2_y}}{\sigma^4_Y} - \frac{2\sigma_{s_{xy}s^2_y}}{\sigma_{XY}\sigma^2_Y} + 0(n^{-2})$$

Since

$$|\sigma_{s_{xy}s^2_y}| \leq \sigma_{s_{xy}} \sigma_{s^2_y}$$

and

$$\sigma_{s_{xy}} = 0(n^{-1/2}), \quad \sigma_{s^2_y} = 0(n^{-1/2})$$

we have

$$\frac{E(b-\beta)^2}{\beta^2} \leq \left(\frac{\sigma_{s_{xy}}}{\sigma_{XY}} + \frac{\sigma_{s^2_y}}{\sigma^2_Y}\right)^2 + 0(n^{-2}) = 0(n^{-1})$$

Thus

$$E(b-\beta)^2 = 0(n^{-1})$$

and this analysis would indicate that

$$E(\hat{x}_2'' - X)(Y - y')(b - \beta) \text{ is } 0(n^{-3/2})$$

However, since products and ratios of arithmetic means must have integer orders, it follows that this term is at least of order n^{-2}, and this term will be small relative to $\sigma^2_{\hat{x}_2''}$ for sufficiently large n.

Also

$$E(b-\beta)^2(Y-y')^2 \leq [E(b-\beta)^4 E(Y-y')^4]^{1/2}$$
$$= [0(n^{-2})0(n^{-2})]^{1/2}$$
$$= 0(n^{-2})$$

and we have the second term of the remainder (Eq. 1.7):

$$E(b-\beta)^2(Y-y')^2 \text{ of order } 0(n^{-2})$$

Then

$$\sigma^2_{x''_2} = \sigma^2_{\hat{x}''_2} + 0(n^{-2}) \tag{1.8}$$

and the entire remainder term will be small relative to $\sigma^2_{\hat{x}''_2}$ for sufficiently large n.

2. A consistent estimate of the regression coefficient, β (Vol. I, Ch. 11, Eq. 2.6, 2.15, and 2.16). Let the regression coefficient of x' on y' be defined as follows:

$$\beta = \frac{E(x' - X)(y' - Y)}{E(y' - Y)^2} \tag{2.1}$$

or for stratified sampling as:

$$\beta = \frac{E\sum_h^L (x'_h - X_h)(y'_h - Y_h)}{E\sum_h^L (y'_h - Y_h)^2} \tag{2.2}$$

Then, a consistent estimate of β is given by the ratio of consistent estimates of the numerator and of the denominator of Eq. 2.1 or 2.2.

In the special case where

$$x'_h = \frac{\sum_i^{m_h} \dfrac{X_{hi}}{P_{hi}}}{m_h}$$

and

$$y'_h = \frac{\sum_i^{m_h} \dfrac{Y_{hi}}{P_{hi}}}{m_h}$$

then

$$\sum_h^L \frac{\sum_i^{m_h} \left(\dfrac{X_{hi}}{P_{hi}} - x'_h\right)\left(\dfrac{Y_{hi}}{P_{hi}} - y'_h\right)}{m_h(m_h - 1)} \tag{2.3}$$

is an unbiased and consistent estimate of the numerator of Eq. 2.2, and

$$\sum_h^L \frac{\sum_i^{m_h} \left(\dfrac{Y_{hi}}{P_{hi}} - y'_h\right)^2}{m_h(m_h - 1)} \tag{2.4}$$

is an unbiased and consistent estimate of the denominator of Eq. 2.2.

Therefore b is a consistent estimate of β, where

$$b = \frac{\text{Eq. 2.3}}{\text{Eq. 2.4}} \tag{2.5}$$

3. Variance and optimum allocation for double sampling with regression estimates (Vol. I, Ch. 11, Eq. 3.1–3.5). *To prove:* Suppose that an estimate of $\bar{X}$ is wanted from a sample, that $\bar{y}$ is the average value of some related characteristic obtained at low unit costs from a large sample of size n selected by simple random sampling, and that $\bar{x}'$ is the sample average of the X's obtained from a smaller sample of size n', also obtained by simple random sampling, which is a subsample of the larger sample, of size n. Similarly, $\bar{y}'$ is the sample mean of the Y's for this same subsample. Then an estimate of $\bar{X}$, using the regression estimate with simple random sampling, is

$$\bar{x}'' = \bar{x}' + b(\bar{y} - \bar{y}') \qquad \text{(3.1 or I–11.3.1)}$$

where b is the estimate of β, the coefficient of regression of x on y, i.e.,

$$b = \frac{\sum^{n'}(x_i - \bar{x}')(y_i - \bar{y}')}{\sum^{n'}(y_i - \bar{y}')^2} \qquad \text{(3.2 or I–11.3.2)}$$

a. The variance. The variance of $\bar{x}''$ is given approximately by

$$\sigma^2_{\bar{x}''} \doteq \frac{S^2_X}{n'}\left\{1 - \rho^2\left(1 - \frac{n'}{n}\right)\right\} \qquad \text{(3.3 or I–11.3.3)}$$

where ρ is the coefficient of correlation between X_i and Y_i.

Proof. By Sec. 1, Ch. 11, for sufficiently large n', the variance of estimate 3.1 is given approximately by the variance of

$$\bar{x}'' = \bar{x}' + \beta(\bar{y} - \bar{y}') \qquad (3.4)$$

The expected value of estimate 3.4 is

$$E\{E(\bar{x}''|[n])\} = \bar{X} \qquad (3.5)$$

and hence estimate 3.4 is unbiased.

The variance of estimate 3.4 is derived from Theorem 15 of Ch. 3 (p. 65), which, as applied to this problem, states that

$$\sigma^2_{\bar{x}''} = \sigma^2_{E(\bar{x}''|[n])} + E\sigma^2_{\bar{x}''|[n]} \qquad (3.6)$$

The variance of $\bar{x}''$ is equal to the variance of the expected value of $\bar{x}''$ for a fixed sample of n units plus the expected value of the conditional variance of $\bar{x}''$ for a fixed sample of n units. The first term on the right-hand side of Eq. 3.6 is the variance of $\bar{x} = E(\bar{x}''|[n])$ and is, assuming sampling with replacement,

$$\sigma^2_{E(\bar{x}''|[n])} = \frac{S^2_X}{n} \qquad (3.7)$$

The second term of Eq. 3.6 is

$$E\sigma^2_{\bar{x}''|[n]} = E\frac{n-n'}{nn'}(s_X^2 + \beta^2 s_Y^2 - 2\beta s_{XY})$$

$$= \frac{n-n'}{nn'}(S_X^2 + \beta^2 S_Y^2 - 2\beta S_{XY})$$

$$= \frac{n-n'}{nn'} S_X^2(1-\rho^2) \tag{3.8}$$

Then Eq. 3.6 becomes

$$\sigma^2_{\bar{x}''} = \frac{S_X^2}{n'}\left[1 - \rho^2\left(1 - \frac{n'}{n}\right)\right] \tag{3.3}$$

which is approximately equal to the variance of Eq. 3.1.

b. Optimum allocation for a simple cost function. Assume a simple cost function:

$$C = C_1 n + C_2 n' \tag{3.9}$$

where C_1 is the unit cost of including an element in the large sample, C_2 is the unit cost of including an element in the small sample, and C is the total cost, varying with the number of elements in the sample; C_1 is assumed to be considerably smaller than C_2 for the design to have practical significance.

To prove: The optimum values of n and n' such that the variance, Eq. 3.3, is minimized subject to the fixed cost, C, are:

$$\text{opt. } n = \frac{C}{C_1 + C_2\sqrt{\dfrac{1-\rho^2}{\rho^2}\dfrac{C_1}{C_2}}} \tag{3.10 or I–11.3.4}$$

and

$$\text{opt. } n' = n\sqrt{\frac{1-\rho^2}{\rho^2}\frac{C_1}{C_2}} \tag{3.11 or I–11.3.5}$$

Proof. Define the Lagrangian function:

$$F(n, n', \lambda) = \frac{S_X^2}{n'}\left[1 - \rho^2\left(1 - \frac{n'}{n}\right)\right] + \lambda(C_1 n + C_2 n' - C)$$

An expression for n in terms of $\sqrt{\lambda}$ is obtained from setting $\partial F/\partial n = 0$, and an expression for n' in terms of $\sqrt{\lambda}$ from $\partial F/\partial n' = 0$. Substituting these expressions for n and n' in Eq. 3.9 yields the expression for $\sqrt{\lambda}$ independent of n and n'. From this it follows that n and n' are given by Eq. 3.10 and 3.11.

4. Condition for cost and variance of single and double sampling designs to be equivalent (Vol. I, Ch. 11, Eq. 3.6). *To prove:* The optimum double sampling design with the regression estimate will have the same cost and variance as for a single (simple random) sample design with a simple unbiased estimate when

$$\rho^2 = \frac{4C_1C_2}{(C_1 + C_2)^2} \qquad \text{(4.1 or I–11.3.6)}$$

Proof. The variance of the double sampling design having the minimum variance subject to a fixed cost, C, is expressed by substituting the optimum values of n and n' from Eq. 3.10 and 3.11 into the expression for the variance, Eq. 3.3. Under these conditions 3.3 becomes:

$$\sigma^2_{\bar{x}''}\,(\text{opt.}) = \frac{S^2_X}{C}\,[\rho\sqrt{C_1} + \sqrt{(1-\rho^2)C_2}]^2 \qquad (4.2)$$

where the cost function is presumed to be

$$C = C_1 n + C_2 n' = C_1 n + C_2 n \sqrt{\frac{1-\rho^2}{\rho^2}\frac{C_1}{C_2}} \qquad (4.3)$$

The variance of a single (simple random) sample of, say, m units is

$$\sigma^2_{\bar{x}'} = \frac{S^2_X}{m} \qquad (4.4)$$

The cost of such a single sample is assumed to be

$$C = C_2 m \qquad (4.5)$$

From Eq. 4.4 and 4.5, we have

$$\sigma^2_{\bar{x}'} = \frac{S^2_X}{C}\,C_2 \qquad (4.6)$$

We have now to find the condition under which the variances 4.2 and 4.6 are equal when their total costs 4.3 and 4.5 are equal. It follows that the variances will be equal with equal total costs when Eq. 4.1 holds. If we require now that only the costs of these two designs be the same and determine the condition for which the optimum double sampling design has a smaller variance than the single sample, we have from Eq. 4.2 and 4.6

$$\sigma^2_{\bar{x}'} - \sigma^2_{\bar{x}''} = \frac{S^2_X}{C}\,C_2 - \frac{S^2_X}{C}\,[\rho\sqrt{C_1} + \sqrt{(1-\rho^2)C_2}]^2 > 0$$

It follows that this inequality holds for

$$\rho^2 > \frac{4C_1C_2}{(C_1 + C_2)^2}$$

Therefore, for values of ρ that satisfy this inequality, the double sampling design shows a gain over the single sampling design.

5. Variance and optimum allocation for double sampling with stratification (Vol. I, Ch. 11, Eq. 3.8, 3.9, 3.11–3.16). It is desired to estimate a total characteristic, X, from a population of N units, where N_1 of these units have a particular characteristic, Z, and the remaining $N_2 = N - N_1$ units have the characteristic W (i.e., they are non-Z's), and it is assumed that stratification into the Z and W groups would be advantageous. (For example, the units might be farms or business establishments, and the characteristic Z might apply to those establishments larger than some specified size; or the distinction between the Z's and W's might be that the Z's responded to a mailed questionnaire within a specified period of time, and the W's did not, etc.)

Using a double sampling approach, an initial large simple random sample of n units is drawn from the population of N. In the n units, n_1 are found to possess attribute Z, and from the remaining n_2 units having attribute W a simple random subsample of n_2' units is selected. The total sample size is $n_1 + n_2'$.

a. The estimate. To prove: If x_1 is the aggregate of a characteristic for the n_1 units in the sample with attribute Z and x_2 for the n_2' units ultimately included in the sample with attribute W, then an unbiased estimate of X is

$$x' = \frac{x_1}{f} + \frac{k}{f}x_2 \qquad \text{(5.1 or I–11.3.8)}$$

where $f = n/N$ is the sampling fraction for the initial large sample of units, and $k = n_2/n_2'$ is the reciprocal of the subsampling fraction.

Proof. Estimate 5.1 may also be written

$$x' = \frac{N}{n}\sum_i^{n_1} x_{1i} + \frac{N}{n}\frac{n_2}{n_2'}\sum_i^{n'_2} x_{2i}$$

where x_{1i} represents the characteristic for the ith unit in the sample with attribute Z, and x_{2i} for the ith unit in the subsample with attribute W.

The conditional expected value of the estimate 5.1 for a fixed sample size of n_2' units from a fixed set of n_2 units is

$$E(x'|n, n_2') = \frac{N}{n}\sum_i^{n_1} x_{1i} + \frac{N}{n}\sum_i^{n_2} x_{2i}$$

$$= \frac{N}{n}\sum_i^{n} X_i' \qquad (5.2)$$

where X_i' represents the characteristic in the ith unit in the initial sample regardless of whether it has attribute Z or W. The expected value of Eq. 5.2 is

$$E[E(x'|n_2, n_2')] = Ex' = X$$

and estimate 5.1 is therefore unbiased.

b. Variance. To prove: The variance of estimate 5.1 is given by

$$\sigma_{x'}^2 = N^2 \frac{N-n}{Nn} S^2 + \frac{N}{n}(k-1)N_2 S_2^2 \quad \text{(5.3 or I–11.3.9)}$$

where

$$S^2 = \frac{\sum_i^N (X_i - \bar{X})^2}{N-1}$$

with

$$\bar{X} = \frac{\sum_i^N X_i}{N}$$

and

$$S_2^2 = \frac{\sum_i^{N_2} (X_{2i} - \bar{X}_2)^2}{N_2 - 1}$$

with

$$\bar{X}_2 = \frac{\sum_i^{N_2} X_{2i}}{N_2}$$

Proof. The variance, Eq. 5.3, is derived by use of Theorem 15 of Ch. 3 (p. 65), which, as applied to this problem, states

$$\sigma_{x'}^2 = \sigma_{E(x'|n_2, n'_2)}^2 + E\sigma_{x'|n_2, n'_2}^2 \tag{5.4}$$

the variance of x' is equal to the variance of the expected value of x' for a fixed number, n_2', of units from a fixed sample of n_2 units having attribute W plus the expected value of the conditional variance of x' for a fixed number, n_2', of units from a fixed sample of n_2 units having attribute W.

The first term on the right of Eq. 5.4 is merely the variance of estimate 5.2, which is

$$N^2 \frac{N-n}{Nn} S^2 \tag{5.5}$$

The conditional variance of 5.1 for a fixed number, n_2', of units from a fixed set of n_2 units having W is

$$\sigma_{x'|n_2, n'_2}^2 = \frac{n_2(k-1)}{f^2} \frac{\sum_i^{n_2} (x_{2i} - \bar{x}_2')^2}{n_2 - 1} \tag{5.6}$$

where

$$\bar{x}_2' = \frac{\sum_i^{n_2} x_{2i}}{n_2}$$

The expected value of Eq. 5.6 for a fixed number, n_2, of W units is

$$\frac{n_2(k-1)}{f^2} S_2^2 \tag{5.7}$$

There is still the condition on formula 5.7, however, that it is the variance for a fixed sample size of n_2, so that the expected value of 5.7, which is the second term on the right of Eq. 5.4, is

$$E\sigma^2_{x'|n_2, n'_2} = \frac{k-1}{f^2} \frac{n}{N} N_2 S_2^2$$

$$= \frac{N}{n} (k-1) N_2 S_2^2 \tag{5.8}$$

From Eq. 5.4, 5.5, and 5.8, we have the variance of estimate 5.1 as given in Eq. 5.3.

c. Optimum allocation of sample sizes for minimum cost subject to a fixed variance.

(1) The following simple cost relationship is assumed:

$$C = C_0 n + C_1 n_1 + C_2 n_2' \qquad \text{(5.9 or I–11.3.11)}$$

where C is the total cost of the survey less any fixed overhead costs that do not vary with the allocation of the sample.

C_0 is the unit cost of selecting and examining a unit included in the large sample and determining whether it is included at the full rate or not.

C_1 is the additional cost per unit for the units that are included at the full large sample rate but not in the subsample.

C_2 is the additional cost per unit of the subsampled units actually included in the subsample.

Let us now determine the optimum allocation for a fixed error and minimum expected cost. The expected cost is given by the expected value of Eq. 5.9, i.e.,

$$EC = n\left(C_0 + C_1 P_1 + \frac{C_2 P_2}{k}\right)$$

where P_1 is the proportion of all units that have attribute Z; $P_2 = 1 - P_1$.

If we specify that the standard error of estimate 5.1 be equal to ε, the values of n and k which minimize the expected cost are

$$n = \hat{n}\left\{1 + (k-1)P_2\frac{S_2^2}{S^2}\right\} \qquad \text{(5.10 or I–11.3.12)}$$

$$k = \sqrt{\frac{S^2 - P_2S_2^2}{P_1S_2^2}\frac{C_2}{C_1 + (C_0/P_1)}} \qquad \text{(5.11 or I–11.3.13)}$$

where

$$\hat{n} = \frac{NS^2}{S^2 + (\varepsilon^2/N)} \qquad \text{(5.12 or I–11.3.14)}$$

is the size of sample that would be required to achieve the specified accuracy with a simple random sample.

This is seen by constructing the function

$$F(n, k, \lambda) = n\left[C_0 + C_1P_1 + \frac{C_2P_2}{k}\right] + \lambda\left[N^2\frac{N-n}{Nn}S^2 + \frac{N}{n}(k-1)N_2S_2^2 - \varepsilon^2\right]$$

Setting $\partial F/\partial n$ and $\partial F/\partial k$ equal to zero and eliminating λ and n, the expression 5.11 for k follows. An expression for n in terms of k follows from setting $\partial F/\partial\lambda = 0$. This expression is easily shown to be Eq. 5.10, where the value of $\hat{n}$, given in Eq. 5.12, satisfies the relationship

$$\varepsilon^2 = N^2\frac{N-\hat{n}}{N\hat{n}}S^2 \qquad (5.13)$$

Exercise 5.1. Assume that we have a population consisting of L strata with M_h first-stage units in the hth stratum, and N_{hi} second-stage units in the hith first-stage unit. Assume that a simple random sample of m_h first-stage units is selected from the M_h units in the hth stratum, and n_{hi} second-stage units are selected from the hith first-stage unit in the sample. Assume further that $\frac{m_h}{M_h}\frac{n_{hi}}{N_{hi}} = f_h$, i.e., a uniform sampling fraction is used in the hth stratum, and that

$$\frac{n_{hi}}{N_{hi}} = \frac{E\bar{n}_h}{\bar{N}_h}$$

where

$$\bar{n}_h = \frac{\sum^{m_h} n_{hi}}{m_h}$$

and

$$\bar{N}_h = \frac{\sum^{M_h} N_{hi}}{M_h}, \quad \text{and} \quad N_h = \sum^{M_h} N_{hi}$$

Let

$$\sum_h^L \sum_i^{m_h} n_{hi} = n$$

the total number of second-stage units in the sample, represent the initial sample of a double sampling design. Assume that n_1 of the n units in the sample are from class 1, i.e., have some specified characteristic (such as being large farms or large stores or respondents to a mailed questionnaire) and are retained in the sample, and that the remaining $n - n_1 = n_2$ cases in the initial sample are in class 2, i.e., do not have the specified characteristic, and that we draw a simple random sample of 1 in k from those remaining n_2 cases.

Now, in the hth stratum, let (dropping the subscript h)

$$x' = \frac{1}{f} \sum_i^m \sum_j^{n_i} x_{1ij} + \frac{k}{f} \sum_i^m \sum_j^{n'_{2i}} x_{2ij}$$

where $x_{1ij} = X_{1ij}$ if the ijth selection is a member of class 1,
$= 0$ otherwise.

$x_{2ij} = X_{2ij}$ if the ijth selection is a member of class 2,
$= 0$ otherwise.

n_i is the number of elements in the ith first-stage unit in the sample.

n'_{2i} is the number of elements subsampled from the n_{2i} elements in class 2, with $\sum_i^m n_{2i} = n_2$.

N_{2i} is the number of elements in class 2 in the ith first-stage unit.

Show that

$$\sigma^2_{x'} = E\sigma^2_{x'|[1,2]} + \sigma^2_{E(x'|[1,2])}$$

where the first term of the right-hand expression is the expectation of the conditional variance of x' for a fixed set of m first-stage units and a fixed set of n_i observations in the ith sample first-stage unit and is equal to

$$E\sigma^2_{x'|[1,2]} = \frac{M}{m} \frac{k-1}{f} E \sum_i^m N_{2i} \hat{s}^2_{x_2}$$

where

$$\hat{s}^2_{x_2} = \frac{\sum_i^m \sum_j^{N_{2i}} \left(X_{2ij} - \frac{\sum_i^m \sum_j^{N_{2i}} X_{2ij}}{\sum_i^m N_{2i}} \right)^2}{\sum_i^m N_{2i} - 1}$$

and the second term of the right-hand expression is the variance of the conditional expected value of x' for a fixed set of m first-stage units and a fixed set of n_i observations in the ith sample first-stage unit and is equal to

$$\sigma^2_{E(x'|[1,2])} = M^2 \frac{M-m}{Mm} S^2_b + N^2 \frac{\bar{N} - E\bar{n}}{\bar{N} m E\bar{n}} S^2_w$$

where

$$S_b^2 = \frac{\sum_i^M (X_i - \bar{X})^2}{M - 1}$$

and

$$S_w^2 = \frac{1}{N} \sum_i^M \frac{N_i}{N_i - 1} \sum_j^{N_i} (X_{ij} - \bar{\bar{X}}_i)^2$$

where X_{ij} is the value of X for the jth element in the ith first-stage unit, and $X_i = \sum_j^{N_i} X_{ij}$ is the value of X for the ith first-stage unit,

$$\bar{\bar{X}}_i = \frac{\sum_j^{N_i} X_{ij}}{N_i}, \quad \text{and} \quad \bar{X} = \frac{\sum^M X_i}{M}$$

★6. Estimate and variance of Latin-square design (Vol. I, Ch. 11, Eq. 4.2–4.8).* Suppose that a population consisting of $M = L^2\bar{M}$ units is classified into L "columns" of $L\bar{M}$ units each, and that each column is classified into L "rows" of $\bar{M}$ units each. Let the L^2 parts into which the whole population is thus classified be called "cells," so that each cell consists of $\bar{M}$ units. Select a cell at random from the first column. From the second column select at random any cell except that in the row selected from the first column. Continue in this way, selecting at random in the rth column a cell from any of the $L - r + 1$ rows not selected for columns 1, 2, · · ·, $(r - 1)$. In each selected cell, choose at random one of the $\bar{M}$ units. There will then be $m = L$ units in the sample.

Let x_{abi} be the value of the ith unit selected from the cell in the ath row in column b, and consider the estimate of the total

$$x' = \frac{1}{f} \sum_a^L \sum_b^1 \sum_i^1 x_{abi} = \frac{1}{f} x \qquad \text{(6.1 or I–11.4.1)}$$

where $f = m/M$ is the sampling fraction, and x is the aggregate value of a characteristic for the units in the sample.

To prove: x' is an unbiased estimate of

$$X = \sum_a^L \sum_b^L \sum_i^{\bar{M}} X_{abi}$$

and the variance of x' is

$$\sigma_{x'}^2 = \frac{1}{m - 1} (\tilde{\sigma}^2 - \tilde{\sigma}_1^2 - \tilde{\sigma}_2^2) + \frac{\tilde{\sigma}_w^2}{m} \qquad \text{(6.2 or I–11.4.3)}$$

---

* This result is due to Jerome Cornfield and W. Duane Evans.

★ May be deferred.

where

$$\tilde{\sigma}^2 = L^2 \sum_a^L \sum_b^L (X_{ab} - \bar{X})^2 = L^4\sigma^2$$

$$\tilde{\sigma}_1^2 = L^3 \sum_a^L (\bar{X}_{a\cdot} - \bar{X})^2 = L^4\sigma_1^2$$

$$\tilde{\sigma}_2^2 = L^3 \sum_b^L (\bar{X}_{\cdot b} - \bar{X})^2 = L^4\sigma_2^2$$

$$\tilde{\sigma}_w^2 = L^2\bar{M} \sum_a^L \sum_b^L \sum_i^{\bar{M}} (X_{abi} - \bar{\bar{X}}_{ab})^2 = L^4\bar{M}^2\sigma_w^2$$

with

$$X_{ab} = \sum_i^{\bar{M}} X_{abi}, \quad \bar{\bar{X}}_{ab} = X_{ab}/\bar{M}, \quad X_{a\cdot} = \sum_b^L X_{ab}, \quad \bar{X}_{a\cdot} = X_{a\cdot}/L$$

$$X_{\cdot b} = \sum_a^L X_{ab}, \quad \bar{X}_{\cdot b} = X_{\cdot b}/L, \quad X = \sum_a^L X_{a\cdot} = \sum_b^L X_{\cdot b}$$

and

$$\bar{X} = X/L^2$$

Proof. (*a*) Since $1/f = L\bar{M}$,

$$Ex' = L\bar{M}E \sum_a^L \sum_b^1 \sum_i^1 x_{abi} = L\bar{M} \sum_a^L \sum_b^1 \sum_i^1 Ex_{abi}$$

$$= L\bar{M} \sum_a^L \sum_b^1 \sum_i^1 \frac{1}{L^2\bar{M}} \sum_a^L \sum_b^L \sum_i^{\bar{M}} X_{abi}$$

$$= \sum_a^L \sum_b^L \sum_i^{\bar{M}} X_{abi} = X$$

Thus, x' is an unbiased estimate of X.

(*b*) We now consider the variance of x'. We may write

$$\sigma_{x'}^2 = E\sigma_{x'|c}^2 + \sigma_{E(x'|c)}^2 \tag{6.3}$$

where $\sigma_{x'|c}^2$ denotes the conditional variance of x' for a fixed selection of cells and $E(x'|c)$ denotes the conditional expectation of x' for a fixed selection of cells.

We first note that

$$E(x'|c) = L\bar{M} \sum_a^L \sum_b^1 E \sum_i^1 (x_{abi}|c)$$

$$= L\bar{M} \sum_a^L \sum_b^1 \frac{1}{\bar{M}} \sum_i^{\bar{M}} X_{abi}$$

$$= L \sum_a^L \sum_b^1 X_{ab}$$

where X_{ab} denotes the population total for the cell selected in the ath column.

To evaluate $\sigma^2_{E(x'|c)}$ we write

$$E\{E(x'|c) - Ex'\}^2 = E\left(L\sum_a^L\sum_b^1 X_{ab} - X\right)^2$$

$$= L^2E\left(\sum_a^L\sum_b^1 X_{ab}\right)^2 - X^2 \qquad (6.4)$$

Now consider

$$E\left(\sum_a^L\sum_b^1 X_{ab}\right)^2 = E\left[\sum_a^L\left(\sum_b^1 X_{ab}\right)^2 + \sum_{a\neq c}^L\left(\sum_b^1 X_{ab}\right)\left(\sum_d^1 X_{cd}\right)\right] \qquad b \neq d$$

$$= \sum_a^L \frac{1}{L}\sum_b^L X_{ab}^2 + \sum_{a\neq c}^L \frac{1}{L(L-1)}\sum_{b\neq d}^L X_{ab}X_{cd}$$

$$= \sum_a^L \frac{1}{L}\sum_b^L X_{ab}^2 + \sum_{a\neq c}^L \frac{1}{L(L-1)}\left[\left(\sum_b^L X_{ab}\right)\left(\sum_b^L X_{cb}\right) - \sum_b^L X_{ab}X_{cb}\right]$$

$$= \frac{1}{L}\sum_a^L\sum_b^L X_{ab}^2 + \frac{1}{L(L-1)}\sum_{a\neq c}^L\left(X_{a\cdot}X_{c\cdot} - \sum_b^L X_{ab}X_{cb}\right)$$

$$= \frac{1}{L}\sum_a^L\sum_b^L X_{ab}^2 + \frac{1}{L(L-1)}\left[\left(\sum_a^L X_{a\cdot}\right)^2 - \sum_a^L X_{a\cdot}^2\right]$$

$$- \frac{1}{L(L-1)}\sum_b^L\left[\left(\sum_a^L X_{ab}\right)^2 - \sum_a^L X_{ab}^2\right]$$

$$= \frac{1}{L-1}\sum_a^L\sum_b^L X_{ab}^2 - \frac{L}{L-1}\left(\sum_a^L \bar{X}_{a\cdot}^2 + \sum_b^L \bar{X}_{\cdot b}^2\right) + \frac{L^3\bar{X}^2}{L-1} \qquad (6.5)$$

and Eq. 6.4 becomes

$$\sigma^2_{E(x'|c)} = \frac{L^4}{L-1}\left[\left(\sum_a^L\sum_b^L \frac{X_{ab}^2}{L^2} - \bar{X}^2\right) - \left(\sum_a^L \frac{\bar{X}_{a\cdot}^2}{L} - \bar{X}^2\right) - \left(\sum_b^L \frac{\bar{X}_{\cdot b}^2}{L} - \bar{X}^2\right)\right]$$

$$= \frac{L^4}{L-1}(\sigma^2 - \sigma_1^2 - \sigma_2^2) \qquad (6.6)$$

where σ^2 is the variance among the cell totals.

σ_1^2 is the variance among the means per cell from the columns.

σ_2^2 is the variance among the means per cell from the rows.

Now consider

$$\sigma^2_{x'|c} = E\{[x' - E(x'|c)]^2|c\} \qquad (6.7)$$

For a sample of one unit per cell

$$\sigma^2_{x'|c} = L^2\bar{M}^2 \sum_a^L \sum_b^1 \sigma^2_{ab}$$

where

$$\sigma^2_{ab} = \frac{1}{\bar{\bar{M}}} \sum_i^{\bar{\bar{M}}} (X_{abi} - \bar{\bar{X}}_{ab})^2$$

and is the variance within the cell selected in the ath column. Now

$$E\sigma^2_{x'|c} = L^2\bar{M}^2 \sum_a^L E \sum_b^1 \sigma^2_{ab}$$

$$= L^2\bar{M}^2 \sum_a^L \frac{1}{L} \sum_b^L \sigma^2_{ab} \tag{6.8}$$

Combining these results into Eq. 6.3, we have

$$\sigma^2_{x'} = L^4 \left[\frac{1}{L-1} (\sigma^2 - \sigma^2_1 - \sigma^2_2) + \frac{1}{L} \bar{M}^2 \sigma^2_w\right]$$

where

$$\sigma^2_w = \frac{1}{L^2} \sum_a^L \sum_b^L \sigma^2_{ab} \tag{6.9 or I-11.4.7}$$

is the average variance within cells.

It therefore follows from the definitions of m, $\tilde{\sigma}^2$, $\tilde{\sigma}^2_1$, $\tilde{\sigma}^2_2$ given above that

$$\sigma^2_{x'} = \frac{1}{m-1} (\tilde{\sigma}^2 - \tilde{\sigma}^2_1 - \tilde{\sigma}^2_2) + \frac{\tilde{\sigma}^2_w}{m} \tag{6.2}$$

It should be noted that this approach easily generalizes to other experimental designs.

***7. The optimum allocation of sample and the optimum weights for estimating a ratio from a stratified sample** (Vol. I, Ch. 11, Sec. 6). *To prove:* Consider a population divided into two strata and let

$$r = w_1 \frac{x'_1}{y'_1} + w_2 \frac{x'_2}{y'_2} \tag{7.1}$$

where

$$x'_1 = \frac{N_1}{n_1} \sum_i^{n_1} x_i, \quad y'_1 = \frac{N_1}{n_1} \sum_i^{n_1} y_i$$

are estimated aggregates for stratum I, and

$$x'_2 = \frac{N_2}{n_2} \sum_i^{n_2} x_i, \quad y'_2 = \frac{N_2}{n_2} \sum_i^{n_2} y_i$$

* May be deferred.

are estimated aggregates for stratum II, and are unbiased estimates of X_1, Y_1, X_2, and Y_2, respectively; the n_1 constitute a simple random sample from the N_1 elements of stratum I, the n_2 elements constitute a simple random sample from the N_2 elements of stratum II, and the selections are made independently from the two strata. Also, w_1 and w_2 are weights to be applied to the stratum estimates, where $w_1 + w_2 = 1$. Then

$$E(r - R)^2 \doteq w_1^2 \frac{N_1 - n_1}{N_1 n_1} S_1^2 + w_2^2 \frac{N_2 - n_2}{N_2 n_2} S_2^2 + (R_1 - R_2)^2 \left(w_1 - \frac{Y_1}{Y}\right)^2 \quad (7.2)$$

where

$$R_1 = \frac{X_1}{Y_1} \quad \text{and} \quad R_2 = \frac{X_2}{Y_2}$$

and

$$\frac{Y_1}{Y} R_1 + \left(1 - \frac{Y_1}{Y}\right) R_2 = R = \frac{X}{Y}$$

$$S_1^2 = R_1^2(V_{1X}^2 + V_{1Y}^2 - 2V_{1XY}) \quad (7.3)$$

$$S_2^2 = R_2^2(V_{2X}^2 + V_{2Y}^2 - 2V_{2XY}) \quad (7.4)$$

and V_{1X}^2, V_{1Y}^2, and V_{1XY} are the rel-variances and covariance in the first stratum of the X_{1i} and Y_{1i}, and the terms are similarly defined in the second stratum. Moreover, the values of w_1, w_2, n_1, n_2 which minimize estimate 7.1 subject to the condition that $w_1 + w_2 = 1$ and $n_1 + n_2 = n$ are determined by

$$w_1 = \frac{S_2^2 - S_1 S_2 + n\left[\frac{Y_1}{Y}(R_1 - R_2)^2 - \frac{S_2^2}{N_2}\right]}{(S_2 - S_1)^2 + n\left[(R_1 - R_2)^2 - \frac{S_1^2}{N_1} - \frac{S_2^2}{N_2}\right]} \quad \text{(7.5 or I–11.6.3)}$$

$$n_1 = \frac{S_2^2 - S_1 S_2 + n\left[\frac{Y_1}{Y}(R_1 - R_2)^2 - \frac{S_2^2}{N_2}\right]}{\left[\frac{Y_1}{Y}(R_1 - R_2)^2 - \frac{S_2^2}{N_2}\right] + \frac{S_2}{S_1}\left[(R_1 - R_2)^2\left(1 - \frac{Y_1}{Y}\right) - \frac{S_1^2}{N_1}\right]}$$

(7.6 or I–11.6.4)

The optimum value of w_2 is obtained by subtracting w_1 in Eq. 7.5 from 1 and of n_2 by subtracting the value of n_1 in Eq. 7.6 from n.

Proof. For sufficiently large values of n_1 and n_2,

$$E(r - R)^2 \doteq E[w_1^2(r_1 - R_1)^2 + w_2^2(r_2 - R_2)^2] + [w_1(R_1 - R) + w_2(R_2 - R)]^2 \quad (7.7)$$

where

$$r_1 = \frac{x_1'}{y_1'} \quad \text{and} \quad r_2 = \frac{x_2'}{y_2'}$$

By substituting the values for

$$E(r_1 - R_1)^2 \doteq \frac{N_1 - n_1}{N_1 n_1} S_1^2$$

and

$$E(r_2 - R_2)^2 \doteq \frac{N_2 - n_2}{N_2 n_2} S_2^2$$

where S_1^2 is given by Eq. 7.3 and S_2^2 by Eq. 7.4,* and $\frac{Y_1}{Y} R_1 + \left(1 - \frac{Y_1}{Y}\right) R_2$ for R and $1 - w_1$ for w_2 in expression 7.7 above, we obtain

$$F(n_1, w_1) \doteq w_1^2 \left(\frac{S_1^2}{n_1} - \frac{S_1^2}{N_1}\right) + (1 - w_1)^2 \left(\frac{S_2^2}{n - n_1} - \frac{S_2^2}{N_2}\right) + \left(w_1 - \frac{Y_1}{Y}\right)^2 (R_1 - R_2)^2 \qquad (7.8)$$

If we set $\partial F/\partial n_1 = 0$ and $\partial F/\partial w_1 = 0$ and solve for n_1 and w_1, we obtain Eq. 7.5 and Eq. 7.6.

Exercises

7.1. Find the optimum values of w_1 and w_2 for fixed values of n_1, n_2. The solution to this problem is appropriate when the sample allocation is predetermined.

7.2. Show that as $(R_1 - R_2)^2$ increases, and if the other terms are not substantially affected, the optimum value of w_1 approaches Y_1/Y.

7.3. Show that

$$E(R_1 - R_2)^2 \doteq \frac{S_1^2}{N_1} + \frac{S_2^2}{N_2}$$

if it is assumed that the first stratum was made up by drawing a random sample of N_1 establishments from some very large population for which the ratio $X_1/Y_1 = R$, and in which the variance defined by Eq. 7.3 was equal to S_1^2, and if the second stratum was made up in an analogous manner from another population in which $X_2/Y_2 = R$ and S_2^2 was the variance defined by Eq. 7.4.

7.4. Show that if

$$(R_1 - R_2)^2 \doteq \frac{S_1^2}{N_1} + \frac{S_2^2}{N_2} \quad \text{(see Ex. 7.3)}$$

$$w_1 \text{ (opt.)} = \frac{S_2^2 - S_1 S_2 + \frac{n}{Y}(\bar{Y}_1 S_1^2 - \bar{Y}_2 S_2^2)}{(S_2 - S_1)^2}$$

* From Vol. II, Ch. 5, Sec. 3.

Under these conditions, and if the first stratum is made up of large establishments and the second stratum of small establishments, so that $\bar{Y}_1 = Y_1/N_1$ is larger than $\bar{Y}_2 = Y_2/N_2$, show that w_1 (opt.) increases as $\bar{Y}_1$ increases relative to $\bar{Y}_2$, and, as a practical matter, for $\bar{Y}_1$ large enough relative to $\bar{Y}_2$ (and for the other terms constant), the optimum values are then

$$w_1 = 1, \quad w_2 = 0$$
$$n_1 = n, \quad n_2 = 0$$

See Vol. I, Ch. 11, Sec. 6, for a discussion of this case.

8. Sampling on two occasions (Vol. I, Ch. 11, Sec. 7). Consider a population of N units in existence over a period of time, where N is large relative to the sizes of samples to be drawn or sampling with replacement is assumed.* Suppose that on the first of two occasions a simple random sample of n units is selected. Retain a simple random sample of Pn of these units for the second occasion, and supplement these by a simple random sample of Qn independently selected units, where $P + Q = 1$. Thus, the second sample is also of size n. Let

$\bar{x}'$ = mean per unit for the first period, for the Pn units that are common to the two samples.

$\bar{x}''$ = mean per unit for the first period, for the Qn units that are in the first sample only.

$\bar{y}'$ = mean per unit for the second period, for the Pn units that are common to the two samples.

$\bar{y}''$ = mean per unit for the second period, for the Qn units that are in the second sample only.

a. An estimate of the mean. We wish to estimate $\bar{Y}$, the mean for the second period, by a linear estimate of the form

$$\bar{y} = a\bar{x}'' + b\bar{x}' + c\bar{y}' + d\bar{y}''$$

Since $E\bar{x}'' = E\bar{x}' = \bar{X}$ and $E\bar{y}'' = E\bar{y}' = \bar{Y}$, we find that

$$E\bar{y} = (a + b)\bar{X} + (c + d)\bar{Y}$$

If we now require that $\bar{y}$ be an unbiased estimate of $\bar{Y}$, we must have

$$a + b = 0, \quad c + d = 1$$

so that

$$\bar{y} = a(\bar{x}'' - \bar{x}') + c\bar{y}' + (1 - c)\bar{y}'' \qquad (8.1)$$

* N large relative to n (or sampling with replacement) is assumed for simplicity and as an approximation. The results can be extended to the case where n/N is large relative to 1 and the sampling is without replacement.

The variance of $\bar{y}$ is

$$\sigma_{\bar{y}}^2 = a^2 \left(\frac{1}{Q} + \frac{1}{P}\right) \frac{\sigma_X^2}{n} + \frac{c^2}{P} \frac{\sigma_Y^2}{n} + \frac{(1-c)^2}{Q} \frac{\sigma_Y^2}{n} - \frac{2ac}{P} \frac{\rho\sigma_X\sigma_Y}{n} \tag{8.2}$$

where σ_X^2 is the population variance of an individual observation in the first period, σ_Y^2 is the population variance in the second period, and ρ is the correlation between the first and second periods for an observation on the same sampling unit.

We wish to choose values of a and c that minimize $\sigma_{\bar{y}}^2$. Equating to zero the derivatives of $\sigma_{\bar{y}}^2$ with respect to a and c, it follows that the optimum values are

$$a = \frac{\rho PQ}{1 - Q^2\rho^2} \frac{\sigma_Y}{\sigma_X} \tag{8.3}$$

$$c = \frac{P}{1 - Q^2\rho^2} \tag{8.4}$$

Thus, the estimate with optimum values for a and c may be written

$$\bar{y}_w = \frac{\rho PQ}{1 - Q^2\rho^2} \frac{\sigma_Y}{\sigma_X} (\bar{x}'' - \bar{x}') + \frac{P}{1 - Q^2\rho^2} \bar{y}' + \frac{Q(1 - Q\rho^2)}{1 - Q^2\rho^2} \bar{y}'' \tag{8.5}$$

and its variance is

$$\sigma_{\bar{y}_w}^2 = \frac{\sigma_Y^2}{n} \frac{1 - \rho^2 Q}{1 - \rho^2 Q^2} \tag{8.6 or I–11.7.3}$$

Equating to zero the derivative of $\sigma_{\bar{y}_w}^2$ with respect to Q, we find that for a fixed sample size n the variance of $\bar{y}_w$ will have its minimum value if we choose

$$Q = \frac{1 - \sqrt{1 - \rho^2}}{\rho^2} \tag{8.7}$$

Note that, if $\sigma_X = \sigma_Y$, the estimate given by Eq. 8.5 is somewhat simplified, but its variance is unchanged. Note, also, that an estimate for the first occasion is given by Eq. 8.5, simply by interchanging X's and Y's if the estimate for the first occasion can await a time until data for both occasions are available.

b. Estimates of the change. One possible obvious estimate of the change $\bar{Y} - \bar{X}$ is

$$\Delta = P(\bar{y}' - \bar{x}') + Q(\bar{y}'' - \bar{x}'') \tag{8.8}$$

whose variance is

$$\sigma_\Delta^2 = \frac{1}{n} (\sigma_X^2 + \sigma_Y^2 - 2P\rho\sigma_X\sigma_Y) \tag{8.9}$$

If we consider the more general linear estimate of the change of the form

$$a\bar{x}'' + b\bar{x}' + c\bar{y}' + d\bar{y}''$$

subject to the condition that this provide an unbiased estimate of $\bar{Y} - \bar{X}$, we find that we must take $a + b = -1$, $c + d = 1$. Following the same procedure as for the estimate of the mean in the second period, we find that the estimate that minimizes the variance is

$$\Delta_w = \frac{Q(1 - Q\rho^2)}{1 - Q^2\rho^2}(\bar{y}'' - \bar{x}'') + \frac{P}{1 - Q^2\rho^2}(\bar{y}' - \bar{x}') + \frac{PQ\rho}{1 - Q^2\rho^2}\left[(\bar{x}'' - \bar{x}')\frac{\sigma_y}{\sigma_x} - (\bar{y}'' - \bar{y}')\frac{\sigma_x}{\sigma_y}\right] \quad (8.10)$$

In the special case that $\sigma_x = \sigma_y$, the estimate is greatly simplified to the form

$$\Delta_w = \frac{P}{1 - Q\rho}(\bar{y}' - \bar{x}') + \frac{Q(1 - \rho)}{1 - Q\rho}(\bar{y}'' - \bar{x}'') \quad (8.11)$$

The variance of Δ_w (Eq. 8.11) is

$$\sigma^2_{\Delta_w} = \frac{2(1 - \rho)\sigma^2}{n(1 - Q\rho)} \quad (8.12)$$

and its development is left as an exercise.

Note that the estimate of change given by Eq. 8.10 is exactly the estimate that would be obtained if both $\bar{X}$ and $\bar{Y}$ were estimated from Eq. 8.5 and the difference $\bar{y}_w - \bar{x}_w$ were computed as the estimate of $\bar{Y} - \bar{X}$.

It is also to be noted that for $\rho > 0$ Eq. 8.12 is a minimum for $Q = 0$, i.e., the variance will be minimized if the units on both occasions are identical and ρ is positive.

c. Estimate of the sum of the means. By the same approach, we obtain as the optimum estimate of $\bar{Y} + \bar{X}$ the statistic

$$z_w = \frac{Q(1 - Q\rho^2)}{1 - Q^2\rho^2}(\bar{x}'' + \bar{y}'') + \frac{P}{1 - Q^2\rho^2}(\bar{x}' + \bar{y}') + \frac{PQ\rho}{1 - Q^2\rho^2}\left[(\bar{x}'' - \bar{x}')\frac{\sigma_y}{\sigma_x} + (\bar{y}'' - \bar{y}')\frac{\sigma_x}{\sigma_y}\right]$$

In the special case that $\sigma_x = \sigma_y$, the optimum estimate is again greatly simplified, so that it may be written

$$z_w = \frac{P}{1 + Q\rho}(\bar{x}' + \bar{y}') + \frac{Q(1 + \rho)}{1 + Q\rho}(\bar{x}'' + \bar{y}'')$$

The variance of z_w is readily obtained.

Note that $z_w = \bar{x}_w + \bar{y}_w$, where $\bar{y}_w$ is given by Eq. 8.5, and $\bar{x}_w$ is given by Eq. 8.5 with X's and Y's interchanged.

d. Joint estimates of change and the means. It has been seen that, if the required timing of the survey estimates is such that the results from the samples for both occasions can be used in preparing estimates for each occasion, then the use of Eq. 8.5 to obtain estimates of the mean for each occasion also results in an estimate of the sum and of the difference that is the best linear estimate (i.e., smallest variance) that can be made from the data from the two samples. Often, however, estimates must be made for the first occasion before sample results from the second occasion are available; the initial estimate must be made from the first sample only, and it may not be feasible to revise this initial estimate. Thus, suppose that we have estimated the mean on the first occasion as

$$P\bar{x}' + Q\bar{x}''$$

Suppose then that we wish to estimate both the mean on the second occasion and the change from the first to the second occasion in such a way that the estimated change is the difference between the estimated means. Let us denote the estimated mean on the second occasion by $\bar{y}$ and the estimated change by Δ, and require that these are to have the forms:

$$\bar{y} = a\bar{x}' + b\bar{x}'' + c\bar{y}' + d\bar{y}''$$

$$\Delta = e\bar{x}' + f\bar{x}'' + g\bar{y}' + h\bar{y}''$$

where the coefficients are constants. We have already required that

$$\Delta = \bar{y} - \bar{x}$$

If we require further that $E\bar{y} = \bar{Y}$, $\bar{y}$ and Δ may be written

$$\bar{y} = (e + P)\bar{x}' - (e + P)\bar{x}'' + c\bar{y}' + (1 - c)\bar{y}''$$

$$\Delta = e\bar{x}' - (e + 1)\bar{x}'' + c\bar{y}' + (1 - c)\bar{y}''$$

It may be useful to determine the constants e and c so as to minimize a linear function of the variances of $\bar{y}$ and Δ. Without loss of generality, we may minimize

$$\sigma_\Delta^2 + w\sigma_{\bar{y}}^2$$

where w is a specified positive number. The solution to this problem is straightforward and yields

$$e = \frac{P}{1 - Q^2\rho^2}\left(\frac{w}{w + 1}\,Q^2\rho^2 - Q\rho\,\frac{\sigma_y}{\sigma_x} - 1\right)$$

$$c = \frac{P}{1 - Q^2\rho^2}\left(\frac{Q\rho}{w + 1}\,\frac{\sigma_x}{\sigma_y} + 1\right)$$

In the special case that $\sigma_x = \sigma_y$ and $w = 1$ (i.e., we wish to minimize $\sigma_\Delta^2 + \sigma_{\bar{y}}^2$), the estimates are thus

$$\bar{y} = \frac{PQ\rho(2 + Q\rho)}{2(1 - Q^2\rho^2)}(\bar{x}'' - \bar{x}') + \frac{P(2 + Q\rho)}{2(1 - Q^2\rho^2)}(\bar{y}' - \bar{y}'') + \bar{y}''$$

$$\Delta = \frac{P(2 + 2Q\rho - Q^2\rho^2)}{2(1 - Q^2\rho^2)}(\bar{x}'' - \bar{x}') + \frac{P(2 + Q\rho)}{2(1 - Q^2\rho^2)}(\bar{y}' - \bar{y}'') + \bar{y}'' - \bar{x}''$$

★9. Sampling for a time series* (Vol. I, Ch. 11, Sec. 7). *a. The sample model.* From a universe of size N, twelve independently selected samples are chosen at random, each of size n. One of the twelve is enumerated in the first month of *each* calendar year, a second in the second month of each calendar year, a third in the third month, and so forth, the twelfth being enumerated in the twelfth month of each calendar year.

During the enumeration, each member of the sample reports both sales for the current month and also sales for the month previous to the current month. After each enumeration a simple unbiased estimate x_u is made of the total sales for the current month, and from the same enumeration a simple unbiased estimate of total sales is also made for the previous month, $u - 1$. Let x_u and y_{u-1}, respectively, represent these estimates as obtained on the uth enumeration.

b. A composite estimate and its variance.

(1) *The estimate.* An estimate, x_u''', of the total sales for the latest month, u, is given by

$$x_u''' = K(x_{u-1}''' + x_u - y_{u-1}) + K_2 x_u \tag{9.1}$$

where $K + K_2 = 1$ with $0 \leq K \leq 1$.

The estimate may also be written as

$$x_u''' = Kx_{u-1}''' + (x_u - Ky_{u-1}) \tag{9.2}$$

In the special case where $K = 1$, an estimate analogous to a chained estimate results, while $K = 0$ makes the estimate the simple unbiased estimate.

(2) *Variance of the monthly total,* x_u'''. To obtain this variance, we shall first express the estimate for the latest month, u, in terms of all the simple unbiased estimates of totals which have been made.

Since, in general,

$$x_i''' = Kx_{i-1}''' + (x_i - Ky_{i-1})$$

we can write, after multiplying both sides of this equation by K^{u-i},

$$K^{u-i}x_i''' = K^{u-i+1}x_{i-1}''' + K^{u-i}(x_i - Ky_{i-1}) \tag{9.3}$$

---

* By Max A. Bershad, Bureau of the Census.

★ May be deferred.

Substituting $z_i = x_i - Ky_{i-1}$, and summing both sides of 9.3 from the first month ($i = 1$) through the latest month ($i = u$), we have

$$\sum_{i=1}^{u} K^{u-i}x_i''' = \sum_{i=1}^{u} K^{u-i+1}x_{i-1}''' + \sum_{i=1}^{u} K^{u-i}z_i$$

$$= \sum_{i=0}^{u-1} K^{u-i}x_i''' + \sum_{i=1}^{u} K^{u-i}z_i \tag{9.4}$$

Cancelling the similar terms on both sides of Eq. 9.4,

$$x_u''' = K^u x_0''' + \sum_{i=1}^{u} K^{u-i}z_i \tag{9.5}$$

But if we take, as the initial estimate (for the first month) in this time series, $x_1''' = Kx_0''' + z_1 = x_1$, Eq. 9.5 becomes

$$x_u''' = K^{u-1}x_1 + \sum_{i=2}^{u} K^{u-i}z_i \tag{9.6}$$

In the following it is assumed that the variances of the estimates x_i and y_i are equal and are the same from month to month (i.e., $\sigma_{x_i}^2 = \sigma_{y_i}^2 = \sigma_x^2$). Similarly, it will be assumed that the monthly correlations between x_i and y_{i-1} are all equal to ρ. Remembering that all the z's are independent of one another except those for months which are a year apart (or multiples of a year) and taking

$$\sigma_{z_i}^2 = \sigma_{x_i}^2 + K^2\sigma_{y_{i-1}}^2 - 2K\sigma_{x_i, y_{i-1}}$$

as

$$= (1 - 2K\rho + K^2)\sigma_x^2 \tag{9.7}$$

$$\sigma_{z_i, z_{i-12r}} = \sigma_{x_i, x_{i-12r}} + K^2\sigma_{y_{i-1}, y_{i-1-12r}} - K\sigma_{x_i, y_{i-1-12r}} - K\sigma_{x_{i-12r}, y_{i-1}}$$

as

$$= (1 - K)^2\rho_{12r}\sigma_x^2 \qquad \text{for integral } r \tag{9.8}$$

when $\qquad \sigma_{x_i, x_{i-12r}} = \sigma_{y_{i-1}, y_{i-1-12r}} = \sigma_{x_i, y_{i-1-12r}} = \sigma_{x_{i-12r}, y_{i-1}} = \rho_{12r}\sigma_x^2$

and

$$\sigma_{x_1}^2 = \sigma_x^2 \tag{9.9}$$

we have from Eq. 9.6, ignoring the yearly correlations with x_1,

$$\sigma_{x_u'''}^2 = K^{2(u-1)}\sigma_x^2 + \sigma_z^2 \sum_{i=2}^{u} K^{2(u-i)} + 2\sum_{r=1} \sigma_{z, z-12r} \sum_{i=12r+2}^{u} K^{u-i}K^{u+12r-i} \tag{9.10}$$

Substituting $j = u - i$, we have

$$\sigma_{x_u'''}^2 = K^{2(u-1)}\sigma_x^2 + \sigma_z^2 \sum_{j=0}^{u-2} K^{2j} + 2\sum_{r=1} \sigma_{z, z-12r} \sum_{j=0}^{u-2-12r} K^jK^{j+12r} \tag{9.11}$$

Performing the indicated summations, and substituting Eq. 9.7 and 9.8 in 9.11, we have

$$\sigma^2_{x'''_u} = K^{2(u-1)}\sigma^2_x + \sigma^2_x(1 - 2K\rho + K^2)\frac{1 - K^{2(u-1)}}{1 - K^2}$$

$$+ 2(1-K)^2\sigma^2_x \sum_{r=1} K^{12r}\rho_{12r}\frac{1 - K^{2(u-1-12r)}}{1 - K^2} \tag{9.12}$$

When u is large and when terms involving the twelfth or higher powers of K can be ignored, Eq. 9.12 becomes

$$\sigma^2_{x'''_u} \doteq \sigma^2_x\left(\frac{1 - 2K\rho + K^2}{1 - K^2}\right) \tag{9.13}$$

To find the value of K that makes $\sigma^2_{x'''_u}$ a minimum, the derivative of $\sigma^2_{x'''_u}$ with respect to K is taken and set equal to zero. It is found that

$$K = \frac{1 - \sqrt{1 - \rho^2}}{\rho} \tag{9.14}$$

minimizes $\sigma^2_{x'''_u}$; and, substituting this value in 9.13, we find that at its minimum

$$\sigma^2_{x'''_u} \doteq \sigma^2_x\sqrt{1 - \rho^2} \tag{9.15}$$

Denote as x''_u the special case of x'''_u which is obtained when $K = 1$. Then from Eq. 9.2

$$x''_u = x''_{u-1} + x_u - y_{u-1} \tag{9.16}$$

Since Eq. 9.13 does not apply when K is equal to 1, we must refer to Eq. 9.12. Substituting $K = 1$ in Eq. 9.12 and remembering that the

$$\lim_{K\to 1}\frac{1 - K^{2(u-1)}}{1 - K^2} = u - 1$$

we have, for $u \le 12$,

$$\sigma^2_{x''_u} = \sigma^2_x[1 + 2(1-\rho)(u-1)] \tag{9.17}$$

In the special case of $K = 0$, $K_2 = 1$, x'''_u becomes the simple unbiased estimate x_u with variance σ^2_x.

(3) *Variance of the month-to-month change*, $x'''_u - x'''_{u-1}$. From Eq. 9.6, we may write

$$\Delta''' = x'''_u - x'''_{u-1} = K^{u-2}(K-1)x_1 + z_u + \sum_2^{u-1} z_i(K^{u-i} - K^{u-1-i}) \tag{9.18}$$

$$= K^{u-2}(K-1)x_1 + z_u + \sum_{j=1}^{u-2}(K^j - K^{j-1})z_{u-j} \tag{9.19}$$

$$\sigma^2_{\Delta'''} = \sigma^2_x K^{2(u-2)}(1-K)^2 + \sigma^2_x(1-2K\rho+K^2)\left[1+\left(\frac{1-K}{K}\right)^2 \sum_1^{(u-2)} K^{2j}\right]$$

$$+ 2\left(\frac{1-K}{K}\right)^2 (1-K)^2\sigma^2_x \left[\sum_{r=1} K^{12r}\rho_{12r} \sum_1^{u-12r-2} K^{2j}\right]$$

$$- 2\frac{(1-K)^3}{K}\sigma^2_x \sum_{r=1} K^{12r}\rho_{12r} \qquad (9.20)$$

Equation 9.20 ignores yearly correlations involving x_1. Performing the indicated summations, we have

$$\sigma^2_{\Delta'''} = \sigma^2_x K^{2(u-2)}(1-K)^2 + \sigma^2_x(1-2K\rho+K^2)\left[\frac{2-K^{2(u-2)}(1-K)}{1+K}\right]$$

$$- 2\frac{(1-K)^3}{K(1+K)}\sigma^2_x \sum_{r=1} K^{12r}\rho_{12r}\,[1+K^{2(u-12r-2)+1}] \qquad (9.21)$$

When u is large and when terms involving the twelfth or higher order of K can be ignored, Eq. 9.21 becomes

$$\sigma^2_{\Delta'''} \doteq 2\sigma^2_x\left[\frac{1-2K\rho+K^2}{1+K}\right] \qquad (9.22)$$

Comparison with Eq. 9.13 shows that

$$\sigma^2_{\Delta'''} \doteq 2(1-K)\sigma^2_{x'''_u} \qquad (9.23)$$

In the special case where $K = 1$, the variance of the difference $x''_u - x''_{u-1}$ becomes on substitution in Eq. 9.21

$$\sigma^2_{\Delta''} = 2\sigma^2_x(1-\rho) \qquad (9.23a)$$

In the special case where $K = 0$, the variance of the difference between two independent simple unbiased estimates becomes

$$\sigma^2_{\Delta} = 2\sigma^2_x$$

(4) *Variance of the total of* 12 *months*, $x''' = \sum_{i=u-11}^{i=u} x'''_i$. Since by Eq. 9.6

$$x'''_u = K^{u-1}x_1 + \sum_{j=0}^{u-2} K^j z_{u-j} \qquad (9.6a)$$

$$x''' = \sum_{j=0}^{11} z_{u-j}\left(\frac{1-K^{j+1}}{1-K}\right) + \frac{1-K^{12}}{1-K}K^{-11}\sum_{j=12}^{u-2} z_{u-j}K^j + \frac{1-K^{12}}{1-K}K^{u-12}x_1 \qquad (9.24)$$

Then

$$\sigma^2_{x'''} = \sigma^2_z \left[\sum_0^{11} \left(\frac{1 - K^{j+1}}{1 - K} \right)^2 + \sum_{12}^{u-2} \left(\frac{1 - K^{12}}{1 - K} \right)^2 K^{2(j-11)} \right]$$

$$+ \sigma^2_{x_1} \left[\frac{1 - K^{12}}{1 - K} \right]^2 K^{2(u-12)}$$

$$+ 2 \sum_{r=1} \sigma_{z,\, z-12r} \left[\sum_{j=0}^{11} \left(\frac{1 - K^{12}}{1 - K} \right) \left(\frac{1 - K^{j+1}}{1 - K} \right) (K^{j+12r-11}) \right.$$

$$\left. + \sum_{j=12}^{u-2-12r} \left(\frac{1 - K^{12}}{1 - K} \right)^2 K^{j-11} K^{j+12r-11} \right] \qquad (9.25)$$

The above ignores yearly correlations with the initial value, x_1. Performing the indicated summations and substituting Eq. 9.7, 9.8, and 9.9 in 9.25, simplifying, and then dropping terms in K of the twelfth order or higher, assuming u to be large, Eq. 9.25 becomes

$$\sigma^2_{x'''} \doteq \sigma^2_x \left[\left(\frac{1 + K^2 - 2\rho K}{(1 - K)^2} \right) \left(12 - \frac{2K}{1 - K^2} \right) + 2 \frac{K}{1 - K^2} \rho_{12} \right] \qquad (9.26)$$

In the special case of $K = 1$ and $u = 12$, there are no yearly correlations, and the variance of the annual total is derived by utilizing only the first term in the coefficient of σ^2_z and the $\sigma^2_{x_1}$ term in Eq. 9.25.

After evaluating the indeterminate forms, and, because x_1 is one of the first 12 terms, summing the coefficient of σ^2_z only through 10, we have

$$\sigma^2_{x''} = \sigma^2_z \sum_{j=0}^{12-2} (j + 1)^2 + 12^2 \sigma^2_{x_1}$$

$$= 2(1 - \rho)\sigma^2_x \frac{(12 - 1)12[2(12) - 1]}{6} + 12^2 \sigma^2_{x_1}$$

$$= 12^2 \sigma^2_x \left\{ 1 + (1 - \rho) \frac{(12 - 1)[2(12) - 1]}{3(12)} \right\} \qquad (9.27)$$

In the special case where $K = 0$

$$\sigma^2_{x'} = \sigma^2_z \sum_{j=0}^{12-2} (1) + \sigma^2_{x_1} = 12\sigma^2_x = 12^2 \frac{\sigma^2_x}{12} \qquad (9.28)$$

(5) *Variance of the month-to-month-a-year-ago change*, $x'''_u - x'''_{u-12}$. From Eq. 9.5 or 9.6*a*, we may write

$$x'''_u - x'''_{u-12} = \sum_{j=0}^{11} z_{u-j} K^j - \sum_{j=12}^{u-2} z_{u-j} K^{j-12}(1 - K^{12}) - x_1 K^{u-13}(1 - K^{12}) \qquad (9.29)$$

The variance of $x'''_u - x'''_{u-12}$ is equal to

$$\sigma_z^2 \left[\sum_{j=0}^{11} K^{2j} + \left(\frac{K^{12}-1}{K^{12}}\right)^2 \sum_{12}^{u-2} K^{2j} \right] + \sigma_{x_1}^2 K^{2(u-13)}(K^{12}-1)^2$$

$$+ 2 \sum_{r=1} \sigma_{z,\, z-12r} \left[\sum_{j=0}^{11} K^j \left(\frac{K^{12}-1}{K^{12}}\right) K^{j+12r} \right.$$

$$\left. + \sum_{j=12}^{u-2-12r} \left(\frac{K^{12}-1}{K^{12}}\right)^2 K^j K^{j+12r} \right] \tag{9.30}$$

Ignoring yearly correlations with the initial value x_1, performing the indicated summations, dropping terms of high order of K, assuming large u, and simplifying, we obtain

$$\sigma^2_{x'''_u - x'''_{u-12}} \doteq \sigma_z^2 \frac{2}{1-K^2} + 2\sigma_{z,\, z-12}\left(-\frac{1}{1-K^2}\right) \tag{9.31}$$

Substituting Eq. 9.7 and 9.8 in 9.31 yields

$$\sigma^2_{x'''_u - x'''_{u-12}} \doteq \frac{\sigma_x^2}{1-K^2}\,[2(1-2\rho K + K^2) - 2\rho_{12}(1-K)^2] \tag{9.32}$$

For the special case where $K = 0$, Eq. 9.32 becomes

$$2(1-\rho_{12})\sigma_x^2 \tag{9.33}$$

c. An analogous composite estimate and its variance. Analogous to the composite estimate

$${}_A x'''_u = K(x'''_{u-1} + x_u - y_{u-1}) + K_2 x_u \tag{9.1}$$

is the estimate

$${}_B x'''_u = K x'''_{u-1} \frac{x_u}{y_{u-1}} + K_2 x_u \tag{9.34}$$

The variance of ${}_A x'''_u$ can be written as

$${}_A\sigma_u^2 = K^2[(\sigma^2_{x'''_{u-1}} + \sigma^2_{x_u} + \sigma^2_{y_{u-1}}) + 2(\sigma_{x'''_{u-1},\, x_u} - \sigma_{x'''_{u-1},\, y_{u-1}} - \sigma_{x_u,\, y_{u-1}})]$$

$$+ K_2^2[\sigma^2_{x_u}]$$

$$+ 2KK_2[\sigma_{x'''_{u-1},\, x_u} + \sigma_{x_u,\, x_u} - \sigma_{y_{u-1},\, x_u}] \tag{9.35}$$

The variance of ${}_B x'''_u$ can be written as

$$\sigma^2_{{}_B x'''_u} \doteq \left(\frac{Ex'''_{u-1}Ex_u}{Ey_{u-1}}\right)^2 K^2[(V^2_{x'''_{u-1}} + V^2_{x_u} + V^2_{y_{u-1}})$$

$$+ 2(V_{x'''_{u-1},\, x_u} - V_{x'''_{u-1},\, y_{u-1}} - V_{x_u,\, y_{u-1}})]$$

$$+ (Ex_u)^2 K_2^2 [V^2_{x_u}]$$

$$+ \frac{Ex'''_{u-1}(Ex_u)^2}{Ey_{u-1}} 2KK_2(V_{x'''_{u-1},\, x_u} + V_{x_u,\, x_u} - V_{y_{u-1},\, x_u}) \tag{9.36}$$

Now, if Ex_u''', Ex_{u-1}''', Ex_u, and Ey_{u-1} are approximately equal in level, and if both sides of Eq. 9.36 are divided by the square of this level, the resulting equation will be 9.35 with V's in place of σ's.

For ready reference, the following table shows the equation numbers of the variances of estimates based on ${}_Ax_u'''$, which are given in this section, and the corresponding equation numbers of the analogous rel-variances based on ${}_Bx_u'''$, which are given in Vol. I.

Estimate	Equation number of the variance of the specified estimate	
	In Vol. I, based on ${}_Bx_u'''$	In this volume, based on ${}_Ax_u'''$
Monthly level	7.23	9.17
Annual level	7.25	9.27
Monthly change	7.27	9.23*a*
Monthly level	7.30	9.13
Monthly change	7.33	9.22
Annual level	7.35	9.26
Month-a-year-ago change	7.36	9.32

Remark 1. It will be readily noticed from Eq. 9.1 or 9.34 that the composite estimate, x_u, does not involve y_u, an unbiased estimate for the same month. The reason, of course, is that the observations necessary to make this estimate are obtained during the $(u + 1)$st enumeration, at which time observations for both x_{u+1} and y_u are obtained simultaneously at very little additional cost for obtaining y_u. Consequently, as a general rule, y_u is obtained one month too late to be used in making the estimate x_u''' for the uth month.

However, in many circumstances, it is possible to obtain the observations for y_u in time to be used for the uth estimate. (If one were willing to pay for obtaining y_u and x_{u+1} separately rather than together, this could always be done.) When y_u can be obtained in time, an improved estimate for monthly level over Eq. 9.1 is

$$x_u'''' = K[x_{u-1}'''' - y_{u-1}] + \frac{K}{\rho} x_u + \left(1 - \frac{K}{\rho}\right) y_u \tag{9.37}$$

where as before, y_u as well as x_u is an unbiased estimate of the sales for the uth month and $K = \dfrac{1 - \sqrt{1 - \rho^2}}{\rho}$.

The different variances of x_u'''' can be derived by methods similar to those used for x_u'''. The variance of the monthly level of sales, for example, is smaller than that of x_u''' and will be found to be

$$\sigma^2_{x_u''''} = \sigma^2_x \frac{K}{\rho} \sqrt{1 - \rho^2} \tag{9.38}$$

(These latter results coincide with the results given by H. D. Patterson in "Sampling on Successive Occasions with Partial Replacement of Units," *J. Roy. Stat. Soc.*, Series B, **12** (1950), 241–255, for 50 per cent overlapping sample.)

Remark 2. If the sample model is altered so that the sample for any month is independent of that for any other month (i.e., the samples for the months of one year are not repeated in subsequent years), then it can be shown that the best linear unbiased estimate x_u''' for month u (u being large) of the form

$$x_u''' = a_0 x_u + a_1 x_{u-1} + a_2 x_{u-2} + \cdots + a_{u-1} x_1 - b_1 y_{u-1} - b_2 y_{u-2} - \cdots - a_{u-1} y_1$$

is the composite estimate (9.1).

Similarly, the best linear unbiased estimate x_u'''' for month u (u being large) of the form

$$x_u'''' = a_0 x_u + a_1 x_{u-1} + a_2 x_{u-2} + \cdots + a_{u-1} x_1 + b_0 y_u + b_1 y_{u-1} + b_2 y_{u-2} + \cdots + b_{u-1} y_1$$

is the composite estimate 9.37.

REFERENCES

(1) W. G. Cochran, "Sampling Theory When the Sampling Units Are of Unequal Size," *J. Amer. Stat. Assn.*, **37** (1942), 199–212.

(2) J. Neyman, "Contributions to the Theory of Sampling Human Populations," *J. Amer. Stat. Assn.*, **33** (1938), 101–116.

(3) Chameli Bose, "Notes on the Sampling Error in the Method of Double-Sampling," *Sankhya*, **6** (1943), 329–330.

(4) M. H. Hansen and W. N. Hurwitz, "The Problem of Non-Response in Sample Surveys," *J. Amer. Stat. Assn.*, **41** (1946), 517–529.

(5) R. J. Jessen, "Statistical Investigation of a Sample Survey for Obtaining Farm Facts," *Iowa Agr. Exp. Stat. Res. Bull.* 304, 1942.

(6) H. D. Patterson, "Sampling on Successive Occasions with Partial Replacement of Units," *J. Roy. Stat. Soc.*, Series B, **12** (1950), 241–255.

(7) W. G. Madow and L. H. Madow, "On the Theory of Systematic Sampling, I," *Annals Math. Stat.*, **15** (1944), 1–24.

(8) W. G. Madow, "On the Theory of Systematic Sampling, II," *Annals Math. Stat.*, **20** (1949), 333–354.

(9) F. Yates, "Systematic Sampling," *Phil. Trans. Roy. Soc.*, Series A, **241** (1948), 345–377.

(10) W. G. Cochran, "Relative Accuracy of Systematic and Stratified Random Samples for a Certain Class of Populations," *Annals Math. Stat.*, **17** (1946), 164–177.

(11) W. G. Cochran, *Sampling Techniques*, John Wiley & Sons, New York, 1953, Chapters 7, 8, and 12.

CHAPTER 12

Response Errors in Surveys*

1. Role of nonsampling errors in determining survey design. As discussed in Ch. 2 of Vol. I, the nonsampling errors in a survey involving original collection of data may often be a more serious problem than the sampling errors. Many of the limitations placed on our choice of sample designs arise out of response error rather than sampling error considerations (more precisely, arise from the joint consideration of response and sampling errors). As an example, one basic limitation imposed by the Census Bureau in designing its Current Population Survey (Case Study B in Ch. 12 of Vol. I) was that there be a full-time supervisor for each primary sampling unit. This limitation had a very substantial influence upon the cost equation for the CPS and, in consequence, upon the ultimate decision regarding the number of psu's to be used. The decision to have a full-time supervisor for each psu was not based on sampling considerations but on the belief that close supervision of the interviewing process would reduce nonsampling errors. Fewer supervisors would, in fact, have allowed the use of more psu's and a reduction in the sampling error. Implicitly, such a decision assumed that the reduction in response error achieved by increased supervision outweighed any increases in sampling error which might result. Actually, very few data are available for determining whether, in fact, decisions made on the nonsampling features of survey design contribute to an improvement in the over-all accuracy and value of a survey.

The paucity of dependable data on response errors is unquestionably the greatest present obstacle to sound survey design. In survey after survey, losses in sampling efficiency are taken on the basis of quite dubious assumptions about the magnitudes and distributions of response errors. Frequently, this point is obscured by the implicit (practically "unconscious") nature of survey designers' assumptions regarding response error.

For a demonstration of the relationship between response and sampling errors in survey design, the student is referred to an article,† "A Case

* This chapter represents a minor revision of a paper by Morris H. Hansen, William N. Hurwitz, Eli S. Marks, and W. Parker Mauldin (9).

† Marks, Mauldin, Nisselson (7).

History in Survey Design: The Post-Enumeration Survey of the 1950 Census." This article outlines the numerous decisions which had to be made and indicates the mixture of opinion and habit which had to be relied on in making such important decisions as those on questionnaire design, interviewer selection, length and type of interview, and training and supervision of interviewers.

Although work on the measurement of response errors is relatively new, several excellent analyses of sources and types of error are available. A summary of the main sources of response errors is included in Ch. 2 of Vol. I. For more extensive discussions of this topic the student is referred to papers by Deming (3), Marks and Mauldin (6), Marks, Mauldin, and Nisselson (7), and Ackoff and Pritzker (1). Mahalanobis (5) has developed several important techniques for measuring and controlling response errors, particularly those arising from the interviewer.

Most of this chapter is devoted to the explicit formulation of a mathematical model for "response errors."* An essential preliminary to such a formulation is a determination of some of the important requirements that a mathematical model should meet in order to make it conform reasonably well to actual survey conditions. One important feature of all survey designs is the estimating procedure. The processes of sampling, data collection, coding, and tabulating introduce "errors" into survey results. These errors may be affected by the choice of an estimating procedure. The present chapter does not involve consideration of the relationship between response errors and choice of estimating procedures.

2. Some requirements on a mathematical model for response errors. In defining "error" we start with an "estimate" and a "value estimated." The "estimate" is some value determined from the survey data and, for any particular survey, is a definite number but varies from survey to survey. In many surveys, the "value estimated" is not defined explicitly, and the problem of survey design is complicated by vagueness regarding what is being measured. However, if the aim is orderly planning of a survey rather than catch-as-catch-can methods, it is essential that the "value estimated" be defined precisely.

Estimating an average or aggregate. A common type of "value estimated" in social surveys is one which is an average or an aggregate of the values for the individual elements that make up the population. Each element of the population has attached to it some value of a variable,

---

* In the broadest sense, "response error" as used here includes both processing and data collection errors. The present chapter is oriented primarily towards data collection errors, but most of the discussion is directly applicable to the control of processing errors.

and we want to know the average or the aggregate for some or all these values. For the present, we shall consider only the case in which we are estimating an average or aggregate for all the population elements, or an average or aggregate for a subgroup of the population, where the members of the subgroup are identified as such without error.

In making a sample survey to estimate a population aggregate or average, we observe the values of some of the population elements and derive the estimate from these observed values. The fact that we have selected for observation some but not all the elements ordinarily introduces some error (sampling error). In addition, we frequently find that there are response errors in the individual observations. Thus, even if we were to observe all the elements of the population (i.e., take a census), we would usually have an error in our estimate of the population average or aggregate.

It should be noted that errors of *non*response play a peculiar role. Failure to secure a response can be considered a sampling bias on the assumption that the "nonresponse" elements have a zero chance of inclusion in a sample. Failure to secure a response can also be considered a response error, since any estimating procedure involves assigning values to the nonresponse elements either implicitly or explicitly; e.g., estimating the population average on the basis of the respondents alone is equivalent to assigning the average of nonresponses a value equal to the estimated average.

The concept of "individual true values." In defining the "value to be estimated" we shall define a true value for each of the individuals who make up the population and define the value to be estimated as an average or aggregate of these individual true values. The individual true value will be conceived of as a characteristic of the individual quite independent of the survey conditions which affect the individual response. Thus, age is usually defined as a time interval between two events, and this definition is quite independent of how we determine an individual's age. It should be remembered, however, that the number you get when you ask a person his age is not necessarily the true value for the age as defined. The respondent may not know his "true" age. Sometimes he does not know exactly the age of his wife or others for whom he may report. Even if he does know the correct answer, he may misunderstand the question or become confused in "recall," or he may purposely give an incorrect answer.

Difficulties of ascertaining individual true values. For some variables (e.g., age or sex) a survey may get the true values for a large proportion of the individuals. For other variables (e.g., income, brand preference, or purchases) the true values may be obtained for a much smaller proportion of the population. A survey rarely gets the true values for all the individuals regardless of the characteristic measured. Frequently, by a sufficient expenditure of well-directed effort we can come much nearer to the true value. For example, in many countries the determination of age could involve an examination of birth or baptismal certifications, or of primary school records if no birth certificate exists, or of the first census in which the individual was listed if neither birth certificate nor primary school record exists. Exhaustive record searches might give the true age for most individuals, although there would obviously be persons for whom we could find no records and other individuals whose records were in error. The searches would, of course, be relatively expensive compared with methods ordinarily used for determining age.

Criteria for a definition of true value. There are many cases in which we might encounter tremendous difficulty in defining a "true" value (entirely apart from the problem of determining the value once we have defined it). What, for example, is a person's "true intelligence," "true attitude toward revision of the Taft-Hartley Act," or "true brand preference for cigarettes"? No definitive answer can be given to these questions. We would suggest, however, three criteria for the definition of "true value":

(1) The true value must be *uniquely* defined.
(2) The true value must be defined in such manner that the purposes of the survey are met.
(3) Where it is possible to do so consistently with the first two criteria, the true value should be defined in terms of operations which can actually be carried through (even though it might be difficult or expensive to perform the operations).

It is possible to define true value in such manner that a survey is subject to no (or negligible) response error. It will be useful to consider such definitions of "true value" in the light of the criteria listed above. For example, we could define a person's "attitude toward revision of the Taft-Hartley Act" as the alternative (in a set of six alternatives) that he first selects after an accredited interviewer for the survey has asked him: "What do you think of the Taft-Hartley Act?" We could define a person's birthplace as the answer recorded for him by an interviewer who is instructed to ask: "In what state or foreign country were you born?" These definitions meet (or, with a little expansion, can be made to meet) two of the three criteria: i.e., they are unique and are defined in terms of operations which can be carried through. In most cases, however, they will not be acceptable as "true values." There might, perhaps, be survey directors who would accept these definitions as the things they really want to measure, but most consumers of data are after something less dependent on the particular interview conditions (even though results of this type may be quite acceptable as approximations to the true value). We may want to know how a person is likely to act toward a Congressman who favored or opposed the Taft-Hartley Act, not what his casual reply is to a rather vague question asked by a person whose motives and sponsorship may generate a very complex reaction in the respondent. We may want to know where a person was actually born, not what gets recorded as his birthplace when the interviewer fails to ask the question properly, or the respondent misunderstands the question, or the interviewer misinterprets the answer.

Use of an expected response value to approximate the true value. In the examples cited (and in many other cases) it may be impossible to define a true value which meets all of the three criteria listed. Often, however, we can define a value which meets the first two criteria and can at least define an operation whose "expected value" will give a satisfactory approximation to the true value. An example is a study done by the Bureau of the Census. After the 1950 Census of Population was completed by the large number of personnel hired as enumerators, carefully selected and highly trained interviewers recanvassed a sample of areas, taking with them a record of the original enumeration, looking carefully for persons missed in the original enumeration, and checking a sample of those persons who were enumerated in the area to make sure that they should have been enumerated. The individuals who did the recanvass were (in general) well-trained, conscientious, and thoroughly familiar with the rules that prescribe which persons are to be enumerated in a given enumeration district. The recanvass procedure did not, of course, insure a "perfect" measurement for each individual, but it came nearer to doing so than the procedure used originally.

Consider interviewing each individual a large number of times under exactly the same conditions as the recanvass. This would yield a population of responses for all individuals. We might draw a sample of individuals and then a sample of one of the possible responses from each of the individuals in the sample. In practice, the conditions for subsequent interviews might change because of the conditioning effect of earlier interviews, but we can conceive of a set of independent recanvass interviews of a respondent and can regard the particular interview made on the recanvass as a sample from this set. The expected value of an estimate from this sample could be regarded as approximating the true value. For a reasonably large set of such observations, the estimates made from the recanvass would then be close to the "true population count."

The concept of an individual response error. The term *individual response error* will be used here to denote the difference between an individual observation and the true value for the individual. For example, the survey might want age as of last birthday as a difference in whole years between date of birth and some specified date (say April 1, 1950). If one of the persons covered by the survey was born April 1, 1897, but in 1950 is reported as 50 years old, the "individual response error" would be 3 years.

A less obvious case of response error is the failure to report an individual in a census of population (or in a sample survey used to estimate total population). Here the "true" value (the value the census is trying to obtain) is 1 (1 person), the value obtained for this individual is 0, and the response error is 1. Since the direction of error may be important, it would be better to call this an error of -1. Similarly, counting the same individual twice would be an error of $+1$.

Variance and bias of response errors. As here defined, an "individual response" is the value obtained on a particular observation (e.g., the result obtained in a specified measurement or interview by a specified interviewer with a specified respondent at a given time). Under slightly different conditions, therefore, the value of the individual response might be different. Thus, the individual response is influenced by the conditions of the observation or interview or written response.

The variability of individual responses has often been treated in terms of random variation. Although this approach has certain defects, we shall adopt it for purposes of the present analysis. Consequently, the response error of a particular individual in a given survey will be thought of as having an expected value (the individual response bias) and a random component of variation around that expected value. Similarly, the aggregate or average of a set of responses for different individuals will have a response bias and a response variance which will be determined by the response biases and variances for the population of individuals.

Essential conditions of a survey. To say that an individual response is a random variable is not, however, sufficient—we must define somewhat more precisely the universe of individual responses involved. For this purpose we shall consider all responses obtainable under certain "essential" conditions. In general, these conditions are "specified" (either implicitly or explicitly) by the survey design. As a minimum a survey design must specify the subject of inquiry, the method of obtaining information (interview, mail inquiry, direct observation, etc.), and the method of recording the information (checking a box, entering a figure, writing a description of the response, etc.). These specifications may be general or specific.

Particular surveys may involve additional specifications, e.g., that the survey be taken during a particular period. There are also "essential" conditions of a survey which arise implicitly as necessary consequences of the explicitly specified conditions. For example, if we specify that a survey of individual income received during 1949 be taken during April 1950, there is implicit in that specification a certain "recall" situation for each respondent and a relationship of this "recall" situation to income-tax-filing activities. If we also specify that responses be obtained by interview, the fact that the survey is to be done in April 1950 implicitly specifies a certain condition of the labor market and this may impose restrictions on the type of interviewer obtainable. The compensation paid and training given to interviewers, the wording of questions to be asked, and the sponsorship of the survey are frequently a part of the specified survey conditions, and these specifications determine, in turn, other conditions which will distinguish this response situation from other response situations.

On the other hand, there are usually present, at the time of any response, conditions which may affect that response but which are neither specified survey conditions nor the direct consequences of specified survey conditions. If the survey design specifies the types of interviewers, the sponsorship of the study, the compensation offered, and the hiring procedures used, these specifications may make it certain that John Jones will be interviewed by one of a certain class of individuals (e.g., persons over 30 years of age who have had at least 2 years of high school education and some experience as interviewers for other surveys), but the exact identity of the interviewer may still vary within the limits of the specified class. The survey design may instruct the interviewer to ask certain questions, but it cannot insure that the questions will always be asked in exactly the same way. The survey design may specify a certain approach to respondents, but it will not specify how that approach will be received by a respondent who happens to be interrupted while she is doing the family laundry.

In general, the survey specifications (explicit or implicit) restrict the range of response variation but by no means eliminate variation completely. Under some conditions the range of variation may be narrow; under others it may be wide. Similarly, the response errors may be compensating in character or they may be more or less systematic in direction, thus creating a response bias. The expected value of the response errors and the random component of variation around that expected value may be regarded as determined by the essential survey conditions.

In practice, some of the essential conditions of a survey will be difficult to separate from the unessential ones, but the fact that some are essential and others are of an accidental character needs to be recognized. Basically, the "essential conditions" of a survey are those variables which we are consciously trying to keep uniform over all cases covered. The "uniformity" may be in a rule rather than being absolute (e.g., in a study of sex behavior, we might require that female interviewers interview female respondents and male interviewers interview males), but the important point is that we deliberately attempt to bring these conditions "under control" (or are forced by circumstances to accept a uniformity in certain conditions). Often the problem of improving survey design will be to identify and deal with some of the more important essential conditions.

In contrast to the essential conditions of the survey, the "random errors" are controlled not by the introduction of uniform rules and procedures but by taking several units—several clusters, several elements for the variables discussed in earlier chapters, several interviewers for the "interviewer variance" discussed in this chapter, or several coders or punchers when we are dealing with "coding or punching variance." The use of "scores" based on several questions in attitude surveys and other psychological measurements is another example of the control of random variation by increasing the number of units (in this case the number of questions).

Correlation of response errors when interviewers are used. It would be convenient to assume that in any particular survey the random component of the response error for one individual is uncorrelated with the random component of the response error for another individual. Unfortunately, such an assumption does not accord with known facts about response variation. In particular a mathematical model which postulates independent responses of all individuals will not fit a survey which uses interviewers unless the interviewer is assumed to have no influence on the response. If we were to assign at random a different interviewer to each individual, the effect of the interviewer on the response would be uncorrelated for any two obtained responses. Ordinarily, however, a given interviewer obtains and records the responses for a number of individuals, and often we have reason to believe that the errors made by a particular interviewer are correlated. Even casual observation of an interviewer at work reveals the presence of interviewing patterns distinctive to that interviewer. In an inquiry about labor force status an interviewer, who implies by his manner that he does not expect to find a housewife with children gainfully employed, may tend to record fewer employed women than an interviewer who seems to insist that every adult should be gainfully employed.

The present analysis uses a mathematical model which assumes that responses are uncorrelated if they are obtained for different individuals by different interviewers. However, there may be correlation between responses even when both the individual and the interviewer are different. For example, the presence of a common supervisor or participation in the same training class may result in correlated errors for two different interviewers (unless these common influences are specified as essential conditions). Correlation between responses obtained by different interviewers may also be introduced in processing; e.g., the same clerk may make similar errors in coding both responses. We shall assume that these correlations are small and can be neglected, although the model could be extended to include them.

*Specification of a mathematical model.** The discussion thus far presented leads to a mathematical model for the analysis of response errors in which we have:

(*a*) A population of N individuals and a population of M interviewers, both of which will, for convenience, be assumed to be large.

(*b*) Associated with each individual, a true value.

(*c*) A set of essential survey conditions which determine for a particular individual and interviewer the expected value of a random variable.

(*d*) Zero correlation between the random components of responses for two different individuals with two different interviewers.

(*e*) The order of interviewing respondents by an interviewer either randomly determined or not affecting the responses.

In many surveys interviewers are available to interview only certain classes of the population and only in certain geographic areas. We shall, therefore, conceive of our interviewers as divided into L groups with M_h interviewers in the hth group who are available to interview a particular N_h individuals and no others. Where all interviewers are available to interview all individuals, $L = 1$, $M_h = M$, $N_h = N$.

3. The effect of interviewers on the variance of sample estimates. *Effect of response errors on estimates of sample variance.* One major advantage possessed by probability sampling as compared with the other types of sampling is the possibility of estimating the sampling error from the sample. In situations where the sample results are uniquely determined by the act of selecting the sample individuals (i.e., given the fact that the ith individual is in the sample, there is one and only one

---

* See also Sec. 7*a* of this chapter.

value to be ascribed to the ith individual) there is, of course, no question of our ability to estimate sampling error from a reasonably large sample.

When the individual responses are subject to error, we shall see that, with appropriate methods, the sampling variance of a statistic such as a mean or total will reflect the response variation as well as the error due to including only a sample of individuals. Appropriate analysis of the sources of error will point to the methods for minimizing the total variance. However, although the use of probability sampling will insure that the variance of the individual true values will be appropriately reflected in the variance of a sample estimate, the accurate reflection of response variance will depend on the applicability of whatever mathematical model is assumed.

The *response bias* of a statistic such as an estimated mean or total will not be reflected in the variance of a sample statistic, although its effect, if it can be estimated, will be reflected in the mean square error and its influence on accuracy thus taken into account. Response bias is not *per se* a "sampling" problem, i.e., bias arising from response errors is ordinarily independent of the sample design and is, in fact, of the same magnitude for a study involving a complete canvass of the population as it is for a sample survey if both the complete canvass and the sample survey are taken under the same essential conditions. Postponing to a later section the consideration of response bias, we shall examine first the other component of survey error, i.e., the variance of a sample estimate, and shall examine particularly the contribution of the interviewer to this variance.

The design of a survey to evaluate response variance due to interviewers. In evaluating sampling variance we must consider the particular technique of drawing a sample and making an estimate from this sample. In studies which involve the use of interviewers we must consider also some specified technique for selecting the interviewers and assigning them to the various individuals included in the sample.

Actually, survey practice in the making of interviewer assignments is far from standard. A common pattern is to group the units selected in the sample by geographic areas and then to assign the units in a given area to one or more interviewers, making the different interviewers' assignments approximately equal. The sampling units may be individuals or clusters, but in either event, in surveys in which interviewers are used, costs of travel and time required for identification of the sample usually suggest some clustering of the assignments to interviewers. This clustering of interviewer assignments led to the introduction of interviewer groups into the population specifications outlined above. In terms of the specified mathematical model, this practice can be approximated by the following sample design:

(*a*) n of the N individuals in the population are selected at random without restriction.*

(*b*) m_h interviewers are selected at random without restriction from the hth interviewer group to interview those sample individuals selected who are available for interview by this interviewer group. Let $m = \sum_h^L m_h =$ the total number of interviewers selected.

(*c*) An equal number, $\bar{n}$, of individuals is assigned to each of the m interviewers. The $\bar{n}$ individuals assigned to any interviewer are a random subsample of all the sample individuals available for interview by this interviewer group.

The applicability of these conditions to actual surveys will be considered later. The conditions stated apply reasonably well to many surveys.

It should be noted that n_h, the number of sample cases drawn which will be available for interview only by interviewers in the hth group, is a random variable. In designing a survey we could decide to use a fixed number of interviewers from the hth group and adjust the size of assignment given each interviewer. For example, if we were using 2 interviewers for a given group and happened to draw 84 sample cases available to this group, we could give each interviewer 42 cases; if we drew 76 individuals, each interviewer would be assigned 38, etc. Another method of determining interviewer assignments is the one used here, i.e., to fix the *size* of the assignments and let the number of interviewers vary. The restriction that the size of the interviewer assignment be fixed does not represent any great loss of generality, since the variance of most sample estimates will be about the same whether the size of assignment or the number of interviewers in a group is fixed.

The sample estimate and its mean square error.† Assume that a simple random sample of n units is selected from a population of N units. Let $x_{hij} =$ the value obtained for the jth sample unit by the ith sample interviewer in the hth (population) group. Let the sample mean be

$$\bar{x} = \frac{\sum_h^L \sum_i^{m_h} \sum_j^{\bar{n}} x_{hij}}{n} \tag{3.1}$$

* We shall restrict this discussion to simple random sampling. The results can be extended to stratified and cluster sampling.

† See Sec. 7*b* for derivation of the formulas presented in this section.

With the survey design specified, $\bar{x}$ would be used as an estimate of the true population mean $\bar{Y}$. The mean square error of $\bar{x}$ is:

$$\text{MSE}\,\bar{x} = B_X^2 + \sigma_{\bar{x}}^2 \tag{3.2}$$

where $B_X = E\bar{x} - \bar{Y}$.

$E\bar{x}$ = the expected value of $\bar{x}$.

Section 7*b* shows the derivation of an expression for $\sigma_{\bar{x}}^2$ as approximately* equal to:

$$\left.\begin{aligned} \sigma_{\bar{x}}^2 &= \frac{\sigma_X^2 - \sigma_{XI}}{n} + \frac{\sigma_{XI}}{m} \\ &= \frac{\sigma_X^2}{n} + \frac{n-m}{n}\frac{\sigma_{XI}}{m} \end{aligned}\right\} \tag{3.3}$$

Here σ_X^2 represents the "total variance" of individual responses around the mean of all individual responses in the population; i.e., it is the variance over all responses for all individuals to all interviewers in the interviewer group, and over all interviewer groups, and σ_{XI} is the covariance between responses obtained from different individuals by the same interviewer (this covariance being taken within interviewer groups since independent selections of interviewers are made from each interviewer group). If we divide the covariance σ_{XI} by the average variance of responses *within* interviewer groups, we have δ, the intraclass correlation; i.e., δ is the correlation between responses of different individuals for the same interviewer. Thus:

$$\sigma_X^2 = \sigma_{wX}^2 + \sigma_{bX}^2 \tag{3.4}$$

$$\delta = \frac{\sigma_{XI}}{\sigma_{wX}^2} \tag{3.5}$$

where σ_{wX}^2 is the variance of responses within interviewer groups (taken over all responses of every individual to every interviewer in the group).

σ_{bX}^2 is the variance of expected responses for interviewer groups, i.e., between average values for interviewer groups.

The sampling design used is, in effect, to sample "clusters" of responses (the responses obtained by each of the interviewers) and, within sample clusters, to subsample $\bar{n}$ responses (such that no two responses are for the

---

* Assuming that N is large relative to n and that the interviewers used in the hth group are a random sample from a potential infinite supply of such interviewers.

same individual). The similarity to cluster sampling may be more apparent if we express $\sigma_{\bar{x}}^2$ as

$$\sigma_{\bar{x}}^2 = \frac{\sigma_{wX}^2}{n} [1 + \delta(\bar{n} - 1)] + \frac{\sigma_{bX}^2}{n} \tag{3.6}$$

In the above expressions, σ_{bX}^2/n represents the variance arising because individuals were sampled independently of the interviewer groups. If we had only one interviewer group ($L = 1$), $\sigma_{bX}^2 = 0$ and $\sigma_{wX}^2 = \sigma_X^2$, so that

$$\sigma_{\bar{x}}^2 = \frac{\sigma_X^2}{n} [1 + \delta(\bar{n} - 1)] \tag{3.7}$$

and if we sampled individuals within interviewer groups, so that the interviewer groups served as strata, $\sigma_{\bar{x}}^2$ would be given by Eq. 3.7, but with σ_{wX}^2 substituted for σ_X^2.

This formula is identical with that for the variance of a sample mean when we draw m clusters of $\bar{n}$ elements each (or sample m clusters of equal size and subsample $\bar{n}$ elements from each cluster). There are, of course, differences from straight cluster sampling arising from the restriction that we must sample only one response for any individual, but the basic sampling principles are analogous.

*Estimates of variance from the sample.** From the sample we can obtain unbiased estimates of σ_{XI} and σ_X^2. These estimates are, respectively,

$$s_{XI} = \frac{\sum_h^L \frac{m_h}{m_h - 1} \sum_i^{m_h} (\bar{x}_{hi} - \bar{x}_h)^2}{m} - \frac{\sum_h^L \sum_i^{m_h} \sum_j^{\bar{n}} (x_{hij} - \bar{x}_{hi})^2}{n(\bar{n} - 1)} \tag{3.8}$$

$$s_X^2 = \frac{\sum_h^L \sum_i^{m_h} \sum_j^{\bar{n}} (x_{hij} - \bar{x})^2}{n - 1} + \frac{(n - m)}{(n - 1)} \frac{s_{XI}}{m} \tag{3.9}$$

where

$$\bar{x}_{hi} = \frac{\sum_j^{\bar{n}} x_{hij}}{\bar{n}} = \text{average for the } i\text{th sample interviewer in the } h\text{th group} \tag{3.10}$$

$$\bar{x}_h = \frac{\sum_i^{m_h} \sum_j^{\bar{n}} x_{hij}}{m_h \bar{n}} = \frac{\sum_i^{m_h} \bar{x}_{hi}}{m_h} = \text{sample average for the } h\text{th group} \tag{3.11}$$

---

* For derivations of the formulas in this and the following sections see Sec. 7c.

Thus, an unbiased estimate of $\sigma_{\bar{x}}^2$ is

$$s_{\bar{x}}^2 = \frac{\sum_h^L \sum_i^{m_h} \sum_j^{\bar{n}} (x_{hij} - \bar{x})^2}{n(n-1)} + \frac{(n-m)}{(n-1)} \frac{s_{XI}}{m} \tag{3.12}$$

Contribution of interviewers to the variance. It should be noted that the effect of using interviewers is to introduce into the variance of $\bar{x}$ a term involving the intraclass correlation within interviewers' assignments. Except as indicated below, the techniques given in Vol. I and in earlier chapters of Vol. II for estimating the variance of $\bar{x}$ (and of other sample estimates) from a sample disregard the contribution of this intraclass correlation. If we disregard the intraclass correlation within interviewers' assignments, the estimate of the variance of $\bar{x}$ would be the first term of Eq. 3.12. It can be seen that the result will usually be an underestimate in cases where there is a noticeable interviewer contribution to the total variance.

If $m = n$, each interviewer interviews only one unit, and there is, of course, no effect of the intraclass correlation. In this case the formulas for the variance given in earlier chapters include any interviewer contribution to the variance.* On the other hand, the larger the interviewer error, or the more individuals we assign to an interviewer, the more important it is to use Eq. 3.12 to estimate the variance of $\bar{x}$.

Equation 3.12 is useful in indicating the effect of interviewer error upon the variance of a sample mean. Where there is no need for a separate estimate of s_{XI}, Eq. 3.12 can be written as

$$s_{\bar{x}}^2 = \frac{\sum_h^L \frac{n_h - 1}{m_h - 1} \sum_i^{m_h} (\bar{x}_{hi} - \bar{x})^2}{m(n-1)} + \frac{\sum_h^L n_h(\bar{x}_h - \bar{x})^2}{n(n-1)} \tag{3.13}$$

and, if $L = 1$,

$$s_{\bar{x}}^2 = \frac{\sum_i^m (\bar{x}_i - \bar{x})^2}{m(m-1)} \tag{3.14}$$

---

* A study director would not ordinarily assign a single individual to an interviewer, but, with cluster samples, a single cluster might be assigned to one or more interviewers. Where an interviewer or group of interviewers is assigned to a single cluster, the estimated variance between ultimate clusters, by the usual analysis, will include interviewer variance. Thus, with large-cluster sampling, discussed in Chapter 9, it may often be true that one or more interviewers will work in only a single primary sampling unit. In this event, the "ultimate cluster" estimate of the total variance will include the appropriate interviewer contribution to the total variance (see, for example, Sec. 4 and 5, Ch. 9, and Sec. 15 and 28, Ch. 9, Vol. I). Where more than one cluster is assigned to an interviewer, we can estimate the interviewer variance separately by the methods given here.

The covariance, σ_{XI}, reflects the effect on response of the interviewer and the interaction between interviewer and respondent. If the observation is of a type which permits a large effect of the interviewer on the response, the contribution of the term involving σ_{XI} to the variance of $\bar{x}$ may be quite substantial. In many cases, however, σ_{XI} will be negligible. We might, for example, expect a large variance among interviewers for estimates of farm acreage under corn, where the interviewer does the estimating by direct observation without measurement. Where the farmer furnishes information to the interviewer on the number of cattle on a farm, there may be little or no effect of the interviewer on the response and σ_{XI} may be negligible or zero.

Reducing interviewer contribution to variance. Where interviewer contribution to the variance is important, it may be possible to reduce this contribution significantly by training and adequately supervising the interviewers, and this should be the first line of attack. Sometimes, however, training of the interviewers beyond a certain point will have very little effect on interviewer variances. Instead of trying to make additional reductions in interviewer variance by increased training and supervision of the interviewer or using other (and perhaps expensive) techniques to obtain greater interviewer uniformity, we might devote our attention to another method for reducing the effect of interviewer variance on our final estimate. From Eq. 3.3 it will be seen that, for fixed values of σ_{XI}, the effect of interviewer variance on $\sigma_{\bar{x}}^2$ decreases as we increase the number of interviewers. Thus, if cost were not a factor, maximum accuracy with this sample design would be obtained by assigning one individual to each interviewer.*

Determining the optimum number of interviewers. With the ordinary survey which has a fixed total budget, increasing the number of interviewers will increase costs and will require a reduction of expenditure at some other point, e.g., reducing the expenditure per interviewer or per individual or reducing the number of individuals included in the sample. When the cost function is simple, as in Eq. 3.15, optimum values of n and m can be readily determined by joint solution of the cost and variance functions. With more complicated cost functions, the optimum values can be determined in the same way as for subsampling designs (see Chapters 6, 7, and 9).

---

* This statement is subject to the limitation that response bias and variance between interviewers remain fixed. Ordinarily, it will not be possible to make extreme changes in size of interviewer assignment without changing the response bias and interviewer variance, but the analysis is acceptable within reasonable limits of variation.

We shall consider the case in which the cost is given by

$$C = C_1 n + C_2 m \tag{3.15}$$

where C = total budget for the survey, excluding fixed overhead costs.

C_1 = cost per elementary unit included in the sample.

C_2 = cost per interviewer used in the survey.

With this cost function and the variance given by Eq. 3.3, the optimum values of n and m are

$$n = a\sqrt{\frac{\sigma_X^2 - \sigma_{XI}}{C_1}} \tag{3.16}$$

$$m = a\sqrt{\frac{\sigma_{XI}}{C_2}} \tag{3.17}$$

where

$$a = \frac{C}{\sqrt{C_1(\sigma_X^2 - \sigma_{XI})} + \sqrt{C_2\sigma_{XI}}} \tag{3.18}$$

Some illustrations. Assuming that cost functions are known or can be roughly approximated, application of the technique may be illustrated by data from two studies where interviewer assignments were randomized. However, for a satisfactory estimate of interviewer variance, we will need more interviewers than the numbers used in the studies mentioned here. The interviewer variance estimates of Tables 1 and 2 (pp. 297 and 298) are based on a very small number of cases and are, therefore, quite unreliable. They are presented only for purposes of illustration.

The Indian Statistical Institute has pioneered in the design of surveys so as to make possible the evaluation of response variation associated with the interviewer. Methods similar to the survey design described in this section have been used for some time by the Indian Statistical Institute to control and measure the effects of the "human agency." Some of these techniques are described by Mahalanobis (5). One such design was used in an inquiry to determine the economic conditions of factory workers in an industrial area at Jagaddal. The entire area was divided into 5 subareas. Within each subarea 5 independent random samples of structures were selected for interview. Each of the 5 samples was assigned to a different interviewer, but the same 5 interviewers worked in all 5 subareas.

This design is similar to the one described above. In this case all interviewers are (presumably) available to interview the entire population, so that $L = 1$. There is a stratification within interviewers' assignments (the sampling by subareas). Results are presented on a "family" basis, although the sampling unit used was actually a structure and thus might involve a cluster of families. To simplify the use of Mahalanobis' data for illustrative purposes, we shall ignore the stratification and clustering and treat the sample of families as if it were an unrestricted random sample of the population surveyed, the families being the individual members of this population.

Mahalanobis did 3 studies in the Jagaddal area (in 1941, 1942, and 1945), all involving approximately the same design. He also reports a study using a similar design (5 subareas but only 4 interviewers) carried out in the Nagpur in 1942–43 by M. P. Shrivastava. Table 1 shows estimates of σ_X^2 and σ_{XI} for various characteristics made from the results of these surveys, assuming an unrestricted random sampling design. With a suitable cost function, these variance estimates can be used to determine the optimum number of interviewers. Suppose, for example, that C (the total survey budget) was \$2000, that C_1 (the cost per family) was \$2, and that C_2 (the cost per interviewer for training, supervision, travel to the five areas to be enumerated, etc.) was \$80. With these values (and the cost function $C = C_1n + C_2m$) the optimum number of interviewers, m, and the optimum number of families, n, would be those shown in the last 2 columns of Table 1. The analysis would point to the use of somewhere between 5 and 8 interviewers for the Jagaddal study and to about 5 interviewers for the Nagpur study. It should be remembered, however, that the estimates s_{XI} are based on 4 degrees of freedom for the Jagaddal study and only 3 degrees of freedom for the Nagpur study. These estimates are, therefore, subject to a high sampling variance. As a matter of fact, the values reported for s_{XI} are entirely consistent with a zero value for σ_{XI}. This situation points to the need for using the results of more interviewers if we wish to make reliable estimates of interviewer contributions to the variance from the sample.

If the cost per interviewer, C_2, had been taken as \$4 instead of \$80, the optimum number of interviewers for estimating monthly per capita expenditures in Jagaddal would have been 49. In this case the use of only 5 interviewers would mean an 80 per cent increase in the variance of our estimate as compared with the optimum.

A small experiment similar to those of Mahalanobis was conducted in Baltimore by the Bureau of the Census as part of the December 1947 Current Population Survey. In this study, segments (small areas) were selected for interview in the Baltimore area. These segments had an expected size of 6 households. The households in 25 of the segments were divided into 2 sets of alternate households. Two interviewers were assigned to each of the 25 segments and given (at random) 1 of the sets of households for interview. Interviewers A and B shared 6 segments,* interviewers B and C shared 5, interviewers A and C shared 5, and interviewers D and E shared 9.

The situation in this study is approximated reasonably well by the specified mathematical model if we assume that interviewers A, B, and C were drawn from one interviewer group and interviewers D and E from another. The sample design is, of course, different, but the difference requires only minor modifications of the formulas presented above.

Table 1. Some examples of interviewer covariances and of optimum determination of number of interviewers

Study	Characteristic	Estimate of inter-viewer covariance s_{XI}	Estimate of total variance s_X^2	Optimum number* of:	
				Inter-viewers m	Indivi-duals n
Jagaddal, 1942	1. Monthly expenditure (rupees per capita)	1.01	171.0	8	680
	2. Consumption of cereals (pounds per head per month)	.13	100.8	5	800
Nagpur, 1943	3. Total monthly expenditures	.80	399.1	6	760

* Values giving minimum variance subject to the cost restriction that $2n + 80m = C = 2000$.

To determine the optimum allocation of resources for the Baltimore study design, we let n equal the number of segments and assume costs of:

C = total budget = \$400.

C_1 = cost per segment (using one interviewer to cover each segment) = \$6.

C_2 = cost per interviewer = \$7.

Table 2 shows the values of s_X^2 and s_{XI} determined from the Baltimore study data and the optimum values of n and m with the cost function of Eq. 3.15. In the Baltimore study 2 interviewers were assigned to each segment. The optimum values n and m were determined for the case in which only 1 interviewer is assigned to any segment.

* To simplify calculations, 1 of these segments was eliminated at random.

It will be noted that s_{XI} is negative for 3 of the 5 characteristics. Negative values of s_{XI} frequently will be obtained when σ_{XI} is near zero (since s_{XI} is an unbiased estimate of σ_{XI}) and are particularly likely to occur when s_{XI} is based on a small number of degrees of freedom (i.e., relatively few interviewers), making the variance of s_{XI} relatively large. Where s_{XI} is negative, we have taken σ_{XI} as zero in estimating $\sigma_{\bar{x}}^2$ and the optimum values of n and m. In these cases, of course, the optimum requires that m be as small as possible (i.e., $m = 2$, the number of interviewer groups).

Table 2. Variance estimates and optimum values of n and m for the Baltimore study conditions

Item	Characteristic to be estimated				
	Total persons	Persons under 14 years of age	Total employed	Persons employed at nonfarm job for wages or salary	Persons operating own business or profession
Variance and covariance estimates					
s_X^2	64.5	4.98	34.3	44.0	1.51
s_{XI}	1.04	− .68	− .013	1.28	− .14
Optimum values					
n	58	64	64	56	64
m	7	2	2	9	2
$s_{\bar{x}}^2$	1.23	.078	.54	.90	.024
Variance of $\bar{x}$ with					
$n = 64$ and $m = 2$	1.51	.078	.54	1.32	.024
$n = 62$ and $m = 4$	1.28	.080	.55	1.01	.024
$n = 57$ and $m = 8$	1.24	.088	.60	.91	.026
$n = 48$ and $m = 16$	1.39	.104	.71	.96	.031
$n = 31$ and $m = 31$	2.08	.161	1.11	1.42	.049

For 2 characteristics (total persons and persons employed at a nonfarm job for wages and salaries) there is some contribution of interviewer error to the total variance. For these characteristics the optimum is fairly broad, i.e., for m between 4 and 16 the variance of $\bar{x}$ will be within 13 per cent of the optimum.

4. Use of the specified mathematical model in minimizing the effect of both bias and variance. The preceding section indicates a method for determining the optimum under fixed essential conditions. In many cases where it is evident that a particular survey technique is subject to substantial response bias, alternative techniques may be available that will reduce the bias. We must, of course, consider the relative cost of such alternatives.

Choosing a single sampling design. We may have a choice of alternative methods, each with different essential conditions, response bias, and optimum values of n and m. For a fixed total cost we can determine the optimum values of n and m for each such method. Then the optimum *method* among those examined is the one which gives the lowest mean square error. For example, experience in determining farm expenditures by direct questioning of farm operators has shown that the results are often subject to considerable error. Determining farm expenditures by other techniques, such as detailed examination of purchase records, may be more accurate but considerably more expensive. We can determine the optimum for direct questioning and for detailed examination of purchase records, subject to a fixed total budget, and select the method which gives the lower mean square error. The optimum method for one budget level may be different from that for another budget level.

Use of double sampling. In some cases, instead of using a single method, a combination of two methods in a double sampling design may prove more efficient. For example, we could interview a relatively large number of cases (possibly even the entire population) by one of the cheaper (and less accurate) methods and reinterview a subsample by one of the more expensive methods. Such a double sampling approach is likely to be useful in instances where methods with low response bias cost many times as much as methods with higher response bias.

Suppose that our original sample is drawn as described in the previous section and we have sampled n individuals and m interviewers (m_h from the hth group). For this sample, we obtain responses x_{hij} under the essential conditions of the initial survey, which we shall designate as essential conditions X. For the subsample we take (at random) an equal number out of the individuals assigned to each interviewer, giving a subsample of n' individuals. For the subsample we shall use a set of L' interviewer groups (which may or may not be the same as the original interviewer groups). We draw m' interviewers (m'_p from the pth group) in such manner that an equal number of interviews can be given to each interviewer. It will be noted that the interviewers for the subsample are drawn independently of those for the original sample and that m' can be less than, equal to, or greater than m. For the subsample we have the responses x_{hij} obtained by the original interviewers and we also have responses z_{pqj} obtained by the second set of interviewers under essential conditions Z. We may use as an estimate* of the true population mean $\bar{Y}$:

$$\hat{z} = \frac{\bar{x}\bar{z}'}{\bar{x}'} \tag{4.1}$$

where $\bar{x}$ = mean of the x_{hij} values for the entire sample of n individuals.

$\bar{x}'$ = mean of the x_{hij} values for the subsample of n' individuals.

$\bar{z}'$ = mean of the z_{pqj} values for the subsample of n' individuals.

Actually, it might be more efficient from a sampling viewpoint to draw the m' interviewers for the subsample of clusters as a subsample of the m interviewers used for the original sample of clusters. However, the main purpose is to reduce the response bias, and this may mean the use of better-qualified or better-trained interviewers. Consequently, the second set of interviewers may be drawn from a different population of interviewers.

The mean square error with double sampling.† It is assumed that interviews under conditions Z are more expensive than under conditions X and that method Z has a considerably smaller response bias. For the specified mathematical model, the mean square error of z will be approximately

$$\text{MSE}\,\hat{z} = B_Z^2 + \bar{Z}^2\left(\frac{U}{n} + \frac{V}{n'} + \frac{W}{m'}\right) \tag{4.2}$$

where

$$\bar{Z} = E\bar{z}' \doteq E\hat{z}$$

$$\bar{X} = E\bar{x} = E\bar{x}'$$

$$B_Z = \bar{Z} - \bar{Y}$$

$$U = \frac{2\rho_{XZ}\sigma_X\sigma_Z}{\bar{X}\bar{Z}} - \frac{\sigma_X^2 - \sigma_{XI}}{\bar{X}^2} \tag{4.3}$$

$$V = \frac{\sigma_Z^2 - \sigma_{ZI}}{\bar{Z}^2} - U \tag{4.4}$$

$$W = \frac{\sigma_{ZI}}{\bar{Z}^2} \tag{4.5}$$

ρ_{XZ} = the correlation between the expected X and Z values for the same individual

---

* This estimate $\hat{z}$ is, of course, a ratio of random variables and is biased but consistent.

† For the derivation of the formulas presented in this section, see Sec. 7*d*.

An optimum double sampling design. With a combination of two methods there is, in general, a set of optimum values for n, n', and m'. As in the preceding section we shall consider only the case where the cost function is simple and the optimum values can be determined directly. We shall assume the cost function:

$$C = C_1 n + C_2 m + C_3 n' + C_4 m' \tag{4.6}$$

where C_1 = cost per individual under conditions X.

C_2 = cost per interviewer under conditions X.

C_3 = cost per individual under conditions Z.

C_4 = cost per interviewer under conditions Z.

C — total survey budget (excluding any fixed overhead costs).

Since MSE of z does not involve m (to the order of approximation used in Equation 4.2*) but the cost increases with m, the optimum design would call for making m as small as possible. Usually, the minimum number of interviewers will be determined by administrative considerations, i.e., an interviewer can be expected to complete a certain number of interviews a day, and, if the survey results must be available at some specified time, we must give an interviewer no more than the number of cases he can complete within the time period allowed. If, then, we decide that an interviewer shall not do more than t_X interviews, the smallest value we can give to m is $m = n/t_X$, and the optimum values of n, n', and m' are

$$n = a\sqrt{\frac{U}{C_1 + \frac{C_2}{t_X}}} \tag{4.7}$$

$$n' = a\sqrt{\frac{V}{C_3}} \tag{4.8}$$

$$m' = a\sqrt{\frac{W}{C_4}} \tag{4.9}$$

where

$$a = \frac{C}{\sqrt{U\left(C_1 + \frac{C_2}{t_X}\right)} + \sqrt{VC_3} + \sqrt{WC_4}} \tag{4.10}$$

---

* Equation 4.2 is an approximation which ignores terms of the third and higher order. Where m, n, m', or n' is small, some of these terms may be appreciable and the approximation to σ_z^2 may be poor.

Using the optimum values in the equation for MSE of $\hat{z}$ (Eq. 4.2) will permit us to compare a combination of Methods X and Z with either method alone or with other methods and combinations to determine the optimum design.

Estimation of variances and biases. We must, of course, have some idea of the costs and of the values of U, V, W, and B_Z. We can estimate $\bar{X}$ and $\bar{Z}$ from a sample (using $\bar{x}$ as an estimate of $\bar{X}$ and $\hat{z}$ as an estimate of $\bar{Z}$). The variances can be estimated by means of Eq. 3.8 and 3.9, and an unbiased estimate of $\rho_{XZ}\sigma_X\sigma_Z$ is provided by

$$\rho'_{XZ}s_Xs_Z = \frac{\sum_{j}^{n'} x_{hij}z_{pqj} - n'\bar{x}'\bar{z}'}{n' - 1} \tag{4.11}$$

Estimation of the response bias, B_Z, is a more difficult problem since this error involves the unknown true population mean $\bar{Y}$. However, a satisfactory comparison of several methods can sometimes be made in instances where one is justified in assuming a negligible response bias for the method which is considered most accurate and (from previous experience or a pilot sample study) estimating the differences in expected value between this most accurate method and the other methods considered.

For example, if Method Z is one which is subject to negligible response bias, as an estimate of bias for some other method we can use either

$$\bar{b}_X = \bar{x} - \bar{z}' \tag{4.12}$$

$$\hat{b}_X = \bar{x} - \hat{z} \tag{4.13}$$

These estimates are, of course, subject to sampling error, and formulas for the variances are given in Sec. 7*e*.

Illustration of jointly minimizing variance and bias. To illustrate the technique for determining the method which minimizes the mean square error, we shall use a problem which involves estimating the average dollar inventory of a group of retail stores. Let us assume that the population consists of all retail stores in a large city and that our budget for the survey is \$15,000, of which \$2500 has been set aside for fixed overhead (so that $C = 12{,}500$). We shall also assume that the maximum assignment to an interviewer, t_X, is as shown in Table 3. Suppose that pilot studies and previous experience give cost, variance, correlation, and response bias estimates for five different techniques and that we wish to determine which technique (or combination of two techniques) to use. Let us assume that the estimates of unit costs, response bias, and variances for each technique are as shown in Table 3 and the correlations, ρ_{XZ}, for each pair of techniques are as shown in Table 4. We shall take $\bar{Y} = \$100{,}000$.

Table 3. Cost factors, maximum assignments, biases, and variances for a study of retail store inventories

Method	Unit costs (dollars)		Maximum assignment per interviewer t_X	Response bias (thousands of dollars) B_X	Square root of variances and covariances (thousands of dollars)	
	Per store C_1	Per interview C_2			σ_X	$\sqrt{\sigma_{XI}}$
1	$\frac{1}{4}$	25	100	− 11.0	83	25
2	2	50	60	− 6.0	80	15
3	6	100	40	− 2.5	76	10
4	12	150	35	− .8	73	9
5	20	150	35	− .6	71	6

Table 4. Correlation between expected values of individual responses (ρ_{XZ})

Method	Method 1	2	3	4
2	.79			
3	.82	.84		
4	.84	.87	.91	
5	.85	.88	.92	.95

Table 5 shows the mean square error which would be obtained for each method and each combination of methods, using with the single sampling method the values of n and m given by Eq. 3.16 and 3.17 and with the double sampling method the values of n, n', and m' given by Eq. 4.7, 4.8, and 4.9 with $m = n/t_X$.

The optimum, if only a single method is employed, is to use Method 4 with $n = 723$ and $m = 25$. However, double sampling permits a further reduction of 35 per cent in the MSE by using Methods 1 and 5 with $n = 3480$, $m = 35$, $n' = 382$, and $m' = 21$. In many cases double sampling will not give gains of this magnitude over a good single sampling method. It should be noted that the figures in Tables 4 and 5 are hypothetical and are used only to illustrate the methods.

In situations of this type it may frequently be necessary to increase expenditures per unit many times in order to reduce the response bias from 10 per cent to 2 per cent. For example, the A. C. Nielsen Company, which compiles data on sales of commodities by retail stores, has found that it can obtain sufficiently accurate reports on sales only by personally checking physical inventory and purchase invoices. As another example, the Bureau of the Census reinterviewed a sample of respondents, using in the reinterview professional personnel from the Washington office.

Table 5. Comparison of minimum mean square errors for five alternative methods (and combinations of them)

Methods	n	m	n'	m'	Minimum mean square error for the indicated method
1	12,022	380	—	—	123.2
2	3,197	122	—	—	39.8
3	1,350	44	—	—	12.7
4	723	25	—	—	11.1
5	508	16	—	—	12.5
1 and 2	5,242	52	1,869	123	39.7
1 and 3	4,382	44	883	50	11.2
1 and 4	3,582	36	503	31	7.6
1 and 5	3,480	35	382	21	7.3
2 and 3	1,521	25	690	41	13.8
2 and 4	1,340	22	392	27	10.1
2 and 5	1,329	22	302	18	9.5
3 and 4	703	18	260	23	13.1
3 and 5	718	18	201	16	12.0
4 and 5	481	14	129	14	15.7

The average cost per interview was of the order of 7 times the average cost for the original interview, and there was a significant increase in the accuracy of certain items such as coverage of persons. On the other hand, in some cases increases in expenditure may yield only small gains in accuracy or large gains for some items and small gains for others. In the same study by the Bureau of the Census, the per cent distribution into 10-year age groups, for example, was practically identical for both interviews, with none of the 10-year age groups differing by more than $\frac{1}{4}$ of 1 per cent from the original interview to the reinterview. If the primary aim of the survey were to obtain an accurate per cent distribution by age, the more expensive method would not be justified.

Frequently, it is also possible for very small increases in expenditures to produce large gains in accuracy. In one instance, for example, the Bureau of the Census in its Current Population Survey had been getting a large number of persons erroneously reported as not in the labor force. A revision of the questions asked added nearly 2 million of these "missed" persons to the labor force (2). The revision added practically nothing to the cost of the survey.

Thus, there is no "typical" relation between cost and accuracy. Each survey presents its own picture. The method outlined is general in its applicability, although the answers obtained will vary.

It should be noted that the work of determining the "optimum" design can frequently be shortened by eliminating from consideration alternatives which are obviously inefficient. For example, Method 5 in the illustration above involves a cost per store two-thirds greater than that of Method 4, but the response bias for the two methods differs by only a trivial amount. The higher expenditure per unit in Method 5 improves the values for individual stores, but the individual response errors of Method 4 are largely "compensating" in nature. Although the combination of Methods 1 and 5 gives the lowest MSE, the result does not differ appreciably from that for Methods 1 and 4. Thus, consideration of both Method 4 and Method 5 was really unnecessary in selecting an optimum.

5. Effect of uncorrelated and compensating response errors. A consideration of the specified mathematical model leads to the conclusion that response errors that are uncorrelated with each other and compensating in character do not necessarily need any special attention in survey design whenever the purpose is to estimate a mean or total for the total population or for a subgroup when the members of the subgroup are identified without error. Furthermore, in this case, the formulas presented in previous chapters for estimating sampling error reflect the response error properly and no special attention need be given to the presence of response errors. This situation is, however, often assumed to exist without valid evidence. It is not at all uncommon for the results of the survey to be justified on the basis that "some of the errors were positive and some were negative, so that the net effect is undoubtedly close to zero." Of course, it is not possible to assume that just because there are both positive and negative errors their effect is necessarily compensating, and such an assumption can very often lead to erroneous conclusions. Moreover, as we have already seen, if the response errors are correlated with each other (as within the work of a single interviewer), the variance is increased and the chances of errors being compensating are reduced. However, let us consider the situation where there is evidence that the errors are, in fact, compensating and uncorrelated.

The essential points can be seen more easily with the very simple situation in which the response errors are uncorrelated for any two individuals in the population and a random sample of individuals is drawn without restriction and with replacement. The variance of a sample estimate of the true population mean under these conditions is

$$\sigma_{\bar{x}}^2 = \frac{\sigma_X^2}{n} \tag{5.1}$$

Since $X_{ij} = Y_i + R_{ij}$ (where Y_i is the true value of the characteristic for the ith elementary unit of the population, and R_{ij} is the response error on the jth measurement for the ith elementary unit), we can express σ_X^2 as

$$\sigma_X^2 = \sigma_Y^2 + \sigma_R^2 + 2\rho_{YR}\sigma_Y\sigma_R \tag{5.2}$$

where σ_Y^2 = variance of the true values.

σ_R^2 = variance of the response errors, which is composed of the variance of response errors for an individual around the individual's expected response error and the variance in the expected value of response errors between individuals.

ρ_{YR} = correlation between the Y_i and the R_{ij}. Note that $\rho_{YR}\sigma_Y\sigma_R = \rho_{Y\bar{R}}\sigma_Y\sigma_{\bar{R}}$, where $\rho_{Y\bar{R}}$ is the correlation between the Y_i and the $\bar{R}_i$, the expected response error for the ith individual.

Response errors are reflected in usual sample variance. We see from Eq. 5.2 that σ_X^2 reflects any effects of the responses as well as the variance of the true values. Similarly, if we estimate σ_X^2 by

$$s_X^2 = \frac{\sum_i^n (x_{ij} - \bar{x})^2}{(n - 1)}$$

the effect of the response errors will appear in the estimated variance, since the expected value of s_X^2 is equal to σ_X^2. Consequently the estimated variance of the sample mean will include appropriate allowance for the response variation.

Reducing the effect of response variation. If we assume a fixed total budget (after deducting fixed overhead), C, then the number of cases which can be sampled is equal to C divided by the unit cost C_X. Thus, we have

$$\sigma_{\bar{x}}^2 = \frac{C_X}{C}\sigma_X^2 \tag{5.3}$$

Let us compare the results of Method X (which gives the response X_i) with those of Method Z (which gives the response Z_i), assuming that Method Z has a unit cost $C_Z > C_X$, that $\sigma_{R_Z} < \sigma_{R_X}$, and that $\bar{R}_Z = \bar{R}_X$

$= 0$. Then, Method Z is preferable to Method X only if

$$\sigma_{\bar{z}}^2 < \sigma_{\bar{x}}^2 \tag{5.4}$$

or if

$$\frac{\sigma_Z^2}{\sigma_X^2} < \frac{C_X}{C_Z} \tag{5.5}$$

Inequality 5.5 provides a test of the relative efficiencies of two data-collecting techniques which have either no response bias or the same response bias. Let us consider a hypothetical example. Suppose that we have a choice between two methods of estimating a characteristic, both methods using a simple random sample of families:

For Method X	For Method Z
$C_X = \$2$	$C_Z = \$5$
$\sigma_{R_X} = 10$	$\sigma_{R_Z} = 2$
$\rho_{YR_X} = .1$	$\rho_{YR_Z} = .1$

$$\sigma_Y = 10$$

Since both methods have the same response bias, Method Z will be more efficient than Method X only if Inequality 5.5 holds. Using the figures just presented gives

$$\frac{C_X}{C_Z} = .4, \quad \frac{\sigma_Z^2}{\sigma_X^2} = \frac{108}{220} = .49$$

Thus Inequality 5.5 does not hold, and we gain more by putting our funds into the larger sample permitted by the lower unit cost of Method X than by putting them into the reduction of response error permitted by Method Z (even though this reduction in response error is quite substantial). It appears that ordinarily any appreciable increase in expenditure to decrease response variation will be unwarranted if there is no effect on the response bias.

If we are estimating the proportion of the elementary units having a given characteristic, any expenditure to reduce response errors will be wasted whenever the response errors are compensating so that $\bar{Z} = \bar{X}$, since, in this case,

$$\sigma_Z^2 = \sigma_X^2 = \bar{X}(1 - \bar{X}) = \bar{Z}(1 - \bar{Z})$$

As a practical matter, the difficulty is that there is rarely any assurance that response errors are, in fact, compensating, and it is not safe, without extensive investigation, to concentrate on sampling variance and assume that response errors are of no importance. Even if we were to reduce sampling error to negligible proportions by taking a complete census, we might still have a substantial MSE because of response bias.

6. Applicability of the specified mathematical model. The analysis presented above applies, of course, only when the conditions of the specified mathematical model apply. It is important, therefore, that we examine these conditions in terms of the situations actually prevalent in typical surveys. With regard to the selection of individuals or other sampling units for interview, it is feasible to use techniques (e.g., selection dependent upon a table of random numbers) which give random samples. However, a determination that individual responses behave like random variables will require much more experimental evidence than is now available, and will necessarily be subject to some question.

As previously noted, the conditions which determine the response of any individual may be regarded as divided into two groups:

(*a*) Those conditions which are "constant," "controlled," and predetermined for a given individual response, e.g., the questions to be asked, the type of interviewer. We have referred to these as the essential survey conditions.

(*b*) Those conditions which are adventitious and "unpredictable," e.g., the mood of the respondent, a momentary distraction which results in a question being misunderstood.

This division is similar to the division between "assignable (i.e., controllable)" causes and "residual" causes of variation in discussions of quality control. We have treated these two groups of factors in the same way as they are treated in the quality control field. Thus, we consider the "adventitious" factors as giving rise to a random variable, the response obtained for a given individual being one of the values of this variate. The "controlled" causes would determine the expected value of this random variable. They also affect its variance.

It should be noted that the present analysis does not provide for measuring the response variance of a single individual apart from the variance between individuals. Such measurement would be feasible experimentally were it not for the conditioning of the respondent. But, in practice, repeated interviews on a single respondent are not independent. Thus, a direct test of the random nature of individual response variation cannot be made by using the specified mathematical model. We can, however, determine the total variance and, by applying the specified mathematical model in a large number of cases, test its approximation to actual conditions.

A further assumption, which can be accomplished in fact, but ordinarily is not, is that the available sample individuals are assigned to interviewers at random and that, within an interviewer group, the selection of interviewers is independent of the selection of individuals. The survey supervisor will usually try to arrange an interviewer's assignment to minimize travel costs rather than making up random arrangements, and definite steps must be taken to make it possible to measure the variable contribution of the interviewers. More experimental work in this area is needed. The implications of this entire problem need thorough exploration, and the analysis presented in the present chapter can be considered only a step toward a systematic treatment of response error.

It was indicated earlier that the results presented were applicable to the estimation of aggregates or averages for a total population, or for subgroups of the population provided that the assignments to subgroups are made without error. If the identification of whether a unit is or is not a member of the subgroup involves response errors, response bias usually will be present in estimates of subgroup means even when estimates of the population mean are unbiased, unless the errors involved in identifying a unit as a member of the subgroup are independent of each other and of the characteristic to be estimated. The effect of errors of measurement on correlations has sometimes been considered, but the situations dealt with ordinarily have been restricted to the case where errors are independent and "attenuate" the correlation. The effect of correlated errors requires much more attention than it has been given but is beyond the scope of the present discussion.

7. Derivations and proofs. *a. Description of the population.* The mathematical model used in this chapter for the analysis of response errors assumes a population divided into N units. The unit may be an elementary unit or it may be a "cluster" of elementary units (e.g., a household or a group of households living in an area). The N units are divided into L "groups" with N_h units in the hth group. There are M_h interviewers available to interview the units in the hth group (and only these units). On any particular interview of a "unit" by an interviewer a response occurs. It is assumed that this response is a random variable for any interviewer and any respondent; i.e., a given response is considered to be only one of the possible responses which might be obtained from this respondent. Let

P_{hijk} = the probability that the response value X_{hijk} will be obtained if the ith interviewer in the hth group interviews the jth unit.

$P_{hijkuvw}$ = the probability that responses X_{hijk} and X_{huvw} are obtained (in the hth group) if the ith and uth interviewers interview the jth and vth units.

We have then

$$\sum_k P_{hijk} = 1 \tag{7.1}$$

$$E_{hij} X_{hijk} = \bar{X}_{hij} = \sum_k P_{hijk} X_{hijk} \tag{7.2}$$

where the sum is taken over all possible responses of the jth unit to the ith interviewer.

We also have

$$\sum_k \sum_w P_{hijkuvw} = \sum_k P_{hijk} = \sum_w P_{huvw} = 1 \tag{7.3}$$

$$E_{hijuv}(X_{hijk} X_{huvw}) = \sum_k \sum_w P_{hijkuvw} X_{hijk} X_{huvw} \tag{7.4}$$

where the sum is taken over all possible responses of the jth unit to the ith interviewer and of the vth unit to the uth interviewer, and E_{hijuv} is the conditional expected value for fixed interviewers i and u, and fixed respondents j and v. We shall assume that, if $i \neq u$ and $j \neq v$ (i.e., if *both* interviewer and respondent are different), the responses obtained are independent. Thus, if $i \neq u$ and $j \neq v$,

$$P_{hijkuvw} = P_{hijk} P_{huvw} \tag{7.5}$$

$$E_{hijuv} X_{hijk} X_{huvw} = \bar{X}_{hij} \bar{X}_{huv} \tag{7.6}$$

b. Contribution of interviewer errors to the variance of a sample estimate from a simple random sample. Let us suppose that the sample design calls for drawing units from the population as a whole and drawing interviewers independently within each group. These selections are made at random. To each sample interviewer drawn from the hth group we assign, at random, a certain number of the units in the sample from the hth group. We fix in advance:

(1) n = the total number of sample units to be drawn.
(2) $\bar{n}_h$ = the number of sample units to be assigned to each sample interviewer from the hth group.

Since the n units are drawn independently of the groups, the number of units falling in the sample in any group is a random variable. Let us designate the number of units in the sample from the hth group by n_h. Since we fix the number of sample units per interviewer in the hth group at $\bar{n}_h$, the number of sample interviewers to be drawn in this group will be $m_h = n_h/\bar{n}_h$, and m_h will also be a random variable.

Let

x_{hijk} = a particular response obtained for the jth sample unit when interviewed by the ith sample interviewer drawn from the hth (population) group.

$$\bar{x} = \frac{\sum_h^L \sum_i^{m_h} \sum_j^{\bar{n}_h} x_{hijk}}{n} \tag{7.7}$$

The summation for k is omitted because only one response is recorded for any individual in the sample.

Contingent upon drawing a unit from the hth group, the probability that $x_{hijk} = X_{hijk}$ is $\frac{1}{N_h}\frac{1}{M_h}P_{hijk}$, and there are n_h values of x_{hijk}. Thus

$$E\bar{x} = \frac{\sum_h^L E_h(n_h x_{hijk})}{n} \tag{7.8}$$

$$E_h(n_h x_{hijk}) = E_h[n_h E_h(x_{hijk}|n_h)] = E_h(n_h \bar{X}_h) \tag{7.9}$$

where

$$\bar{X}_h = \frac{\sum_i^{M_h} \sum_j^{N_h} \sum_k P_{hijk} X_{hijk}}{M_h N_h} = \frac{\sum_i^{M_h} \sum_j^{N_h} \bar{X}_{hij}}{M_h N_h} \tag{7.10}$$

Therefore

$$E\bar{x} = \frac{\sum_h^L \bar{X}_h E_h(n_h)}{n} \tag{7.11}$$

Since

$$E_h(n_h) = \frac{nN_h}{N}$$

$$E\bar{x} = \frac{\sum_h^L N_h \bar{X}_h}{N} = \frac{\sum_h^L \frac{1}{M_h} \sum_i^{M_h} \sum_j^{N_h} \bar{X}_{hij}}{N} = \bar{X} \tag{7.12}$$

$$\sigma_{\bar{x}}^2 = E\bar{x}^2 - \bar{X}^2 = E\left(\frac{\sum_h^L \sum_i^{m_h} \sum_j^{\bar{n}_h} x_{hijk}}{n} - \bar{X}\right)^2 \tag{7.13}$$

To evaluate

$$\sigma_{\bar{x}}^2 = E\left(\frac{\sum_h^L \sum_i^{m_h} \sum_j^{\bar{n}_h} x_{hijk}}{n} - \bar{X}\right)^2$$

we shall make repeated applications of Theorem 15 of Ch. 3 (p. 65). First, let us assume that condition b_1 is that the sample of households and the sample of interviewers assigned to the households are fixed. Then, from Theorem 15,

$$\sigma_{\bar{x}}^2 = E\sigma_{\bar{x}|b_1}^2 + \sigma_{E(\bar{x}|b_1)}^2 \tag{7.14}$$

Now, since

$$E\left(\frac{\sum_h^L \sum_i^{m_h} \sum_j^{\bar{n}_h} x_{hijk}}{n} \mid b_1\right) = \frac{\sum_h^L \sum_i^{m_h} \sum_j^{\bar{n}_h} \bar{X}_{hij}}{n} \tag{7.15}$$

it follows that

$$\sigma^2_{\bar{x}|b_1} = E\left[\left(\frac{\sum_h^L \sum_i^{m_h} \sum_j^{\bar{n}_h} x_{hijk}}{n} - \frac{\sum_h^L \sum_i^{m_h} \sum_j^{\bar{n}_h} \bar{X}_{hij}}{n}\right)^2 \mid b_1\right]$$

$$= \frac{1}{n^2} \sum_h^L \sum_i^{m_h} E_{hi}\left\{\left[\sum_j^{\bar{n}_h} (x_{hijk} - \bar{X}_{hij})\right]^2 \mid b_1\right\}$$

$$= \frac{1}{n^2} \sum_h^L \sum_i^{m_h} \sum_j^{\bar{n}_h} E_{hij}(x_{hijk} - \bar{X}_{hij})^2$$

$$+ \frac{1}{n^2} \sum_h^L \sum_i^{m_h} \sum_{j \neq v}^{\bar{n}_h} E_{hij,v}(x_{hijk} - \bar{X}_{hij})(x_{hivw} - \bar{X}_{hiv}) \tag{7.16}$$

where E_{hi} represents the conditional expected value for the interviewers fixed, E_{hij} represents the conditional expected value for ith interviewer in the sample and jth household in the sample, $E_{hij,v}$ represents the conditional expected value with the jth and vth sample households fixed. Then, $E\sigma^2_{\bar{x}|b_1}$, where $\sigma^2_{\bar{x}|b_1}$ is given by Eq. 7.16, is

$$E\sigma^2_{\bar{x}|b_1} = \frac{1}{n^2} E \sum_h^L \frac{m_h}{M_h} \frac{\bar{n}_h}{N_h} \sum_i^{M_h} \sum_j^{N_h} E_{hij}(x_{hijk} - \bar{X}_{hij})^2$$

$$+ \frac{1}{n^2} E \sum_h^L \frac{m_h}{M_h} \frac{\bar{n}_h}{N_h} \frac{\bar{n}_h - 1}{N_h - 1} \sum_i^{M_h} \sum_{j \neq v}^{N_h} E_{hij,v}(x_{hijk} - \bar{X}_{hij})(x_{hivw} - \bar{X}_{hiv}) \tag{7.17}$$

and since

$$Em_h\bar{n}_h = En_h = n\frac{N_h}{N}$$

we have

$$E\sigma^2_{\bar{x}|b_1} = \frac{1}{nN} \sum_h^L \frac{1}{M_h} \sum_i^{M_h} \sum_j^{N_h} E_{hij}(x_{hijk} - \bar{X}_{hij})^2$$

$$+ \frac{1}{nN} \sum_h^L \frac{\bar{n}_h - 1}{N_h - 1} \frac{1}{M_h} \sum_i^{M_h} \sum_{j \neq v}^{N_h} E_{hij,v}(x_{hijk} - \bar{X}_{hij})(x_{hivw} - \bar{X}_{hiv}) \tag{7.18}$$

To complete the evaluation of $\sigma^2_{\bar{x}}$, we must evaluate $\sigma^2_{E(\bar{x}|b_1)}$, the second term of Eq. 7.14.

Now $E(\bar{x}|b_1)$ is given by Eq. 7.15 and

$$\sigma^2_{E(\bar{x}|b_1)} = E\left(\frac{\sum_h^L \sum_i^{m_h} \sum_j^{\bar{n}_h} \bar{X}_{hij}}{n} - \bar{X}\right)^2 \tag{7.19}$$

We can again apply Theorem 15, Ch. 3 (p. 65), to Eq. 7.19 with the condition b_2 that the interviewers in the sample are fixed. First, let

$$\bar{x}' = \frac{\sum_h^L \sum_i^{m_h} \sum_j^{\bar{n}_h} \bar{X}_{hij}}{n}$$

Then

$$\sigma^2_{E(\bar{x}|b_1)} = \sigma^2_{\bar{x}'} \tag{7.20}$$

Now, from Theorem 15,

$$\sigma^2_{\bar{x}'} = E\sigma^2_{\bar{x}'|b_2} + \sigma^2_{E(\bar{x}'|b_2)} \tag{7.21}$$

Since

$$E(\bar{x}'|b_2) = \frac{\sum_h^L \sum_i^{m_h} \bar{n}_h \bar{X}_{hi}}{n} \tag{7.22}$$

where

$$\bar{X}_{hi} = \frac{\sum_j^{N_h} \bar{X}_{hij}}{N_h}$$

then

$$\sigma^2_{\bar{x}'|b_2} = E\left[\left(\frac{\sum_h^L \sum_i^{m_h} \sum_j^{\bar{n}_h} \bar{X}_{hij}}{n} - \frac{\sum_h^L \sum_i^{m_h} \bar{n}_h \bar{X}_{hi}}{n}\right)^2 \Big| \; b_2\right]$$

$$= \frac{1}{n^2} \sum_h^L \sum_i^{m_h} E_{hi} \left[\sum_j^{\bar{n}_h} (\bar{X}_{hij} - \bar{X}_{hi})\right]^2 \tag{7.23}$$

$$= \frac{1}{n^2} \sum_h^L \sum_i^{m_h} \bar{n}_h \frac{\sum_j^{N_h} (\bar{X}_{hij} - \bar{X}_{hi})^2}{N_h}$$

$$+ \frac{1}{n^2} \sum_h^L \sum_i^{m_h} \frac{\bar{n}_h}{N_h} \frac{\bar{n}_h - 1}{N_h - 1} \sum_{j \neq v}^{N_h} (\bar{X}_{hij} - \bar{X}_{hi})(\bar{X}_{hiv} - \bar{X}_{hi})$$

and since

$$Em_h \bar{n}_h = En_h = n \frac{N_h}{N}$$

we have

$$E\sigma^2_{\bar{x}'|b_2} = \frac{1}{Nn}\sum_h^L \frac{1}{M_h}\sum_i^{M_h}\sum_j^{N_h}(\bar{X}_{hij} - \bar{X}_{hi})^2$$

$$+ \frac{1}{Nn}\sum_h^L \frac{1}{M_h}\frac{\bar{n}_h - 1}{N_h - 1}\sum_i^{M_h}\sum_{j\neq v}^{N_h}(\bar{X}_{hij} - \bar{X}_{hi})(\bar{X}_{hiv} - \bar{X}_{hi}) \quad (7.24)$$

and to complete the evaluation of $\sigma^2_{\bar{x}'}$, we must find $\sigma^2_{E(\bar{x}'|b_2)}$ in Eq. 7.21. From Eq. 7.22

$$E(\bar{x}'|b_2) = \frac{\sum_h^L \sum_i^{m_h} \bar{n}_h \bar{X}_{hi}}{n} \quad (7.25)$$

and

$$\sigma^2_{E(\bar{x}'|b_2)} = E\left(\frac{\sum_h^L \sum_i^{m_h} \bar{n}_h \bar{X}_{hi}}{n} - \bar{X}\right)^2 \quad (7.26)$$

If we now let $E(\bar{x}'|b_2) = \bar{x}''$, we can again apply Theorem 15, with the condition b_3 that the number of interviewers is fixed. Then

$$\sigma^2_{\bar{x}''} = E\sigma^2_{\bar{x}''|b_3} + \sigma^2_{E(\bar{x}''|b_3)} \quad (7.27)$$

Now

$$E(\bar{x}''|b_3) = \frac{\sum_h^L \frac{m_h \bar{n}_h}{M_h}\sum_i^{M_h}\bar{X}_{hi}}{n} = \frac{\sum_h^L n_h \bar{X}_h}{n} \quad (7.28)$$

and

$$\sigma^2_{\bar{x}''|b_3} = E\left[\left(\frac{\sum_h^L \sum_i^{m_h} \bar{n}_h \bar{X}_{hi}}{n} - \frac{\sum_h^L \bar{n}_h \bar{X}_h}{n}\right)^2 \mid b_3\right] \quad (7.29)$$

$$= \frac{1}{n^2}\sum_h^L E_h \bar{n}_h^2 \left[\sum_i^{m_h}(\bar{X}_{hi} - \bar{X}_h)\right]^2$$

$$= \frac{1}{n^2}\sum_h^L \bar{n}_h^2 \frac{m_h}{M_h}\sum_i^{M_h}(\bar{X}_{hi} - \bar{X}_h)^2 \quad (7.30)$$

since $E(\bar{X}_{hi} - \bar{X}_h)(\bar{X}_{hj} - \bar{X}_h) = 0$ when the interviewers are selected independently (i.e., with replacement).

Now, since

$$Em_h\bar{n}_h = En_h = n\frac{N_h}{N}$$

$$E\sigma^2_{\bar{x}''|b_3} = \frac{1}{nN}\sum_h^L \bar{n}_h N_h \frac{\sum_i^{M_h}(\bar{X}_{hi} - \bar{X}_h)^2}{M_h} \quad (7.31)$$

and

$$\sigma^2_{E(\bar{x}''|b_s)} = E\left(\frac{\sum_h^L n_h \bar{X}_h}{n} - \bar{X}\right)^2 = E\left(\frac{\sum_h^L n_h \bar{X}_h}{n} - \frac{\sum_h^L N_h \bar{X}_h}{N}\right)^2 \tag{7.32}$$

$$= E\left[\sum_h^L \bar{X}_h \left(\frac{n_h}{n} - \frac{N_h}{N}\right)\right]^2$$

$$= E\sum_h^L \bar{X}_h^2 \left(\frac{n_h}{n} - \frac{N_h}{N}\right)^2 + E\sum_{h \neq j}^L \bar{X}_h \left(\frac{n_h}{n} - \frac{N_h}{N}\right) \bar{X}_j \left(\frac{n_j}{n} - \frac{N_j}{N}\right) \tag{7.33}$$

Now, to evaluate Eq. 7.33, we have for the first term

$$E\sum_h^L \bar{X}_h^2 \left(\frac{n_h}{n} - \frac{N_h}{N}\right)^2 = \frac{N-n}{(N-1)n}\sum_h^L \bar{X}_h^2 \frac{N_h}{N}\left(1 - \frac{N_h}{N}\right)$$

$$= \frac{N-n}{(N-1)n}\sum_h^L \bar{X}_h^2 \frac{N_h}{N} - \frac{N-n}{(N-1)n}\sum_h^L \bar{X}_h^2 \frac{N_h^2}{N^2}$$

To evaluate the second term of Eq. 7.33, we first let

$$\frac{n_h}{n} - \frac{N_h}{N} = \hat{n}_h, \quad \text{and} \quad \frac{n_j}{n} - \frac{N_j}{N} = \hat{n}_j$$

The second term then is

$$E\sum_{h \neq j}^L \bar{X}_h \hat{n}_h \bar{X}_j \hat{n}_j = \sum_{h \neq j}^L E\bar{X}_h \hat{n}_h \bar{X}_j \hat{n}_j$$

$$= \sum_{h \neq j}^L E(E\bar{X}_h \bar{X}_j \hat{n}_h \hat{n}_j | \hat{n}_j)$$

Since

$$\sum_h^L \hat{n}_h = \sum_j^L \hat{n}_j = 0$$

the second term of Eq. 7.33 is equal to

$$-\sum_{h \neq j}^L E\bar{X}_h \bar{X}_j \hat{n}_j^2 \left(\frac{N_h}{N - N_j}\right) = -\frac{N-n}{(N-1)n}\sum_{h \neq j}^L \bar{X}_h \bar{X}_j \frac{N_j}{N}\left(\frac{N - N_j}{N}\right)\frac{N_h}{N - N_j}$$

$$= -\frac{N-n}{(N-1)n}\sum_{h \neq j}^L \bar{X}_h \bar{X}_j \frac{N_j}{N}\frac{N_h}{N} = -\frac{N-n}{(N-1)n}\left[\frac{\left(\sum_h^L N_h \bar{X}_h\right)^2}{N^2} - \frac{\sum_h^L N_h^2 \bar{X}_h^2}{N^2}\right]$$

$$= -\frac{N-n}{(N-1)n}\bar{X}^2 + \frac{N-n}{(N-1)n}\frac{\sum_h^L N_h^2 \bar{X}_h^2}{N^2}$$

Combining the evaluation of the first term of Eq. 7.33 with that of the second term, we find that Eq. 7.33 is equal to

$$\frac{N-n}{(N-1)n}\left(\frac{\sum_{h}^{L} N_h \bar{X}_h^2}{N} - \bar{X}^2\right)$$

and for N large relative to n, we have

$$\sigma^2_{E(\bar{x}''|b_3)} = \frac{\sum_{h}^{L} N_h(\bar{X}_h - \bar{X})^2}{nN} \tag{7.34}$$

We finally have, collecting terms,

$$\sigma^2_{\bar{x}} = E\sigma^2_{\bar{x}|b_1} + E\sigma^2_{\bar{x}'|b_2} + E\sigma^2_{x''|b_3} + \sigma^2_{E(\bar{x}''|b_3)} \tag{7.35}$$

where the terms in Eq. 7.35 are defined by Eq. 7.18, 7.24, 7.31, and 7.34. Equation 7.35, after considerable algebraic manipulation, reduces to

$$\sigma^2_{\bar{x}} = \frac{1}{n}\sum_{h}^{L}\frac{N_h}{N}\frac{\sum_{i}^{M_h}\sum_{j}^{N_h} E_{hij}(x_{hijk} - \bar{X})^2}{M_h N_h}$$
$$+ \frac{1}{n}\sum_{h}^{L}(\bar{n}_h - 1)\frac{N_h}{N}\frac{\sum_{i}^{M_h}\sum_{j\neq v}^{N_h} E_{hij,v}(x_{hijk} - \bar{X}_h)(x_{hivw} - \bar{X}_h)}{M_h N_h(N_h - 1)} \tag{7.36}$$

Now let

$$\sum_{h}^{L}\frac{N_h}{N}\frac{\sum_{i}^{M_h}\sum_{j}^{N_h} E_{hij}(x_{hijk} - \bar{X})^2}{M_h N_h} = \sigma^2_X \tag{7.37}$$

We note that

$$\sigma_{hXI} = \frac{\sum_{i}^{M_h}\sum_{j\neq v}^{N_h} E_{hij,v}(x_{hijk} - \bar{X}_h)(x_{hivw} - \bar{X}_h)}{M_h N_h(N_h - 1)} \tag{7.38}$$

is the average covariance among the responses for a given interviewer. Then Eq. 7.36 can be written

$$\sigma^2_{\bar{x}} = \frac{1}{n}\sigma^2_X + \frac{1}{n}\frac{\sum_{h}^{L}(\bar{n}_h - 1)N_h\sigma_{hXI}}{N} \tag{7.39}$$

where σ^2_X is given by Eq. 7.37 and σ_{hXI} by Eq. 7.38. Assuming

$$\bar{n}_h = \frac{n}{m} = \bar{n}$$

for all h, Eq. 7.39 becomes

$$\frac{1}{n}(\sigma_X^2 - \sigma_{XI}) + \frac{\sigma_{XI}}{m} \tag{7.40}$$

where

$$\sigma_{XI} = \frac{\sum^{L} N_h \sigma_{hXI}}{N}$$

c. Estimates of variance from sample results. Let

$$s_{hX}^2 = \frac{\sum_i^{m_h}\sum_j^{\bar{n}_h}(x_{hij} - \bar{x}_{hi})^2}{m_h\bar{n}_h} - \frac{\sum_i^{m_h}(\bar{x}_{hi} - \bar{x}_h)^2}{m_h - 1} \tag{7.41}$$

To prove:

$$Es_{hX}^2 = \sigma_{hX}^2 = E(x_{hijk} - \bar{X}_h)^2 \tag{7.42}$$

Proof.

$$E\frac{\sum_i^{m_h}\sum_j^{\bar{n}_h}(x_{hij} - \bar{x}_{hi})^2}{m_h\bar{n}_h} - E\frac{\sum_i^{m_h}\sum_j^{\bar{n}_h}x_{hij}^2}{m_h\bar{n}_h} - E\frac{\sum_i^{m_h}\bar{x}_{hi}^2}{m_h} \tag{7.43}$$

Since

$$E\bar{x}_{hi} = \bar{X}_h \quad \text{and} \quad E\bar{x}_{hi}^2 = m_h\sigma_{\bar{x}_h}^2 + \bar{X}_h^2$$

$$\begin{aligned} \text{Eq. 7.43} &= \sigma_{hX}^2 + \bar{X}_h^2 - m_h\sigma_{\bar{x}_h}^2 - \bar{X}_h^2 \\ &= \sigma_{hX}^2 - m_h\sigma_{\bar{x}_h}^2 \end{aligned} \tag{7.44}$$

Moreover, from Sec. 2, Ch. 6, the expected value of the last term in Eq. 7.41 is

$$E\frac{\sum_i^{m_h}(\bar{x}_{hi} - \bar{x}_h)^2}{m_h - 1} = m_h\sigma_{\bar{x}_h}^2 \tag{7.45}$$

Therefore,

$$Es_{hX}^2 = \sigma_{hX}^2 \tag{7.46}$$

To prove:

$$Es_{hXI} = \sigma_{hXI} \tag{7.47}$$

where

$$s_{hXI} = \frac{\sum_i^{m_h}(\bar{x}_{hi} - \bar{x}_h)^2}{m_h - 1} - \frac{\sum_i^{m_h}\sum_j^{\bar{n}_h}(x_{hij} - \bar{x}_{hi})^2}{m_h\bar{n}_h(\bar{n}_h - 1)} \tag{7.48}$$

Proof. From Sec. 2 of Ch. 6

$$E\frac{\sum_i^{m_h}(\bar{x}_{hi} - \bar{x}_h)^2}{m_h - 1} = m_h\sigma_{\bar{x}_h}^2 \tag{7.49}$$

and

$$\frac{\sum_i^{m_h}\sum_j^{\bar{n}_h}(x_{hij}-\bar{x}_{hi})^2}{m_h\bar{n}_h(\bar{n}_h-1)}=\frac{\sum_i^{m_h}\sum_j^{\bar{n}_h}x_{hij}^2}{m_h\bar{n}_h(\bar{n}_h-1)}-\frac{\sum_i^{m_h}\bar{x}_{hi}^2}{m_h(\bar{n}_h-1)} \tag{7.50}$$

$$E\frac{\sum_i^{m_h}\sum_j^{\bar{n}_h}x_{hij}^2}{m_h\bar{n}_h(\bar{n}_h-1)}=\frac{\sigma_{hX}^2}{\bar{n}_h-1}+\frac{\bar{X}_h^2}{\bar{n}_h-1} \tag{7.51}$$

and

$$E\frac{\sum_i^{m_h}\bar{x}_{hi}^2}{m_h(\bar{n}_h-1)}=\frac{m_h\sigma_{\bar{x}_h}^2}{\bar{n}_h-1}+\frac{\bar{X}_h^2}{\bar{n}_h-1} \tag{7.52}$$

and from Eq. 7.49 and 7.50 we have

$$Es_{hXI}=-\frac{\sigma_{hX}^2}{\bar{n}_h-1}+\frac{m_h\bar{n}_h\sigma_{\bar{x}_h}^2}{\bar{n}_h-1} \tag{7.53}$$

From Eq. 7.39

$$\sigma_{\bar{x}_h}^2=\frac{\sigma_{hX}^2}{m_h\bar{n}_h}+\frac{(\bar{n}_h-1)\sigma_{hXI}}{m_h\bar{n}_h} \tag{7.54}$$

Es_{hXI} = Eq. 7.53, and, substituting Eq. 7.54 for $\sigma_{\bar{x}_h}^2$, we have

$$Es_{hXI}=\sigma_{hXI} \tag{7.55}$$

To prove:

$$Es_{bX}^2=\frac{\sum_h^L N_h(\bar{X}_h-\bar{X})^2}{N} \tag{7.56}$$

where

$$s_{bX}^2=\frac{\sum_h^L n_h(\bar{x}_h-\bar{x})^2}{n-1}-\frac{\sum_h^L\frac{\bar{n}_h(n-n_h)}{m_h-1}\sum_i^{m_h}(\bar{x}_{hi}-\bar{x}_h)^2}{n(n-1)} \tag{7.57}$$

Proof. Now the expected value of the first term in the right-hand side of Eq. 7.57 is

$$E\frac{\sum_h^L n_h(\bar{x}_h-\bar{x})^2}{n-1}=E\frac{\sum_h^L n_h\bar{x}_h^2}{n-1}-\frac{n}{n-1}E\bar{x}^2 \tag{7.58}$$

$$=E\frac{\sum_h^L n_h(\sigma_{\bar{x}_h}^2+\bar{X}_h^2)}{n-1}-\frac{n}{n-1}(\sigma_{\bar{x}}^2+\bar{X}^2) \tag{7.59}$$

$$= \frac{n}{n-1} \sum_h^L \frac{N_h}{N} \sigma_{\bar{x}_h}^2 + \frac{n}{n-1} \frac{\sum_h^L N_h \bar{X}_h^2}{N} - \frac{n}{n-1} \sigma_{\bar{x}}^2 - \frac{n}{n-1} \bar{X}^2 \tag{7.60}$$

$$= \frac{n}{n-1} \sum_h^L \frac{N_h}{N} \sigma_{\bar{x}_h}^2 + \frac{n}{n-1} \frac{\sum_h^L N_h (\bar{X}_h - \bar{X})^2}{N} - \frac{n}{n-1} \sigma_{\bar{x}}^2 \tag{7.61}$$

Since

$$\bar{x} = \frac{\sum_h^L n_h \bar{x}_h}{n}$$

then, by Theorem 15 of Ch. 3 (p. 65),

$$\sigma_{\bar{x}}^2 = E \frac{\sum_h^L n_h^2 \sigma_{\bar{x}_h}^2}{n^2} + E \left(\frac{\sum_h^L n_h \bar{X}_h}{n} - \bar{X} \right)^2 \tag{7.62}$$

Since, as shown above, $En_h = n(N_h/N)$, it follows that

$$E \left(\frac{\sum_h^L n_h \bar{X}_h}{n} - \bar{X} \right)^2 = \frac{\sum_h^L N_h (\bar{X}_h - \bar{X})^2}{Nn} \tag{7.63}$$

We have

$$\sigma_{\bar{x}}^2 = E \frac{\sum_h^L n_h^2 \sigma_{\bar{x}_h}^2}{n^2} + \frac{\sum_h^L N_h (\bar{X}_h - \bar{X})^2}{Nn} \tag{7.64}$$

Therefore

$$E \frac{\sum_h^L n_h (\bar{x}_h - \bar{x})^2}{n-1} = \frac{n}{n-1} \sum_h^L \frac{N_h}{N} \sigma_{\bar{x}_h}^2 + \frac{n}{n-1} \frac{\sum_h^L N_h (\bar{X}_h - \bar{X})^2}{N}$$

$$- \frac{1}{n(n-1)} E \sum_h^L n_h^2 \sigma_{\bar{x}_h}^2 - \frac{\sum_h^L N_h (\bar{X}_h - \bar{X})^2}{N(n-1)} \tag{7.65}$$

Now

$$E \frac{\sum_h^L \bar{n}_h (n - n_h) \frac{\sum_i^{m_h} (\bar{x}_{hi} - \bar{x}_h)^2}{m_h - 1}}{n(n-1)} = E \frac{\sum_h^L n_h (n - n_h) \sigma_{\bar{x}_h}^2}{n(n-1)} \tag{7.66}$$

$$= E \frac{\sum_h^L n_h n \sigma_{\bar{x}_h}^2}{n(n-1)} - E \frac{\sum_h^L n_h^2 \sigma_{\bar{x}_h}^2}{n(n-1)}$$

$$= \frac{n}{n-1} \frac{\sum_h^L N_h \sigma_{\bar{x}_h}^2}{N} - E \frac{\sum_h^L n_h^2 \sigma_{\bar{x}_h}^2}{n(n-1)} \tag{7.67}$$

Hence from Eq. 7.65 and 7.67

$$Es_{bX}^2 = \frac{\sum_h^L N_h(\bar{X}_h - \bar{X})^2}{N} \tag{7.68}$$

Thus, if we let

$$s_{\bar{x}}^2 = \frac{\sum_h^L n_h s_{hX}^2}{n^2} + \frac{\sum_h^L (\bar{n}_h - 1) n_h s_{hXI}}{n^2} + \frac{s_{bX}^2}{n} \tag{7.69}$$

where s_{hX}^2 is given by Eq. 7.41, s_{hXI} by Eq. 7.48, and s_{bX}^2 by Eq. 7.57,

$$Es_{\bar{x}}^2 = \frac{\sum_h^L N_h \sigma_{hX}^2}{nN} + \frac{\sum_h^L (\bar{n}_h - 1) N_h \sigma_{hXI}}{nN} + \frac{\sigma_{bX}^2}{n}$$

$$= \sigma_{\bar{x}}^2 \tag{7.70}$$

$s_{\bar{x}}^2$ can also be written in the forms:

$$s_{\bar{x}}^2 = \frac{\sum_h^L \sum_i^{m_h} \sum_j^{\bar{n}_h} (x_{hij} - \bar{x})^2}{n(n-1)} + \frac{\sum_h^L (\bar{n}_h - 1) n_h s_{hXI}}{n(n-1)} \tag{7.71}$$

and

$$s_{\bar{x}}^2 = \frac{\sum_h^L \frac{n_h(n_h - 1)}{m_h(m_h - 1)} \sum_i^{m_h} (\bar{x}_{hi} - \bar{x})^2}{n(n-1)} + \frac{\sum_h^L n_h(\bar{x}_h - \bar{x})^2}{n(n-1)}$$

When $L = 1$,

$$s_{\bar{x}}^2 = \frac{\sum_i^m (\bar{x}_i - \bar{x})^2}{m(m-1)} \tag{7.72}$$

If $\bar{n}_h = \bar{n} = n/m$,

$$s_{XI} = \frac{\sum_h^L n_h s_{hXI}}{n}$$

$$= \frac{\sum_h^L \frac{m_h}{m_h - 1} \sum_i^{m_h} (\bar{x}_{hi} - \bar{x}_h)^2}{m} - \frac{\sum_h^L \sum_i^{m_h} \sum_j^{\bar{n}} (x_{hij} - \bar{x}_{hi})^2}{n(\bar{n} - 1)} \tag{7.73}$$

$$s_X^2 = \frac{\sum_h^L n_h s_{hX}^2}{n} + s_{bX}^2$$

$$= \frac{\sum_h^L \sum_i^{m_h} \sum_j^{\bar{n}} (x_{hij} - \bar{x}_{hi})^2}{n} + \frac{\sum_h^L \frac{m_h}{m_h - 1} \sum_i^{m_h} (\bar{x}_{hi} - \bar{x}_h)^2}{m}$$

$$+ \frac{\sum_h^L n_h(\bar{x}_h - \bar{x})^2}{n - 1} - \frac{\sum_h^L \bar{n} \frac{n - n_h}{m_h - 1} \sum_i^{m_h} (\bar{x}_{hi} - \bar{x}_h)^2}{n(n - 1)}$$

$$= \frac{\sum_h^L \sum_i^{m_h} \sum_j^{\bar{n}} (x_{hij} - \bar{x})^2}{n - 1} + \frac{\bar{n} - 1}{n - 1} s_{XI} \tag{7.74}$$

$$s_{\bar{x}}^2 = \frac{\sum_h^L \sum_i^{m_h} \sum_j^{\bar{n}} (x_{hij} - \bar{x})^2}{n(n - 1)} + \frac{\bar{n} - 1}{n - 1} s_{XI} \tag{7.75}$$

d. Variance of a mean with a double sampling design. This section considers a case where n individuals are drawn at random, and from these n individuals, $\bar{n}$ individuals are assigned at random to each of m sample interviewers (m_h interviewers from the hth group). For this sample we obtain responses x_{hij} under essential conditions X. From the $\bar{n}$ sample cases assigned to each interviewer we subsample at random $\bar{n}'$ cases, giving a total subsample of $n' = m\bar{n}'$. We also have a set of L' interviewer groups (which may or may not be the same as the original interviewer groups). Of the subsample cases n_p' are available for interview by interviewers in the pth of the L' interviewer groups. We draw $m_p'(= m'n_p'/n'$, where m' is determined in advance) interviewers from the pth interviewer group and assign at random to each of these sample interviewers $\bar{n}''(= n'/m')$ individuals. The second set of interviewers obtains responses z_{pqj} under essential conditions Z from each of the n' individuals in the subsample. We use as an estimate of $\bar{Y}$, the true population mean,

$$\hat{z} = \frac{\bar{x}\bar{z}'}{\bar{x}'} \tag{7.76}$$

where

$$\bar{x} = \frac{\sum_h^L \sum_i^{m_h} \sum_j^{\bar{n}} x_{hij}}{n} \tag{7.77}$$

$$\bar{x}' = \frac{\sum_{h}^{L} \sum_{i}^{m_h} \sum_{j}^{\bar{n}'} x_{hij}}{n'} \tag{7.78}$$

$$\bar{z}' = \frac{\sum_{p}^{L'} \sum_{q}^{m'_p} \sum_{j}^{\bar{n}''} z_{pqj}}{n'}$$

The expected value and variance of $\bar{x}$ have already been derived. The expected values and variances of $\bar{x}'$ and $\bar{z}'$ are identical with the values for a random sample of n' cases drawn without reference to the sample of n cases from which $\bar{x}$ is calculated. Thus

$$E\bar{x}' = \bar{X} = \frac{\sum_{h}^{L} \frac{1}{M_h} \sum_{i}^{M_h} \sum_{j}^{N_h} \bar{X}_{hij}}{N} \tag{7.79}$$

where $\bar{X}_{hij}$ = the expected response value for the jth individual in the population interviewed by the ith interviewer in the hth group under essential conditions X,

and

$$E\bar{z}' = \bar{Z} = \frac{\sum_{p}^{L'} \frac{1}{M'_p} \sum_{q}^{M'_p} \sum_{j}^{N'_p} \bar{Z}_{pqj}}{N} \tag{7.80}$$

where $\bar{Z}_{pqj}$ = the expected response value for the jth individual in the population interviewed by the qth interviewer in the pth group under essential conditions Z.

Using the usual approximations to the expected value and variance of $\hat{z}$ gives

$$E\hat{z} \doteq \frac{(E\bar{x})(E\bar{z}')}{(E\bar{x}')} = \frac{\bar{X}\bar{Z}}{\bar{X}} = \bar{Z} \tag{7.81}$$

$$\sigma_{\hat{z}}^2 \doteq \bar{Z}^2 \left[\frac{\sigma_{\bar{x}}^2 + \sigma_{\bar{x}'}^2 - 2\sigma_{\bar{x}\bar{x}'}}{\bar{X}^2} + \frac{\sigma_{\bar{z}'}^2}{\bar{Z}^2} - \frac{2(\sigma_{\bar{x}'\bar{z}'} - \sigma_{\bar{x}\bar{z}'})}{\bar{X}\bar{Z}}\right] \tag{7.82}$$

Since $\sigma_{\bar{x}\bar{x}'} = \sigma_{\bar{x}}^2$,

$$\sigma_{\hat{z}}^2 = \bar{Z}^2 \left[\frac{\sigma_{\bar{x}'}^2 - \sigma_{\bar{x}}^2}{\bar{X}^2} + \frac{\sigma_{\bar{z}'}^2}{\bar{Z}^2} - \frac{2(\sigma_{\bar{x}'\bar{z}'} - \sigma_{\bar{x}\bar{z}'})}{\bar{X}\bar{Z}}\right] \tag{7.83}$$

and

$$\text{MSE } \hat{z} = (\bar{Z} - \bar{Y})^2 + \sigma_{\hat{z}}^2 \tag{7.84}$$

It will be noted that the bias of $\hat{z}$ is approximately $(\bar{Z} - \bar{Y})$. Thus, if $\bar{Z}$ is closer to $\bar{Y}$ than is $\bar{X}$, $\hat{z}$ may be a better estimate of $\bar{Y}$ than $\bar{x}$, even when the variance of $\hat{z}$ exceeds the variance of $\bar{x}$.

As previously noted, the variances of $\bar{x}'$ and $\bar{z}'$ are the same as the variances when the subsample of n' is drawn independently of the sample of n. Thus

$$\sigma^2_{\bar{x}} = \frac{\sigma^2_X - \sigma_{XI}}{n} + \frac{\sigma_{XI}}{m} \tag{7.85}$$

$$\sigma^2_{\bar{z}'} = \frac{\sigma^2_Z - \sigma_{ZI}}{n'} + \frac{\sigma_{ZI}}{m'} \tag{7.86}$$

$$\sigma^2_{\bar{x}'} = \frac{\sigma^2_X - \sigma_{XI}}{n'} + \frac{\sigma_{XI}}{m} \tag{7.87}$$

We also have

$$\begin{aligned} nn'E\bar{x}\bar{z}' &= n'E(x_{hij}z_{pqj}) + n'(n-1)E(x_{hij}z_{pqv}) \\ &= n'E(x_{hij}z_{pqj}) + n'(n-1)\bar{X}\bar{Z} \end{aligned} \tag{7.88}$$

$$E\bar{x}\bar{z}' = \frac{E(x_{hij}z_{pqj}) - \bar{X}\bar{Z}}{n} + \bar{X}\bar{Z} \tag{7.89}$$

$$\sigma_{\bar{x}\bar{z}'} = \frac{\sigma_{XZ}}{n} \tag{7.90}$$

where

$$\sigma_{XZ} = E(x_{hij}z_{pqj}) - \bar{X}\bar{Z} = \frac{\sum_k^N \bar{X}_k\bar{Z}_k}{N} - \bar{X}\bar{Z} \tag{7.91}$$

and

$$(n')^2E\bar{x}'\bar{z}' = n'Ex_{hij}z_{pqj} + n'(n'-1)Ex_{hij}z_{pqj} \tag{7.92}$$

$$= n'Ex_{hij}z_{pqj} + n'(n'-1)\bar{X}\bar{Z} \tag{7.93}$$

$$\sigma_{\bar{x}'\bar{z}'} = \frac{\sigma_{XZ}}{n'} \tag{7.94}$$

Thus

$$\sigma^2_{\hat{z}} = \bar{Z}^2\left(\frac{U}{n} + \frac{V}{n'} + \frac{W}{m'}\right) \tag{7.95}$$

where

$$U = \frac{2\sigma_{XZ}}{\bar{X}\bar{Z}} - \frac{\sigma^2_X - \sigma_{XI}}{\bar{X}^2} \tag{7.96}$$

$$V = \frac{\sigma^2_Z - \sigma_{ZI}}{\bar{Z}^2} - U \tag{7.97}$$

$$W = \frac{\delta_Z}{\bar{Z}^2} \tag{7.98}$$

Unbiased estimates from the sample of σ_X^2 and σ_{XI} have been derived above. Estimates of σ_Z^2 and σ_{ZI} will have the same form as those for σ_X^2 and σ_{XI}, using the values z_{pqj}. As an unbiased estimate of σ_{XZ} we have

$$s_{XZ} = \frac{\sum_j^{n'} x_{hij} z_{pqj}}{n' - 1} - \frac{n'\bar{x}'\bar{z}'}{n' - 1} \tag{7.99}$$

$$Es_{XZ} = \frac{n'}{n' - 1}(Ex_{hij}z_{pqj} - E\bar{x}'\bar{z}') = \frac{n'}{n' - 1}\left(\sigma_{XZ} + \bar{X}\bar{Z} - \frac{\sigma_{XZ}}{n'} - \bar{X}\bar{Z}\right)$$

$$= \sigma_{XZ} \tag{7.100}$$

e. Variance of estimates of response bias. We have considered the case where samples are drawn as described in Part *d* ("double sampling" design), and it can be assumed that $\bar{Z} \doteq \bar{Y}$. In this situation estimates of the bias of $\bar{x}$ as an estimate of $\bar{Y}$ are

$$\bar{b}_X = \bar{x} - \bar{z}' \tag{7.101}$$

and

$$\hat{b}_X = \bar{x} - \hat{z} \tag{7.102}$$

Thus $\bar{b}_X$ is an unbiased estimate of the response bias $\bar{B}_X$ $(= \bar{X} - \bar{Z})$. The "ratio" estimate $\hat{b}_X$ is a consistent estimate of $\bar{B}_X$. For the variance of $\bar{b}_X$ we have

$$\sigma_{\bar{b}_X}^2 = \sigma_{\bar{x}}^2 + \sigma_{\bar{z}'}^2 - 2\sigma_{\bar{x}\bar{z}'} \tag{7.103}$$

Substituting values obtained above for $\sigma_{\bar{x}}^2$, $\sigma_{\bar{z}'}^2$, and $\sigma_{\bar{x}\bar{z}'}$ gives

$$\sigma_{\bar{b}_X}^2 = \frac{\sigma_X^2 - \sigma_{XI}}{n} + \frac{\sigma_{XI}}{m} + \frac{\sigma_Z^2 - \sigma_{ZI}}{n'} + \frac{\sigma_{ZI}}{m'} - \frac{2\sigma_{XZ}}{n} \tag{7.104}$$

For the variance of $\hat{b}_X$, we have

$$\sigma_{\hat{b}_X}^2 = \sigma_{\bar{x}}^2 + \sigma_{\hat{z}}^2 - 2\sigma_{\bar{x}\hat{z}} \tag{7.105}$$

and

$$\sigma_{\bar{x}\hat{z}} \doteq \sigma_{\bar{x}\bar{z}'} \tag{7.106}$$

Thus

$$\sigma_{\hat{b}_X}^2 \doteq \sigma_{\bar{b}_X}^2 + \sigma_{\hat{z}}^2 - \sigma_{\bar{z}'}^2 \tag{7.107}$$

$$\doteq \sigma_{\bar{b}_X}^2 - \bar{Z}^2\left(\frac{1}{n'} - \frac{1}{n}\right)\left(\frac{2\sigma_{XZ}}{\bar{X}\bar{Z}} - \frac{\sigma_X^2 - \sigma_{XI}}{\bar{X}^2}\right) \tag{7.108}$$

REFERENCES

(1) Russell L. Ackoff and Leon Pritzker, "The Methodology of Survey Research," *Intern. J. Opinion and Attitude Res.*, **5**, No. 3 (1951), 313–334.
(2) Gertrude Bancroft and Emmett H. Welch, "Recent Experience with Problems of Labor Force Measurement," *J. Amer. Stat. Assn.*, **41** (1946), 303–312.
(3) W. Edwards Deming, "On Errors in Surveys," *Amer. Sociol. Rev.*, **9** (1944), 359–369.
(4) Morris H. Hansen and William N. Hurwitz, "The Problem of Non-Response in Sample Surveys," *J. Amer. Stat. Assn.*, **41** (1946), 517–529.
(5) P. C. Mahalanobis, "Recent Experiments in Statistical Sampling in the Indian Statistical Institute," *J. Roy. Stat. Soc.*, **109** (1946), 325–370.
(6) Eli S. Marks and W. Parker Mauldin, "Problems of Response in Enumerative Surveys," *Amer. Sociol. Rev.*, **15** (1950), 649–657.
(7) Eli S. Marks, W. Parker Mauldin, and Harold Nisselson, "A Case History in Survey Design: The Post-Enumeration Survey of the 1950 Census," *J. Amer. Stat. Assn.*, **48** (June 1953).
(8) Gladys L. Palmer, "Factors in the Variability of Response in Enumerative Studies," *J. Amer. Stat. Assn.*, **38** (1943), 143–152.
(9) Morris H. Hansen, William N. Hurwitz, Eli S. Marks, and W. Parker Mauldin, "Response Errors in Surveys," *J. Amer. Stat. Assn.*, **46** (1951), 147–190.

Index